Boundary Layer Flows - Modelling, Computation, and Applications of Laminar, Turbulent Incompressible and Compressible Flows

Edited by Vallampati Ramachandra Prasad, Valter Silva and João Cardoso

Published in London, United Kingdom

Boundary Layer Flows - Modelling, Computation, and Applications of Laminar, Turbulent Incompressible and Compressible Flows
http://dx.doi.org/10.5772/intechopen.100716
Edited by Vallampati Ramachandra Prasad, Valter Silva and João Cardoso

Contributors
Abhishish Chandel, Vijay Shankar, Palle Kiran, S.H. Manjula, Juscelino Batista Leão, Carlos Gavilan, Russell Keanini, Peter Tkacik, Jerry Dahlberg, Tyler Watkins, Franklin Green, Jesse Redford, Sepideh Amiritavasoli, Seyed Jalal Hemmati, Saeed Niazi, Ali Jalali, Adrián R. Wittwer, Acir M. Loredo-Souza, Mario E. De Bortoli, Jorge O. Marighetti, Habib Djourdem, Noureddine Bouteraa, Lamia Thamri, Taoufik Naffouti, Aibing Yu, Zheng Qi, Shibo Kuang, Liangwan Rong, Kejun Dong, Vallampati Ramachandra Prasad, Rahul Boadh, Yogendra Kumar Rajoria

First published in London, United Kingdom, 2023 by IntechOpen
IntechOpen is the global imprint of INTECHOPEN LIMITED, registered in England and Wales, registration number: 11086078, 5 Princes Gate Court, London, SW7 2QJ, United Kingdom

British Library Cataloguing-in-Publication Data
A catalogue record for this book is available from the British Library

Additional hard and PDF copies can be obtained from orders@intechopen.com

Boundary Layer Flows - Modelling, Computation, and Applications of Laminar, Turbulent Incompressible and Compressible Flows
Edited by Vallampati Ramachandra Prasad, Valter Silva and João Cardoso
p. cm.
Print ISBN 978-1-80356-218-6
Online ISBN 978-1-80356-219-3
eBook (PDF) ISBN 978-1-80356-220-9

For EU product safety concerns:
IN TECH d.o.o., Prolaz Marije Krucifikse Kozulić 3, 51000 Rijeka, Croatia,
info@intechopen.com or visit our website at intechopen.com.

Meet the editors

Prof. Vallampati Ramachandra Prasad works in the fields of multi-physical fluid dynamics and numerical simulation. His research investigates complex phenomena of interest in mechanical engineering, applied mathematics, chemical engineering, and material processing. He has made fundamental contributions in magnetic fluid dynamics, thermal radiation heat transfer, fluid modeling, micro polar non-Newtonian hydrodynamics, magnetic induction flows, viscoelastic liquid flows, porous media, and boundary flows. He has published more than 100 journal articles and has presented at numerous conferences. He has written and edited several books on radiation heat transfer, thermo-diffusion and diffusion-thermo effects on boundary layer flows, and viscoelastic flows. He has collaborated with researchers in the United States, India, Spain, Turkey, and Taiwan, among other countries. He is currently a reviewer for more than 100 national and international journals.

Valter Silva is a senior researcher in environment and energy at the University of Aveiro, Portugal. In 2009, he obtained his Ph.D. in Chemical and Biological Engineering at the University of Porto. Since 2012, he has led a research team devoted to the application of experimental and numerical solutions on environmental and energy topics. In the last five years, Dr. Silva has coordinated national and international projects with leading universities across the world, including MIT and Carnegie Mellon, raising approximately $2.5 million in funds. He has supervised more than thirty students. He has participated in more than twenty-five national and international projects in energy and environment and has authored two books, eleven book chapters, and more than sixty papers in international peer-reviewed journals.

João Cardoso obtained a master's degree in Medical Physics from the University of Porto, Portugal, and a Ph.D. in Mechanical Engineering from the University of Lisbon, Portugal. Dr. Cardoso has devoted his research work to numerical fluid mechanics applications set to describe industrial energy processes targeting sustainability. To date, he has published more than twenty peer-reviewed papers in international journals, one book, and four book chapters. He also frequently collaborates as a reviewer for several journals such as *Chemical Engineering Journal*, *Renewable and Sustainable Energy Reviews*, and *Energy Conversion and Management*, among others.

Contents

Preface

This book presents recent advances in boundary layer flows and modern analytical and computational techniques to deepen the understanding of laminar and turbulent flows, which may be compressible or incompressible. It presents interpretations of theoretical and experimental results in boundary layer flows and directions for future research. There are chapters on atmospheric boundary layers, laminar boundary layers, the physics of the transition from laminar to turbulent flow, the turbulent boundary layer and its governing equations in time-averaging form, drag prediction by integral methods, turbulence modeling and differential methods, and current topics and problems in research and industry. The book also includes different numerical methods to solve boundary layer equations, their stability, and convergence. Written by the world's authoritative experts in the field, this book provides the opportunity to explore studies on boundary layer theory and their importance to the nonlinear theory of viscous and electrically conducting flows, the theory of heat and mass transfer, and the dynamics of reactive and multiphase media and aerodynamics. This volume is a useful resource for applied mathematicians, physicists, and engineers working with laminar and turbulent boundary layer flows.

Vallampati Ramachandra Prasad
Department of Mathematics,
School of Advanced Sciences,
Vellore Institute of Technology,
Vellore, Tamil Nadu, India

Valter Silva
Senior Researcher,
University of Aveiro,
Aveiro, Portugal

João Cardoso
Polytechnic Institute of Portalegre,
Instituto Superior Técnico,
University of Lisbon,
Lisbon, Portugal

Section 1

Laminar and Turbulent Flows

Chapter 1

Employing HEC-ResSim 3.1 for Reservoir Operation and Decision Making

Abhishish Chandel, Vijay Shankar and Sumit Jaswal

Abstract

The land of Himachal Pradesh is full of small and big rivers which are perennial. This benefit pushes Himachal Pradesh to build more and more dams to generate electricity, provide better irrigation supply to downstream areas and provide flood protection. To better utilize the huge potential of water, management of such reservoirs is the key issue. For this purpose, HEC-ResSim 3.1 is practiced on Pong Dam situated in western Himachal Pradesh. HEC-ResSim is one of the simulation models that possess single or multi-reservoir simulators and can simulate water resources systems. In this study, reservoir elevation and reservoir storage volume management is the target objective. The presented study was subsidized by daily observed data from 1998 up to 2014 of pool elevation, inflow, and outflow discharge. In addition, geometry and hydraulic data from dam and reservoir were employed to develop the platform to create a simulation using HEC-ResSim. Using the available reservoir data the simulation was performed for the 4 months of 2012. Then simulation results were compared with the real-time recorded data at the site. To validate the results, coefficient of determination for operations like reservoir elevation and reservoir storage was generated through regression plot and found more than 95% accurate. Also, Root Mean Square Error (RMSE) was calculated for both reservoir elevation and reservoir storage simulation and found under an acceptable range. This paper shows the utility of HEC-ResSim 3.1 for reservoir operational management and also throws light on the further scope. Finally, there is a discussion of how useful is HEC-ResSim as a reservoir management tool and integration of HEC-ResSim 3.1 with other hydrologic monitoring systems.

Keywords: HEC-ResSim, simulation, reservoir management, irrigation

1. Introduction

For a better operation of reservoirs, it is extremely necessary to manage the reservoir storage and hence reservoir levels with the seasonal variations throughout the year. Highly variable inflows from the river and water demands from the reservoir make it more challenging to manage reservoir storage volume according to daily demand. In the 1950s, the Corps of Engineers (COE), USA, and the National Weather Service (NWS) jointly developed the Stream-flow Synthesis and Reservoir Regulation

(SSARR) computer model. SSARR was used both as the stream-flow forecasting tool and as the real-time reservoir regulation tool. However, over the past few years, the stream-flow routing algorithms have been migrated to the HEC's ResSim model. Modini [1] studied and described all the challenges and strategies to completely migrate the AUTOREG/SSARR model to HEC-ResSim. An important objective was to ensure that all the provisions of the current Columbia River Treaty Flood Control Operating Plan (FCOP) must be migrated to HEC-ResSim. The main objective of the study was to develop a flexible model to accommodate FCOP strategy changes.

Eichert and Davis [2] generated the HEC-5 model for the study of flood control on the Susquehanna Reservoir System, USA. Model executed a decision support system to overcome the uncertainties of unevenly distributed water resources systems. Hickey et al. [3] used Hec-5 software in reservoir simulation for flood analysis in response to the destructive floods of 1983, 1986, 1995, and 1997. The main targets of study are model development, with a focus on headwater and major terminal reservoirs, and potential improvements to the flood damage reduction system. Emphasis is laid on model development and analyzing the influence of reservoirs in flood hydrology. Kim et al. [4] developed a deterministic optimization model named Coordinated Multiple Reservoir Operating Model (CoMOM) for real-time multi-reservoir operations in the Han River basin in Korea. Matondo and Msibi [5] created a DSS support with three major components, that is, the model input, modeling options, and outputs screens. The output of the DSS comprises the optimal rationing (%), monthly reservoir volume for the desired duration as well as a graphical representation of the reservoir response over the old and new scenario.

Bekele and Knapp [6] coupled storage routing and multi-objective evolutionary algorithms to simulate reservoir release rates of "Shelbyville and Carlyle Lakes" on Kaskaskia River, USA. The resulting coupled model can provide simulations of storage and reservoir pool elevations for the two lakes under varying water use conditions. All the long-term studies proved that the modeling techniques for the management of reservoirs are very helpful and efficient. Todd et al. [7] published studies on reservoir simulation using different techniques. Focuses on the four modeling systems: Reservoir System Simulation (HEC-ResSim), River and Reservoir Operations (RiverWare), River Basin Management Decision Support System (MODSIM), and Water Rights Analysis Package (WRAP). Though fundamentally similar, the four modeling systems differ significantly in their organizational structure, computational algorithms, user interfaces, and data management mechanisms. The Bureau and Tennessee Valley Authority jointly sponsored the development of RiverWare at the Center for Advanced Decision Support for Water and Environmental Systems of the University of Colorado [8, 9]. The Tennessee Valley Authority applied RiverWare in optimizing the daily and hourly operation of the system of multipurpose reservoirs and hydroelectric power plants. The Lower Colorado River Authority also applied RiverWare in daily time step modeling of water supply operations for reservoirs on the Colorado River of Texas [10]. MODSIM is a general-purpose reservoir/river system simulation model based on a network flow linear programming developed at Colorado State University [11, 12]. MODSIM has been used to study several reservoir/river systems in the western United States and throughout the world. The objective function coefficients used in MODSIM are factors entered by the model used to specify relative priorities that govern operating decisions.

The development of WRAP at Texas A&M University began in the late 1980s. WRAP has been greatly expanded since 1997 in conjunction with implementing a statewide Water Availability Modeling (WAM) System [13]. WRAP simulates water resources development, management, regulation, and use in a river basin or multi-basin region under a priority-based water allocation system. In WRAP terminology, a water right is a set of water use requirements, reservoir storage and conveyance facilities, operating rules, and institutional arrangements for managing water resources. Simulation results stored as DSS files accessed with HEC-DSSVue (a program used to manipulate data from HEC-DSS databases) for plotting and other analyses [14, 15]. According to the comparative studies of the basic modeling techniques for the reservoir operational studies, HEC-ResSim is recommended as the most productive and efficient modeling software. In 2004, for evaluation and reservoir management of the Tigris and Euphrates rivers system in Iraq, HEC-ResSim 2.0 was employed by Hanbali [16]. The study included six main reservoirs, three off-stream reservoirs, and seven small reservoirs, and many diversion dams for diverting water from Tigris and Euphrates rivers. HEC-ResSim 2.0 was used for simulation history events especially flood and drought periods.

Babazadeh [17] employed HEC-ResSim for reservoir modeling and stated that the application of simulation models is one of the most efficient ways of analyzing water resources systems. Model verification results indicate that this model can simulate the behavior of the system very well. Modeling resulted in increasing irrigation efficiency by 20% and reducing failures in the system by 12%. McKinney [18] developed a flow model of Lancang Cascade Dams, China, to maximize hydropower production and calibrate the model to match outflow at downstream gauge with the data of most recent year available data. HEC-ResSim came out to be capable to model dams in series. Piman [19] used HEC-ResSim and SWAT simultaneously for the assessment of flow changes from hydropower development and operations in Sekong, Susan, and Srepok Rivers of Mekong Basin, Vietnam. To access the magnitude of potential changes, daily flows were simulated over 20 years using the HEC-ResSim and SWAT models for a range of dam operations and development scenarios. Goodarzi [20] practiced a combination of LINGO and HEC-ResSim models to determine monthly operating rules for the Zayandehrud reservoir system in Iran. The results show that optimizing the operation of the Zayandehrud reservoir system could increase its storage by 88.9% as well as increase the reliability index of regulated water for all downstream demands by more than 10%. Lara [21] employed the HEC-ResSim model on Tucurui Dam, Brazil. It was subsidized by daily observed data from 2001 up to 2006 of pool elevation, inflow, and outflow discharge. HEC-ResSim was established as a powerful tool to support the decision-making of reservoir operations and an interesting alternative for risk management and flood control. Klipsch and Hurst [22] developed the HEC-ResSim 3.1 user's manual which provided great support and learning in employing HEC-ResSim 3.1 for the management of Pong Dam. Along with the software support, HEC-ResSim user's support by socio-networking website played a great role in the execution of the study (HEC-ResSim user's blog). In the present study, the reservoir storage volume and reservoir levels are modeled using Hec-ResSim 3.1 reservoir understudy is "Pong Dam" on Beas River in the state of Himachal Pradesh, India. Reservoir level and reservoir storage's target information are generated which enhance the decision-making capabilities related to reservoir management works. The input data is reservoir inflow and outflow time-series, reservoir physical data, reservoir area-volume-depth relationship, etc. In this paper, HEC-ResSim 3.1 is employed to the reservoir for the generation of a decision support system to regulate the reservoir elevation and reservoir volume which further enhances the reservoir operations (flood control, irrigation water supply, and hydropower generation). The study will help the reservoir management to tackle the future incoming water challenges and also to create simulations to practice the sudden situations like cloud bursts, surprise snowfall, uneven rains, etc. which are very common events over the region. All these parameters demand an improved operational system for the reservoir. The main objectives of the presented study are:

1. To employ HEC-ResSim 3.1 for modeling the real-time situations of the Pong Dam in the state of Himachal Pradesh, India.

2. To carry out a comparative evaluation of simulated elevation and storage targets with the recorded data.

2. Simulation software and data

2.1 HEC-ResSim 3.1 introducing

The simulation software used in the present study is HEC-ResSim (version 3.1) created by the U.S. Army Corp of Engineers—Hydrologic Engineering Center. The software has three main modules: Watershed Setup, Reservoir Network, and Simulation. Res-Sim has a graphical user interface (GUI) and utilizes the HEC Data Storage System (HEC-DSS) for storage and retrieval of input and output time-series data. ResSim is used to simulate reservoir operations including all characteristics of a reservoir and channel routing downstream. The data requirements for HEC-ResSim include the physical and operational characteristics of the dam and reservoir. The physical reservoir data is described through the use of the volume-area and elevation curves (**Figures 1** and **2**).

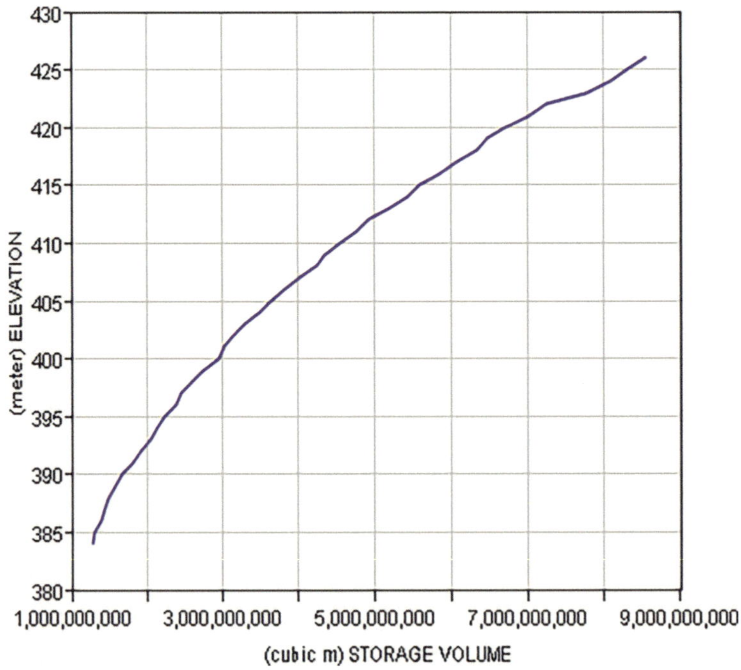

Figure 1.
Elevation-volume curve of Pong reservoir.

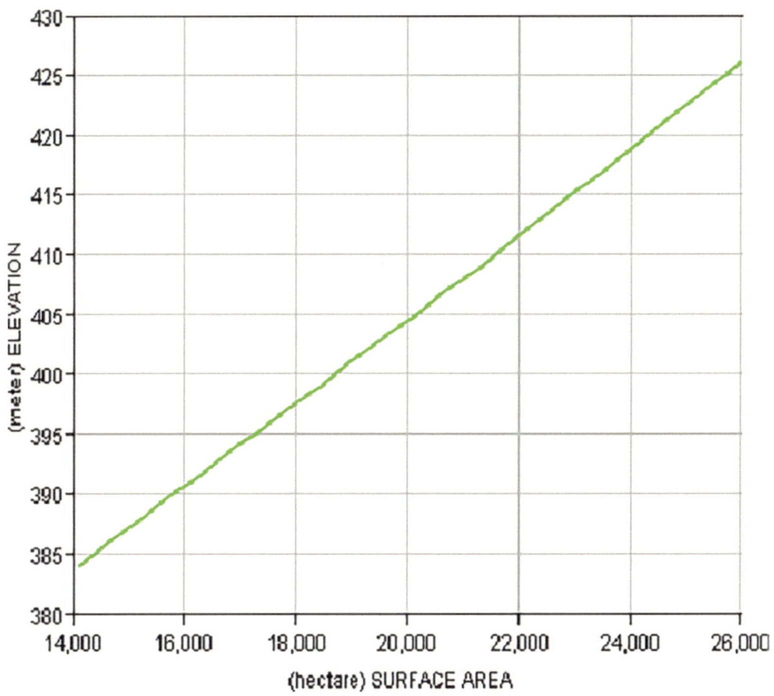

Figure 2.
Elevation-area curve of Pong reservoir.

The physical data of the dam include the type and capacity of each outlet. The operational data includes the zone definitions along with the rules governing the operations in each zone. There are three main management zones or pools, that is, the inactive pool, the conservation pool, and the flood pool.

The model allows the user to define alternatives and run their simulations simultaneously to compare results. Network elements include reservoirs, routing reaches, diversions, and junctions. In ResSim, watersheds include streams, projects (i.e., reservoir, levees), gage locations, impact areas, time-series locations, hydrologic and hydraulic data for the specific area. In the present paper, the reservoir operational rule curves data shown in **Figure 3** are employed to perform the HEC-ResSim simulations and represent the operational patterns of the Pong Dam. "Top of the Dam" shows the physical top-most part of the dam and "flood control curve" shows the max level for emergency releases. Till now maximum pool level achieved by the reservoir is shown by the "maximum pool elevation curve." "Conservation curve" shows the saved usage and operational levels of the reservoir for the past years. Similarly, buffer storage level and inactive pool levels are also described in the **Figure 3**.

Pong Dam modeling is performed in a systematic pattern using HEC-ResSim 3.1. In the first step, the Pong watershed is set up to employ HEC-ResSim 3.1. To set up Pong watershed features are added, stream alignments are drawn, configurations created, and project elements placed into configurations. The next reservoir network is developed over the watershed. In this routing reaches and junctions are added and edited simultaneously. Reservoir data like physical data (pool, dam, and outlet properties) are added. Then operation sets are added by applying zones and rules. Also, reference to the observed data created here. Alternatives by selecting network are

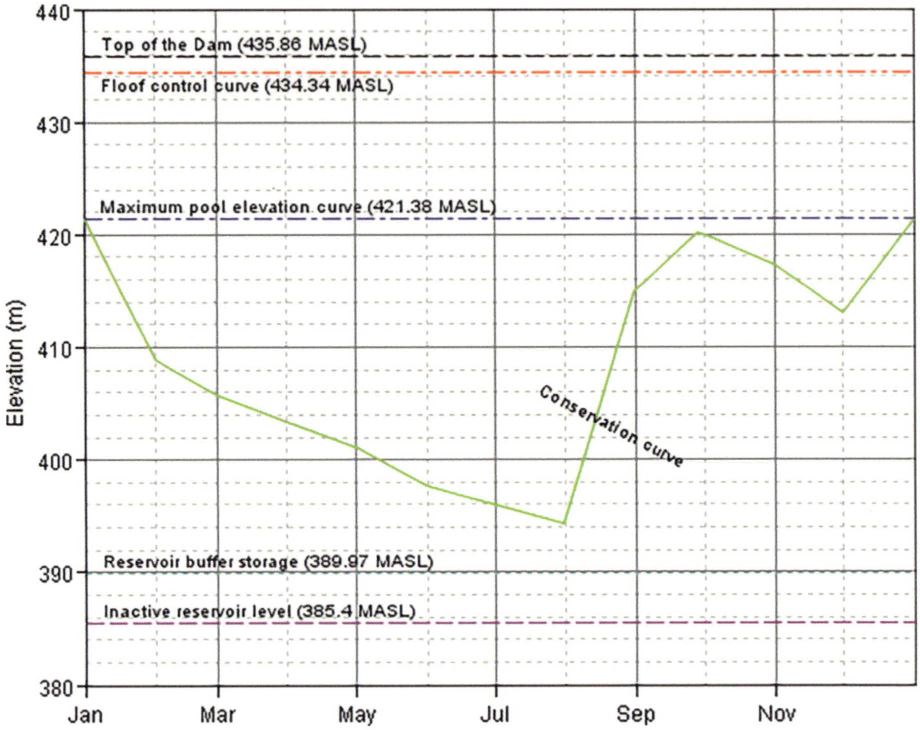

Figure 3.
Observed operational rule curves for Pong Dam.

defined. Run control settings are determined by the operation set for the simulation. The look-back date is defined. Simulation identifies time-series records and observed records from alternatives. To perform simulation a predefined alternative to the current simulation is selected. In the simulation module, the output result is analyzed, simulated results for the reservoir elevation and reservoir volume are extracted and the efficiency of simulated output is computed. The efficiency is computed using two methods: that is, graphical method, that is, to find the coefficient of determination, and statistical method, that is, to find Root Mean Square Error (RMSE).

$$\text{RMSE} = \left(\frac{\sum_{i=0}^{n} (\text{Observed output} - \text{simulated output})^2}{n} \right)^{0.5} \quad (1)$$

where n is the total number of data points.

3. Study location and characteristics

Pong Dam on Beas River is located in the wetland zone of the Shivalik Hills of western Himachal Pradesh, India at 32.0167°N, 76.0833°E. It is the highest earth-fill dam in India. The reservoir is a well-known wildlife sanctuary and one of the 25 international wetland sites declared in India. India. **Figure 4** shows the geographical presence of the Pong Dam in India.

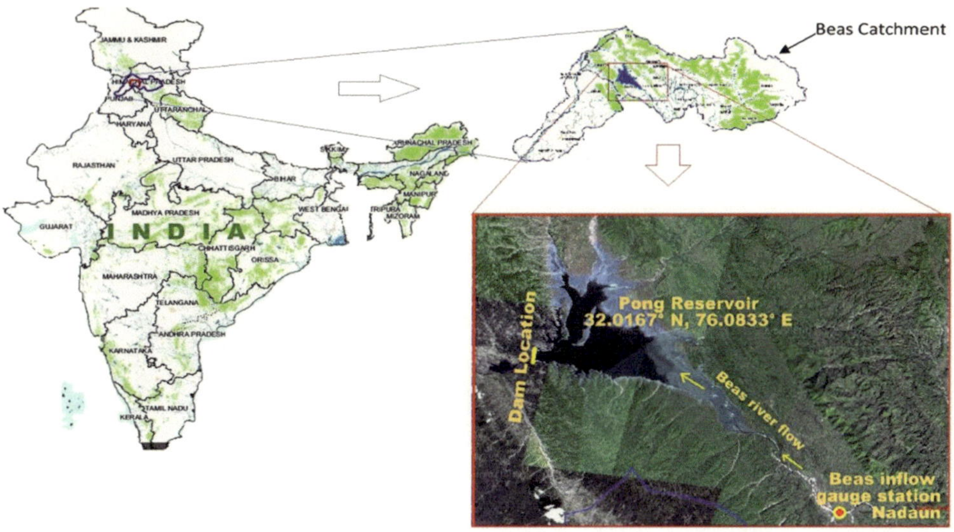

Figure 4.
Location map of the study area.

Beas River is one of the biggest rivers of Himachal Pradesh. The starting point of this permanent river is the snow-covered mountain at Rohtang Pass in Himachal Pradesh, India. Annual volume of water carried by Beas River as measured by monitoring stations in 40 years' time series is 9701.82 MCM. Pong Dam collects water drained from the catchment area of 12,613.998 Sq. km. First of all, operation reservoir volume is 7290 MCM at elevation 426.72 m from sea level and inactive level is 384.5 m above sea level. The length of the dam crest is 1951 m and its height from the river bed is 105.86 m and its crest width is 13.72 m. This dam has six Francis turbines with a capacity of 66 MW each. This is a single-purpose dam designed and constructed to serve irrigation to downstream areas. However, flood control, hydropower, and fisheries are the complimentary

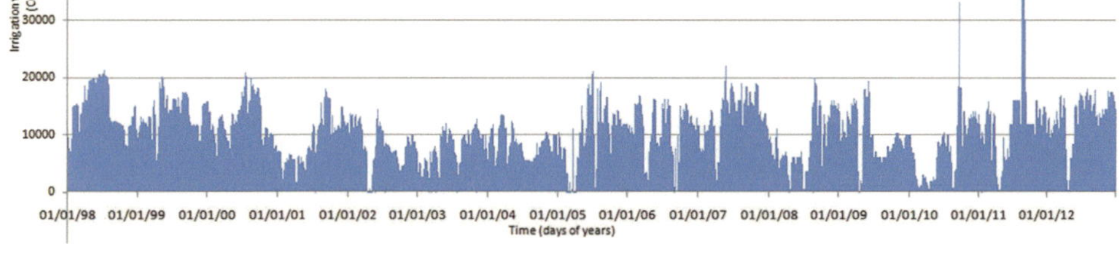

Figure 5.
Irrigation water release from Pong Dam (1998–2012).

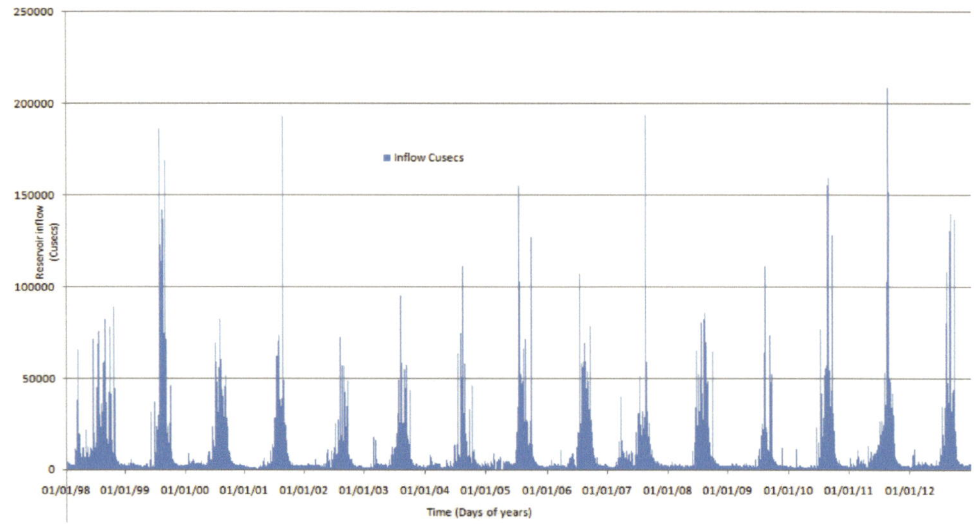

Figure 6.
Pong reservoir inflow (1998–2012).

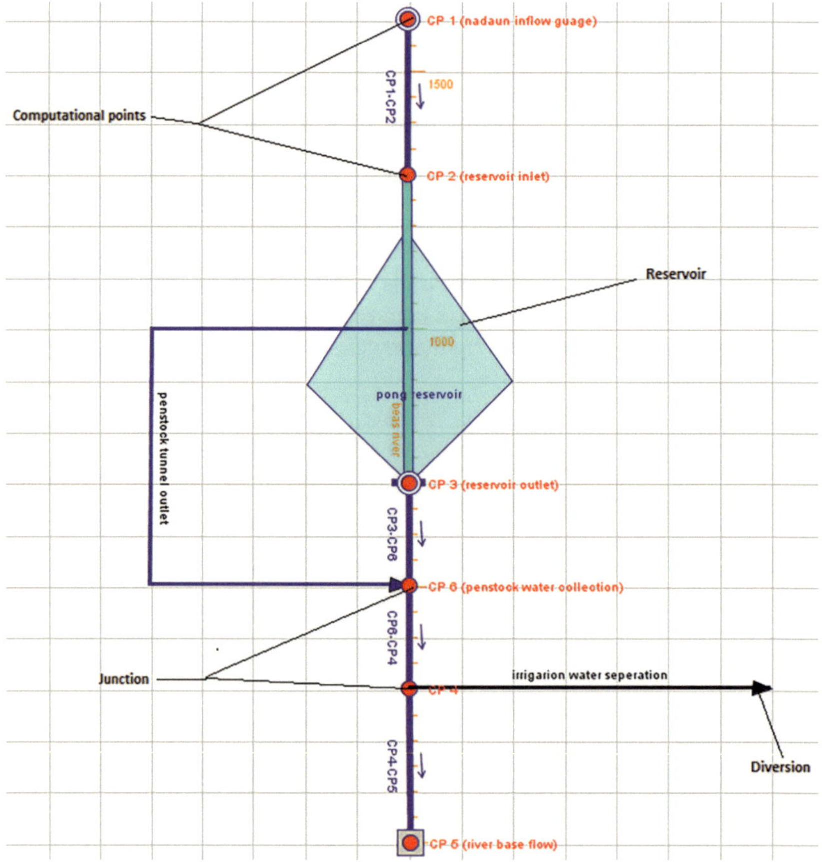

Figure 7.
Model layout of Pong reservoir and component setup in river Beas.

benefits. Based on field survey and published data from Pong Dam Operation Company, that is, Bhakhra Beas Management Board, agricultural sector use of water from dam reservoir is shown in **Figure 5**. Irrigation water release is the major operation of the Pong Dam. Release decisions depend upon many seasonal factors and crop water demands.

Also, there is a daily record for the reservoir inflow from 1998 to 2012 which acts as a very useful parameter for simulation (**Figure 6**). As presented earlier (**Figure 5**) the irrigation water supply is nearly constant but the incoming water to the reservoir is very flashy. Touching extreme peaks during monsoons and consistently very low during the rest of the year. Such flow patterns demand proper management and regulation of water releases from the reservoir.

HEC-ResSim 3.1 is employed to model the pong reservoir's operational setup and comes out as shown in **Figure 7**. It includes a stream, pong reservoir, computational points, penstock tunnel outlet, and irrigation water supply outflow. Penstock outlet is before the spillway and meets to the stream again downstream from where whole water diverted to the irrigation water supply. As hydropower generation is a complimentary operation over the irrigation water supply, the water indent for the powerhouse completely depends upon the irrigation water demand in the downstream fields.

4. Results and discussion

Secondary data of the reservoir (inflow, outflow, reservoir level, storage, and power generation as daily data) are collected from Bhakhra Beas Management Board (BBMB). **Figure 8** shows the behavior of the reservoir as generated by the real-time

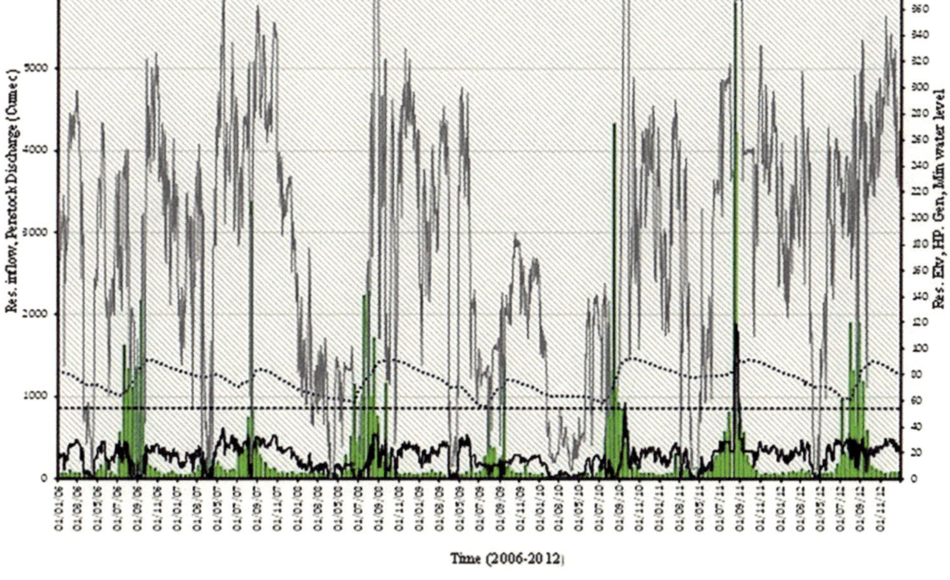

Figure 8.
Observed parameters of Pong storage dam operation.

data records. Reservoir inflow is touching extreme peaks during monsoons, that is, July–September whereas inflow is very low during the rest of the seasons. This makes the River understudy very flashy, hence requiring proper management.

For the study of accuracy and validation of the model, observed data from different years are compared with the respective simulated data. Therefore, input data, output data, and reservoir and dam properties were supplied to the model for variable durations of simulation for different years. Here results are shown for the 4 months of 2012 tagged with the time-series data of the whole year 2012.

4.1 Reservoir elevation operation

When we talk about the head formation for hydropower generation or preparing the reservoir for flood control, the first thing that comes to the manager's mind is the reservoir elevation. If one achieves the target elevation, the further operations of the reservoir become more efficient and safe. Hence if we can predict the future hydrologic condition of the reservoir we can simulate the target reservoir elevation in that situation. Also, we can practice some random expected flood values to check the elevation operation in that situation. In this way, we can prepare our-self for the worst condition even if that situation had never hit before. **Figure 9** shows the comparative plot for the actual reservoir elevations and the simulated reservoir elevations. It is showing results for 4 months of simulation and is utilized for the calculation of RMSE for reservoir elevation. RMSE for the operation of reservoir elevation is 0.78 m. This RMSE value is acceptable and recommends HEC-ResSim 3.1 for such reservoir modeling efforts.

To check the efficiency of the model both the simulated and actual values of the simulation were analyzed in the regression curve (**Figure 10**). Daily observed and simulated data related to reservoir elevation for the 123 days of the simulation period

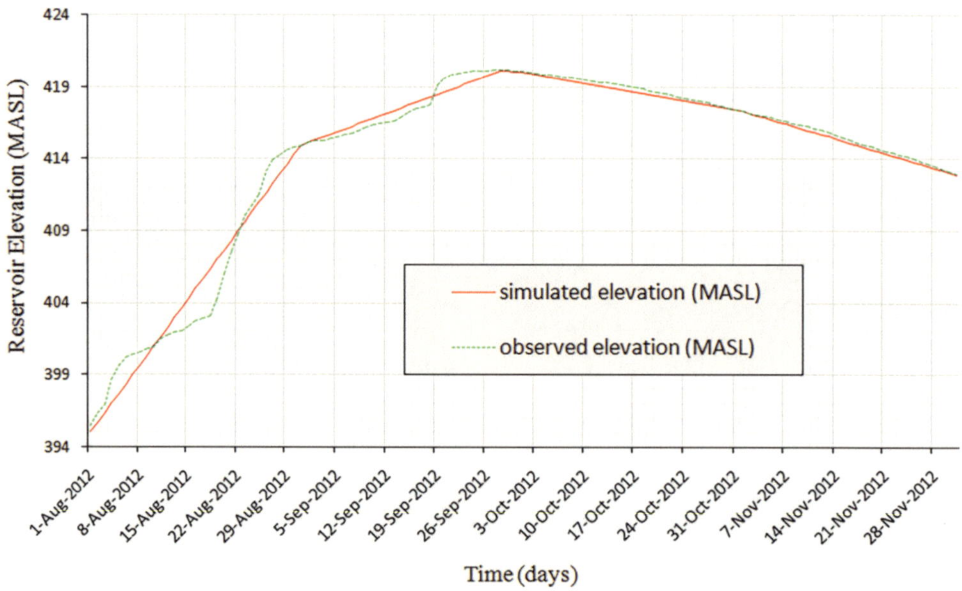

Figure 9.
Comparison of observed reservoir elevation and simulated reservoir elevation for the simulation period of August to November 2012.

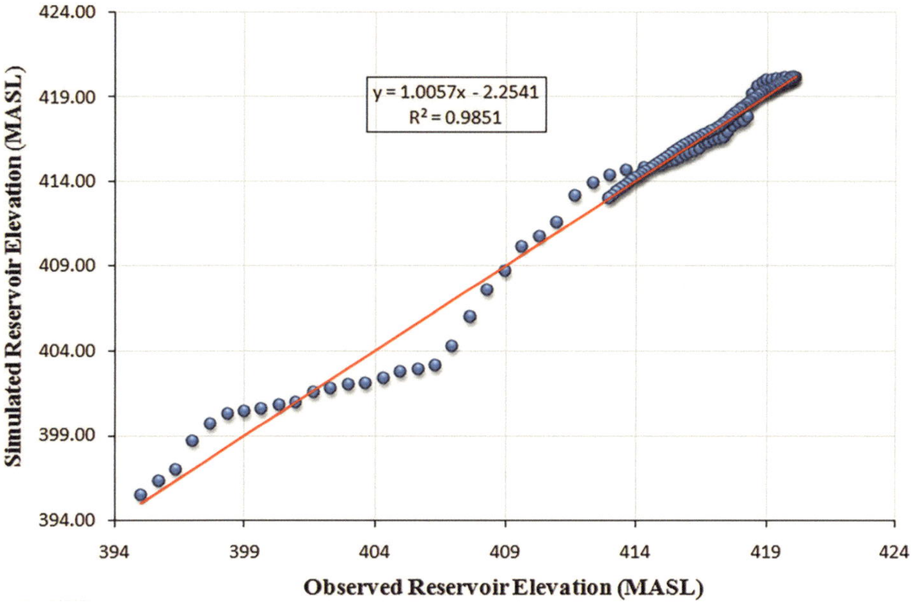

Figure 10.
Regression curve for the observed and simulated reservoir elevations for the simulation period of August to November 2012.

(August–November 2012) plotted in the regression chart to analyze the second efficiency parameter, that is, Coefficient of Determination.

The "Coefficient of Determination" for the plot was evaluated as 0.98, hence we can say that the simulated reservoir elevation values are 98% correct. If we want to employ HEC-ResSim for Pong Dam elevation simulation then it can be a trustful and powerful tool.

4.2 Reservoir storage operation

Reservoir storage is the main point of concern when we have to manage the outflow with the downstream demands like agricultural irrigation, etc. Also for hydropower optimization, the elevation-storage curve is the backbone of the process. If the reservoir is very prone to sedimentation then also the storage volume of the reservoir plays a key role in changing reservoir geometry. So there is a great deal if there is a technique to simulate reservoir storage volume. If one achieves the target storage volume, the further operations of the reservoir become more efficient and safe. Hence if we can predict the future hydrologic condition of the reservoir we can simulate the target reservoir storage in that situation. Also, we can practice some random expected flood values to check the storage operation in that situation. In this way, we can prepare our-self for the worst condition even if that situation had never hit before. **Figure 11** shows the comparative plot for the actual reservoir storage and the simulated reservoir storage. It is showing results for 4 months of simulation and is utilized for the calculation of RMSE for reservoir storage. RMSE was calculated for the reservoir storage values obtained from the simulation duration. RMSE was evaluated as 151.81 MCM for the simulation of reservoir storage. This RMSE value is acceptable and recommends HEC-ResSim for such reservoir modeling efforts.

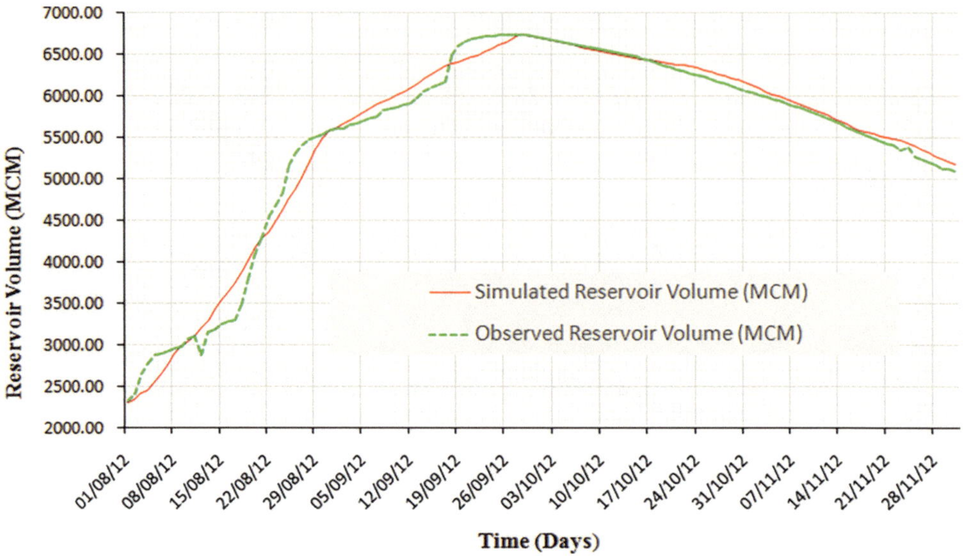

Figure 11.
Comparison of observed reservoir volume and stimulated reservoir volume for the simulation period of August to November 2012.

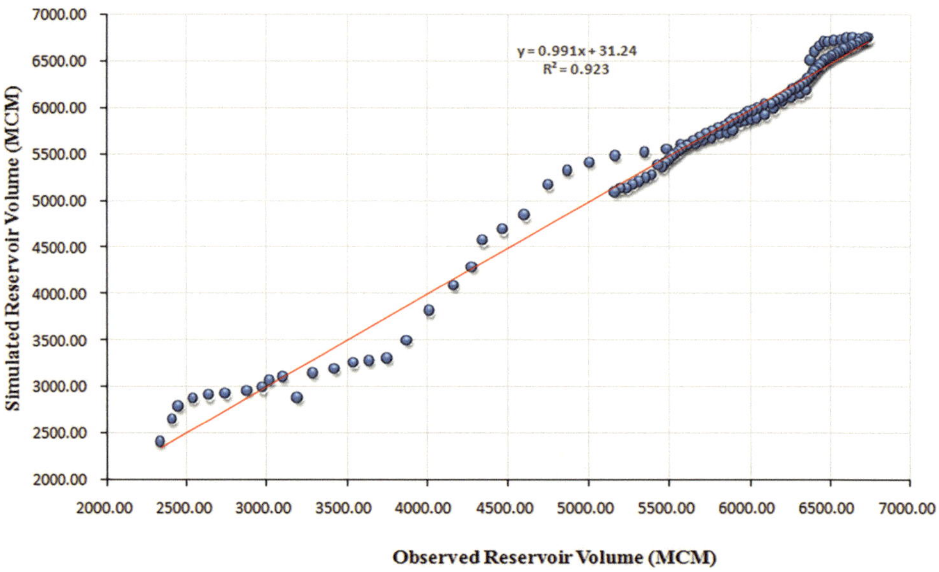

Figure 12.
Regression curve for the observed and simulated reservoir volumes for the simulation period of August to November 2012.

To check the efficiency of the model both the simulated and actual values of simulation are analyzed in the regression curve (**Figure 12**). Daily observed and simulated data related to reservoir storage for the 123 days of the simulation period (August–November 2012) plotted in the regression chart to analyze efficiency parameter, that is, Coefficient of Determination.

The coefficient of regression for the plot was evaluated as 0.92, hence we can say that the simulated reservoir elevation values are 92% correct. If we want to employ HEC-ResSim for Pong Dam storage simulation then it can be a trustful and powerful tool.

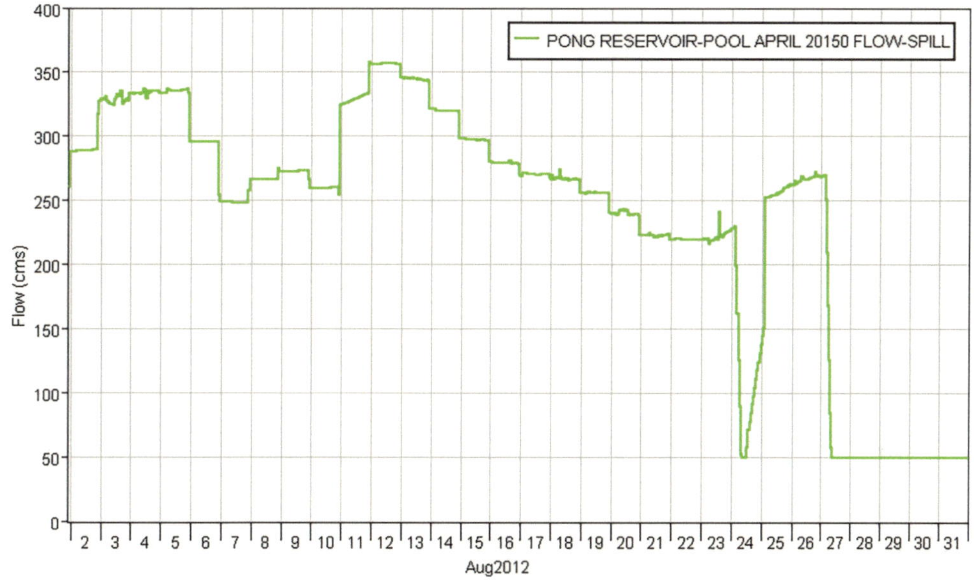

Figure 13.
Simulated and predicted outflow for August 2012.

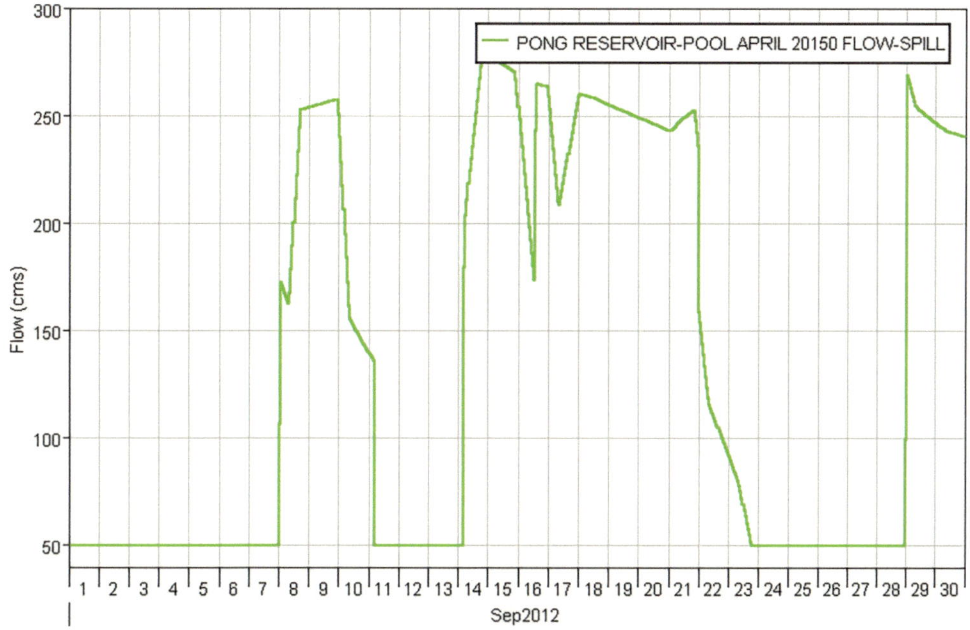

Figure 14.
Simulated and predicted outflow for September 2012.

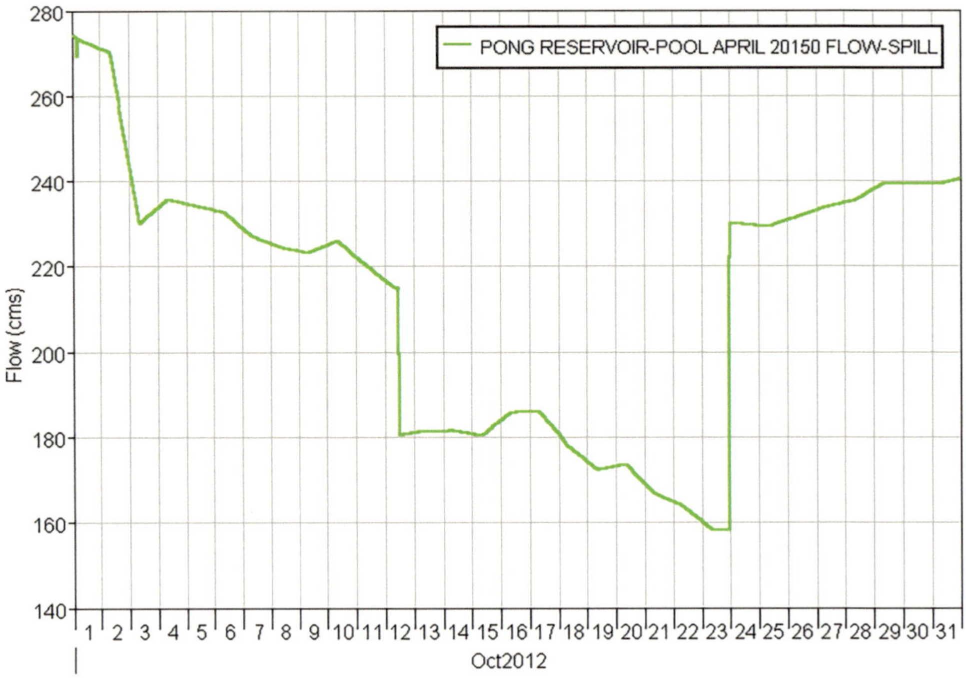

Figure 15.
Simulated and predicted outflow for October 2012.

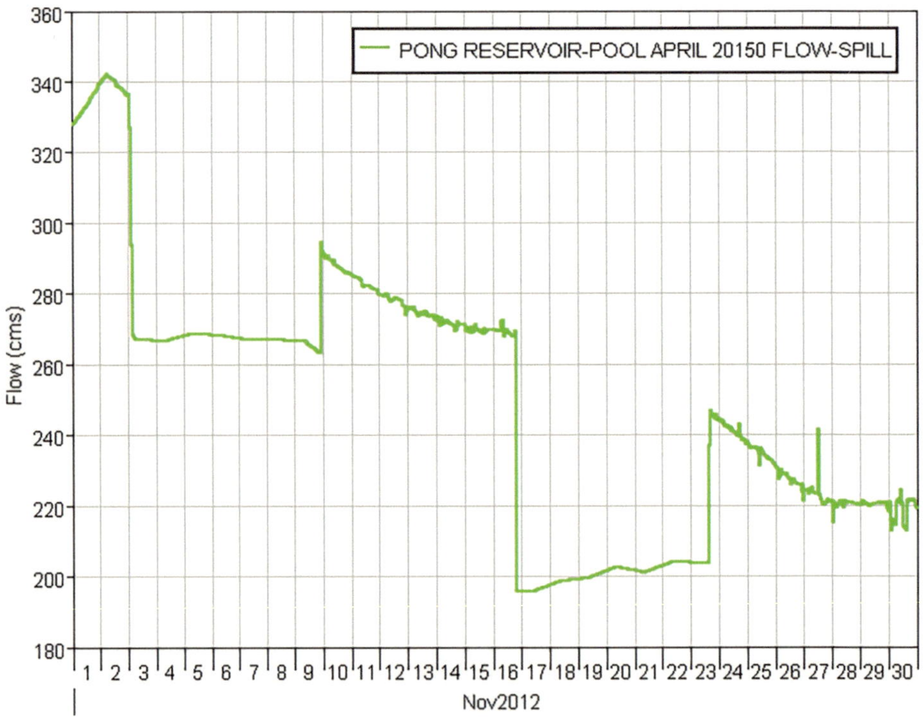

Figure 16.
Simulated and predicted outflow for November 2012.

4.3 Additional simulation results

HEC-ResSim is also capable of predicting unknown flows from the reservoir. In the initial conditions, we have entered irrigation release as an outflow only and kept spill zero. But after simulation, HEC-ResSim found some extra unregulated water in the reservoir which should be thrown out except irrigation release. This aids the management of release decisions even if we do not have any previous record of any such releases. Calculating all the inflow, outflow, elevation, and storage patterns in a synchronized manner and considering rule curves, HEC-ResSim predicted the target water spillage for all 4 months of simulation. **Figures 13-16** show the predicted water spillage from the reservoir for the simulation months of August, September, October, and November, respectively for the year 2012.

5. Conclusion

According to the results presented in this paper, HEC-ResSim is an interesting alternative to reduce the uncertainties of outflow forecasts and support the improvement of the flooding warning program of the Pong Dam. Putting the Pong Dam's hydrological database together with HEC-ResSim, one could reproduce operational aspects of the dam and test different operational scenarios, even in real-time. Since for this model, the availability of data and information falls short but concerning the available data, it was able to prove that Hec-ResSim is a very efficient way to analyze any reservoir for different climatic conditions.

This article attempts to study the performance of the Pong storage dam in actual conditions and simulated conditions using HEC-ResSim and evaluation indices. Results of model validation showed that the model was capable of simulation with suitable accuracy.

HEC-ResSim is a powerful tool, which can support the decision-making of the managers and operators at the Pong Dam. The model presents capabilities to improve the precision of the flooding warnings, reduce dam safety costs, and increase hydropower production. In addition, Hec-ResSim supports a minimum computation time of 15 minutes, which can generate decision support for every 15 minutes of the simulation period. HEC-ResSim is also an interesting alternative for risk management and water control. Moreover, it can further come out to be more precise with the attachment with GIS. It can work a long way with power plants up-to-the turbines level if there is the availability of detailed data regarding it. HEC-ResSim can support the real-time decision, using a real-time integrated database along with a user-friendly interface integrator like HEC-RTS or DELFT-FEWS. HEC-RTS—Real-Time Simulation is another U.S. Army Corps of Engineers tool, which provides support for operational decision-making. HEC-ResSim integrated with HEC-RTS or HEC-HMS allows the water control manager to make short-term (typically a few days or weeks) forecasts of hydrologic conditions at the catchment scale.

List of notations

Hec ResSim	Hydrologic Engineering Center Reservoir Simulation
U.S.	United States
HEC-5	Hydrologic Engineering Center

RMSE	Root Mean Square Error
NFP	Network Flow Programming
WRASIM-M	Warfighter's Simulation-Model
CoMOM	Coordinated Multiple Reservoir Operating Model
DSS	Decision Support System
NSGA-II	Non-dominated Sorting Genetic Algorithm II
COE	Corps of Engineers
NWS	National Weather Service
SSARR	Stream-flow Synthesis and Reservoir Regulation
FCOP	Flood Control Operating Plan
WRAP	Water Rights Analysis Package
USACE	United States Army Corps of Engineers
GUI	Graphical User's Interface
HEC-FIA	Hydrologic Engineering Center-Flood Impact Analysis
HEC-RAS	Hydrologic Engineering Center-River Analysis System
BBMB	Bhakhra Beas Management Board
HEC-DSSVue	HEC-Data Storage System Visual utility engine
WAM	Water Availability Model
HEC-HMS	Hydrologic Engineering Center-Hydrologic Modeling System
MODSIM	River Basin Management Decision Support System
MASL	Meters Above Sea Level
CP	Computational Point

Author details

Abhishish Chandel*, Vijay Shankar and Sumit Jaswal
National Institute of Technology Hamirpur, Hamirpur, India

*Address all correspondence to: abhishishchandel@gmail.com

References

[1] Modini GC. Using HEC-ResSim for Columbia River treaty flood control. In: 2nd Joint Federal Interagency Conference, Las Vegas, NV, June 27-July 1, 2010; 2010

[2] Eichert BS, Davis J. Sizing Flood Control Reservoir System by System Analysis. Hydrologic Engineering Center, US Army Corps Engineers. Technical Paper 44; 1976

[3] Hickey J, Bond M, Patton T, Richardson K, Pugner P. Reservoir simulations of synthetic rain floods for the Sacramento and San Joaquin River Basins. Journal of Water Resources Planning and Management. 2003;**129**(6):443-457

[4] Kim S, Park Y, Kim J. Simulation of the real-time multiple reservoir operating environment with a multiple reservoir operation optimization model. In: Building Partnerships. 2000. pp. 1-10. DOI: 10.1061/40517(2000)148

[5] Matondo J, Msibi K. A decision support system for beneficial use of national dam water resources in Swaziland. In: World Environmental and Water Resources Congress. 2009. pp. 1-14. DOI: 10.1061/41036(342)204

[6] Bekele E, Knapp H. Evolutionary computation for simulating reservoir release rates of Lake Shelbyville and Carlyle Lake. In: World Environmental and Water Resources Congress. 2012. pp. 998-1007. DOI: 10.1061/9780784412312.101

[7] Todd et al. Texas water availability modeling system. Texas Water Journal. 2012;**3**(1):270-279. Available from: texaswaterjournal.org

[8] Zagona EA, Fulp TJ, Shane R, Magee T, Goranflo HM. RiverWare: A generalized tool for complex reservoir system modeling. Journal of the American Water Resources Association. 2001;**37**(4):913-929

[9] Zagona EA, Magee T, Goranflo HM, Fulp T, Frevert DK, Cotter JL. RiverWare. In: Singh VP, Frevert DK, editors. Watershed Models. Boca Raton, FL: CRC Press, Taylor & Francis; 2006. pp. 527-548

[10] Zagona E, Kandl E, Carron J, Bowser S. Water accounting and allocation in RiverWare. In: Proceedings of 2nd Joint Federal Interagency Conference on Sedimentation and Hydrologic Modeling; June 27–July 1 2010; Las Vegas, NV. Boulder, CO: The Center for Advanced Decision Support for Water and Environmental Systems, University of Colorado at Boulder; 2010. Available from: http://cadswes.colorado.edu/PDF/RiverWare/10F_Zagona_03_01_10.pdf

[11] Labadie JW, Larson R. MODSIM 8.1: River Basin Management Decision Support System, User Manual and Documentation. Fort Collins, CO: Colorado State University; 2007. p. 123

[12] Labadie JW. MODSIM: River basin decision support system. In: Singh VP, Frevert DK, editors. Watershed Models. Boca Raton, FL: CRC Press, Taylor and Francis; 2006. pp. 569-592

[13] Wurbs RA. Texas water availability modeling system. Journal of Water Resources Planning and Management. 2005;**131**(4):270-279

[14] HEC. HECDSS User's Guide and Utility Program Manuals. Davis, CA: U.S. Army Corps of Engineers; 1995

[15] HEC. HEC-DSSVue HEC Data Storage System Visual Utility Engine,

User's Manual. Davis, CA: U.S. Army Corps of Engineers; 2009

[16] Hanbali F. HEC-ResSim Reservoir Model for Tigris and Euphrates River Basins in Iraq. Advance in Hydrologic Engineering Center, US Army Corps; 2004

[17] Babazadeh H, Sedghi H, Kaveh F, Jahromi M. Performance evaluation of Jiroft storage dam operation using HEC-ResSim 2.0. In: Eleventh International Water Technology Conference, IWTC11 2007 Sharm El-Sheikh, Egypt; 2007

[18] McKinney DC. HEC-ResSim Model of Model of Lancang Cascade Dams. 2011. Available from: http://www.caee.utexas.edu/prof/mckinney/

[19] Piman T, Cochrane TA, Arias ME, Green A, Dat ND. Assessment of flow changes from hydropower development and operations in Sekong, Sesan, and Srepok Rivers of Mekong Basin. Journal of Water Resources Planning and Management. 2013;**139**(6):723-732

[20] Goodarzi E, Ziaei M, Hosseinipour EZ. Reservoir optimization and simulation modelling: A case study. Topics in Safety, Risk, Reliability and Quality. 2014;**25**:195-215

[21] Lara PG, Bonuma NB, Lopes JD, Luz GM. Reservoir operation employing Hec-ResSim: Case study of Tucurui dam, Brazil. In: 6th International Conference on Flood Management, Sau Paulo, Brazil, September 2014; 2014

[22] Klipsch JD, Hurst MB. HEC-ResSim, Reservoir System Simulation User's Manual, Version 3.1. Davis, CA: U.S. Army Corps of Engineers, Institute of Water Resources, Hydrologic Engineering Center; 2013

Chapter 2

Evaluation SSG-LRR Model on an Homogeneous Turbulence under Inclined Shear for High and Low Stratification: Shear Number Effect

Lamia Thamri and Taoufik Naffouti

Abstract

This chapter develops proposals for an evaluation coupled second order model of SSG-LRR on an homogeneous turbulence submitted to an inclined shear for high and low stratification. The effect of Shear number on thermal and dynamic turbulent fields of the problem is performed for Shear number fixed at 2, 6, 14 and 20. Two values of Froude number equal to 0.35 and 1.29 are adopted for all numerical simulations corresponding to high and low stratification, respectively. For all simulations, value of angle theta is fixed at $\theta = \pi/4$ corresponding to the angle between the shear and the vertical gradient of stratification. SSG-LRR model is adopted to compute turbulent parameters of principal component of anisotropy b_{12}, normalized turbulence dissipation ε/KS and the density flux $\overline{\rho u_1}$. A good agreement is detected by comparison of findings via model of SSG-LRR with the reported results in the literature by Direct Numerical Simulation of Jacobitz (DNSJ). It is found that the variation of Shear number predict a very strong influence on thermal and dynamic turbulent characteristics. Hence, findings with SSG-LRR model prove the existence of an asymptotic equilibrium states for various thermal and dynamic parameters in particularly for a low stratification.

Keywords: coupled models, homogeneous and stratified turbulence, inclined shear, shear number, DNSJ

1. Introduction

Homogeneous sheared and stably stratified turbulence is considered as a fundamental flow relevant to the study of geophysical turbulence and, generally, anisotropic turbulence. Homogeneous and stratified turbulence has received considerable attention in the past few decades. This interest stems from its importance in many industrial and engineering applications. Such applications include geophysical flows, oceanic and atmospheric mixed layers, turbulent thermal plumes, atmospheric circulation, pipes, gas turbines, airplanes, etc. Complex dynamical turbulent processes associated to the stratified turbulence under shear are observed in atmosphere and in industry. This kind of turbulence problem presents an attractive physical field as it plays a key role to improve our understanding of behaviors of geophysical flows [1–4]. Dynamical turbulent processes associated to the stratified turbulence submitted to an inclined shear are encountered in industry and atmosphere. This problem of the turbulence is relevant in many thermal engineering applications in both nature and industry owing to it playing a key role to better describe turbulent characteristics of geophysical flows. Furthermore, the understanding of behaviors of homogeneous and stratified turbulence related to geophysical flows presents a considerable role to develop turbulence theories and performed a comparison of results via different models of LES, SSG (Speziale, Sarkar, and Gatski), k-ε, SL, and DNS [5–9]. So, the exam of the turbulence leads to engineering applications with benefits for society such as the development of high-resolution climate and space weather models.

The available review illustrates numerous investigations on the characterization of thermal and dynamic fields of the turbulence flow using different methods [10–13]. Komori et al. [14] approve that an experimental analysis on a stratified turbulent via vertical shear flow. Experimental studies are performed for a horizontal shear by various authors [15, 16]. Comprehensive studies of homogeneous and stratified turbulence submitted to inclined shear include the experimentally work by Rohr et al. [17] and Piccirillo et al. [18]; they confirmed the importance of the Froude number and reported an influence of additional dimensional parameter such as shear number on evolution of turbulence. Development of LRR (Launder, Reece, and Rodi) model of turbulence in which the Reynolds stresses is studied by Launder et al. [19, 20], they conclude that the numerical solutions of the model equations are presented for a selection of strained homogeneous shear flows and for two-dimensional inhomogeneous shear flows. After that, Jacobitz et al. [21] performed the direct numerical simulations (DNS) to investigate the evolution of turbulence in a uniformly sheared and stably stratified flow. They conclude that the shear number has a strong non-monotone influence on the turbulence evolution. Jacobitz el al. [5] used the direct numerical simulations in order to investigate the evolution of turbulence in a stably stratified fluid forced by nonvertical shear. The flux Richardson number Ri_f depends on the gradient Richardson number Ri but not on the shear inclination angle θ. It is interesting that the normalized turbulence production is strongly influenced by the angle θ, but that the flux Richardson number remains unaffected.

The anisotropy of fluctuating motion in a stably stratified medium with uniform mean shear is studied by Sarkar [22] using the numerical simulation of uniform shear flow. He concludes that the vertical of streamwise velocity is found to dominate the components of turbulent dissipation in both horizontal and vertical shear flows. After that, Bouzaiane et al. [23] studied the evolution of homogeneous stably stratified turbulence submitted to a nonvertical shear using second-order closure models (Craft and Launder CL and Shih and Lumley SL). They improve that a good agreement between the predictions of second-order models and values of DNSJ is generally observed for the principal component of anisotropy b_{12} and a qualitative agreement is observed for the ratios K/E and K_ρ/E of the kinetic and potential energies to the total energy E. Thamri et al. [24] applied three coupled second-order models (of SSG-CL, SSG-SL, and SSG-LRR) in order to simulate the influence of nonvertical shear on homogeneous turbulence and for a low stratification Ri = 0.2. They showed a good accord between results of three second-order models and those of DNSJ.

Basing on the literature, it is found that no combination between Speziale, Sarkar, and Gatski (SSG) model [25] and Launder, Reece, Rodi (LRR) model [19, 20] is improved. The present work is the first systematic investigation related to the shear number effects on thermal and dynamics parameters of a homogeneous turbulence submitted to an inclined shear using second-order coupled models of SSG-LRR, with low and high stratification. For the case of high stratification of the turbulence, the Richardson number is equal to Ri = 2 associated to Froude number equal to 0.35. For low stratification, the Richardson number is fixed at Ri = 0.15 corresponding to Froude number equal to 1.29. In addition, a comparison between our work (SSG-LRR model) and the results of DNSJ [21, 26, 27] is carried out. The paper is organized as follows: The problem is defined in Section 2, which also presents the governing equations and parameters. The methodology is presented in Section 3.1, which describes the numerical method, model setup, boundary conditions, and validation. Results and discussion follow in Section 3.2, and finally, conclusions are drawn in Section 4.

2. Governing equations and numerical approach

All numerical computations are based on the continuity equation of an incompressible fluid, three-dimensional unsteady Navier–Stokes equation with the Boussinesq approximation and the transport equation for the density. The mean velocity $\overline{U} = \left(\overline{U_1}, 0, 0\right)$ has a constant inclined shear rate $S = S.\sin(\theta)$ where $\theta = \pi/4$ is the angle between the shear and the vertical gradient of stratification. The mean density has a constant vertical stratification gradient $S_\rho = \frac{\partial \overline{\rho}}{\partial x_3}$. Using the decomposition of Reynolds, the dependent variables U_i (i-th component of the total velocity), ρ (total density), and P (total pressure) are decomposed into a mean part (denoted by an overbar) and a fluctuating part (denoted by small letters):

$$U_i = \overline{U_i} + u_i, \rho = \overline{\rho} + \rho \text{and} P = \overline{P} + p \tag{1}$$

The decomposition of dependent variables is introduced into equations of motion and the following evolution equations for the fluctuating parts are obtained:

$$\frac{\partial u_i}{\partial x_i} = 0 \tag{2}$$

$$\begin{aligned} &\frac{\partial u_i}{\partial t} + u_k \frac{\partial u_i}{\partial x_k} + \left(S\frac{\sqrt{2}}{2}x_2 + S\frac{\sqrt{2}}{2}x_3\right)\frac{\partial u_i}{\partial x_1} + \left(S\frac{\sqrt{2}}{2}u_2 + S\frac{\sqrt{2}}{2}u_3\right)\delta_{i1} = \\ &\quad -\frac{1}{\rho_0}\frac{\partial p}{\partial x_i} + \nu\frac{\partial^2 u_i}{\partial x_k \partial x_k} - \frac{g}{\rho_0}\rho\delta_{i3} \end{aligned} \tag{3}$$

$$\frac{\partial \rho}{\partial t} + u_k\frac{\partial \rho}{\partial x_k} + \left(S\frac{\sqrt{2}}{2}x_2 + S\frac{\sqrt{2}}{2}x_3\right)\frac{\partial \rho}{\partial x_1} + S_\rho u_3 = \alpha\frac{\partial^2 \rho}{\partial x_k \partial x_k} \tag{4}$$

g is the acceleration due to gravity, ν the kinematic viscosity, α the scalar diffusivity, δ_{ij} is the Kronecker symbol, ρ is the fluctuation of the scalar, and ρ_0 is the density of reference.

$S_\rho = \frac{\partial \overline{\rho}}{\partial x_3}$ is the stable density stratification, and S is the shear rate.

The dimensionless shear number SK/ε is the ratio of the turbulence time scale K/ε to the shear time scale 1/S. Dimensionless Richardson number is the ratio of the turbulence stratification by the turbulence shear. Richardson number is defined as follows:

$$\mathrm{R}i = \frac{N^2}{S^2}$$

The Froude number, which depends on Richardson number, is given as:

$$Fr = \sqrt{\frac{0.5^2}{Ri}}$$

where S is the uniform mean shear, $N^2 = -g(\partial\rho/\partial z)/\rho_0$ is the Brunt–Väisala frequency, g is the acceleration due to gravity, ρ is the fluctuation of the scalar, ρ_0 is the density of reference, and z is the vertical coordinate.

The non-dimensional time related to temporal evolution of physical turbulent parameters, which characterize the considered problem, is defined as: τ = S.t.

The adopted calculation algorithm related to the present problem consists the following steps:

1. Initialization of thermal and dynamic fields of the flow.

2. Declaration of initial conditions of isotropy of findings DNSJ [21, 26, 27]:

$$\left(b_{ij}\right)_0 = 0,0, \quad \left(\frac{\varepsilon}{KS}\right)_0 = 0,5, \quad (F_i)_0 = 0,0, \quad (\eta)_0 = 0,0$$

3. Integration of dimensionless equations of SSG-LRR model.

4. A fourth-order Runge–Kutta method is adopted to resolve various equations of the problem.

5. Calculation of thermal and dynamic turbulent parameters (b_{12}, $\overline{\rho u_1}$, ε/KS).

The set of differential equations governing the transport of the velocity correlation$\overline{u_i u_j}$, transport equation for the density flux $\overline{u_i\rho}$, transport equation for the turbulent kinetic energy $K = \frac{\overline{u_i u_i}}{2}$, and transport equation for the density fluctuation $\overline{\rho^2}$ may be written in the form:

$$\begin{aligned}\frac{d\overline{u_i u_j}}{dt} = &-S\frac{\sqrt{2}}{2}\overline{u_j u_2}\delta_{i1} - S\frac{\sqrt{2}}{2}\overline{u_j u_3}\delta_{i1} - S\frac{\sqrt{2}}{2}\overline{u_i u_2}\delta_{j1} - S\frac{\sqrt{2}}{2}\overline{u_i u_3}\delta_{j1} - \\ &\frac{g}{\rho_0}\left(\overline{u_i\rho}\delta_{j3} + \overline{u_j\rho}\delta_{i3}\right) + \frac{1}{\rho_0}\overline{p\left(\frac{\partial u_i}{\partial x_j} + \frac{\partial u_j}{\partial x_i}\right)} - 2\nu\overline{\frac{\partial u_i}{\partial x_k}\frac{\partial u_j}{\partial x_k}}\end{aligned} \quad (5)$$

$$\begin{aligned}\frac{d\overline{u_i\rho}}{dt} = &-S\frac{\sqrt{2}}{2}\overline{u_2\rho}\delta_{i1} - S\frac{\sqrt{2}}{2}\overline{u_3\rho}\delta_{i1} - S_\rho\overline{u_i u_3} - \frac{g}{\rho_0}\overline{\rho^2}\delta_{i3} + \frac{1}{\rho_0}\overline{p\frac{\partial\rho}{\partial x_i}} - \\ &(\alpha+\nu)\overline{\frac{\partial\rho}{\partial x_k}\frac{\partial u_i}{\partial x_k}}\end{aligned} \quad (6)$$

$$\frac{dK}{dt} = -S\frac{\sqrt{2}}{2}\overline{u_1u_2} - S\frac{\sqrt{2}}{2}\overline{u_1u_3} - \frac{g}{\rho_0}\overline{u_3\rho} - \nu\overline{\frac{\partial u_i}{\partial x_k}\frac{\partial ui}{\partial x_k}} \tag{7}$$

$$\frac{d\overline{\rho^2}}{dt} = -2S_\rho\overline{\rho u_3} + 2\nu\overline{\frac{\partial \rho}{\partial x_k}\frac{\partial \rho}{\partial x_k}} \tag{8}$$

g is the acceleration due to gravity, ν the kinematic viscosity, and α the scalar diffusivity.

$S_\rho = \frac{\partial \overline{\rho}}{\partial x_3}$ is the stable density stratification, S is the shear rate, and δ_{ij} is the Kronecker symbol.ρ is the fluctuation of the scalar, and ρ_0 is the density of reference. where component denoted by an over bar and component denoted by small letters denote mean and fluctuating components of velocity, p is the fluctuation of static pressure about its mean value, and x_i (i = 1, 2, 3) denote Cartesian space coordinates.

$$U_i = \overline{U_i} + u_i, , P = \overline{P} + p\rho = \overline{\rho} + \rho$$

U_i the total values of velocity components, P the total pressure, and ρ the total density.

Nearly every worker obtained a closed system of differential equations. The turbulence modeling remains an important approach retained by several authors [28]. The coupled between second-order modeling (SSG-LRR) is retained here and consists of modeling the pressure-strain $\varphi_{ij}$and pressure–temperature gradient $\varphi_{i\rho}$ correlations.

$$\varphi_{ij} = \frac{1}{\rho_0}\overline{p\left(\frac{\partial u_i}{\partial x_j} + \frac{\partial u_j}{\partial x_i}\right)} = \varphi_{ij}^1 + \varphi_{ij}^2 + \varphi_{ij}^3 \text{ and } \varphi_{i\rho} = \frac{1}{\rho_0}\overline{p\frac{\partial \rho}{\partial x_i}} = \varphi_{i\rho}^1 + \varphi_{i\rho}^2 + \varphi_{i\rho}^3$$

The SSG model [25] has concerned only kinematic turbulence, and it is retained here for these terms, which are written in the following forms:

$$\varphi_{ij} = \varphi_{ij}^1 + \varphi_{ij}^2 \tag{9}$$

$$\varphi_{ij}{}^1 = -C_1 b_{ij} + 3(C_1 - 2)\left(b_{ij}^2 + II_b\frac{\delta_{ij}}{3}\right) \tag{10}$$

$$II_b = b_{kl}b_{lk}, C_1 = 3, 4$$

$$\varphi_{ij}{}^2 = C_2 b_{mn}S_{mn}b_{ij} + \frac{1}{2}\left(C_3 - \sqrt{b_{mn}b_{mn}}C_{33}\right)S_{ij} + \frac{1}{2}C_4\left(b_{ik}S_{kj} + b_{jk}S_{ki} - \frac{2}{3}b_{mn}S_{mn}\delta_{ij}\right) + \frac{1}{2}C_5\left(b_{ik}w_{jk} + b_{jk}w_{ik}\right)$$

$$C_2 = 1, 8 \quad C_3 = 0, 8 \quad C_4 = 1, 25 \quad C_5 = 0, 4 \quad C_{33} = 1, 3$$

The LRR model [19, 20] is retained for scalars field:

$$\varphi_{i\rho}^1 = -C_1'\frac{\varepsilon}{k}\overline{\rho u_i} \tag{11}$$

$$C_1' = 3.2$$

$$\varphi_{i\rho}^2 = 0, 8\overline{\rho u_k}U_{i,k} - 0, 2\overline{\rho u_k}U_{k,i} \tag{12}$$

The contributions 3, φ_{ij}^3, and $\varphi_{i\rho}^3$ are terms due to buoyancy. Lumley et al. [29] and Zeman et al. [30] retained only the classical model for this contribution.

$$\varphi_{ij}^3 = -C_3\left(\beta_j\overline{u_i\rho} + \beta_i\overline{u_j\rho} - \frac{2}{3}\beta_l\overline{u_l\rho}\,\delta_{ij}\right) \tag{13}$$

$$\varphi_{i\rho}^3 = -C_{3\rho}\beta_i\overline{\rho^2} \tag{14}$$

$$C_3 = 0,5, C_{3\rho} = 0,5$$

With the reason of acquisition, the dimensionless equations, basic Eqs. (5)–(8) are deposited in dimensionless form by laying on the dimensionless time St = τ, the kinematic components $b_{ij} = \frac{\overline{u_iu_j}}{2K} - \frac{1}{3}\delta_{ij}$, ε/KS and the scalar component $\eta = \frac{1}{2}\frac{g}{\rho_0 S_\rho}\frac{\overline{\rho^2}}{k} = \frac{k_\rho}{k}$

$$\frac{db_{12}}{d\tau} = -\frac{\sqrt{2}}{2}\left(b_{22}+\frac{1}{3}\right) - \frac{\sqrt{2}}{2}b_{23} + \frac{\varphi_{12}}{2kS} + 2\frac{\sqrt{2}}{2}b_{12}^2 + 2\frac{\sqrt{2}}{2}b_{12}b_{13} + F_3b_{12} + \left(\frac{\varepsilon}{kS}\right)b_{12} \tag{15}$$

$$\frac{db_{22}}{d\tau} = \frac{\varphi_{22}}{2kS} - \frac{\varepsilon}{3kS} + \left(b_{22}+\frac{1}{3}\right)\left(2\frac{\sqrt{2}}{2}b_{12} + 2\frac{\sqrt{2}}{2}b_{13} + F_3 + \frac{\varepsilon}{kS}\right) \tag{16}$$

$$\frac{d\eta}{d\tau} = -F_3 + \eta\left(2\frac{\sqrt{2}}{2}b_{12} + 2\frac{\sqrt{2}}{2}b_{13} + F_3 - \frac{\varepsilon}{kS}\right) \tag{17}$$

$$\begin{aligned}\frac{d}{d\tau}\left(\frac{\varepsilon}{kS}\right) = {}& -2C_{\varepsilon1}\left(\frac{\varepsilon}{kS}\right)b_{13} + (1-C_{\varepsilon2})\left(\frac{\varepsilon}{kS}\right)^2 - C_{\varepsilon1}(1-C_{\varepsilon3})\left(\frac{\varepsilon}{kS}\right)F_3 \\ & +\left(\frac{\varepsilon}{kS}\right)\left(2\frac{\sqrt{2}}{2}b_{12} + 2\frac{\sqrt{2}}{2}b_{13} + F_3\right)\end{aligned} \tag{18}$$

where $F_i = \frac{g}{\rho_0}\frac{\overline{u_i\rho}}{kS}$ is the non-dimensional component of the turbulent scalar flow.

A numerical integration of the differential equations is done. Obtained results are discussed in the following sections.

3. Numerical integration and results

3.1 Validation of the present code based on coupled model of SSG-LRR

In order to make sure on the precision of coupled model of SSG-LRR employed for the computations of the considered problem, the present code taped with Fortran is validated with the published findings of Jacobitz et al. [21, 26, 27] related to direct numerical simulations of the stratified turbulence under shear. However, a comparative analysis of predictions via SSG-LRR model and results of DNSJ is performed (**Figures 1–3**).

A fourth-order Runge–Kutta method is used for integrating a closed system of nonlinear dimensionless differential equations. The initial conditions are taken from the same initial conditions of result of DNSJ [21, 26, 27].

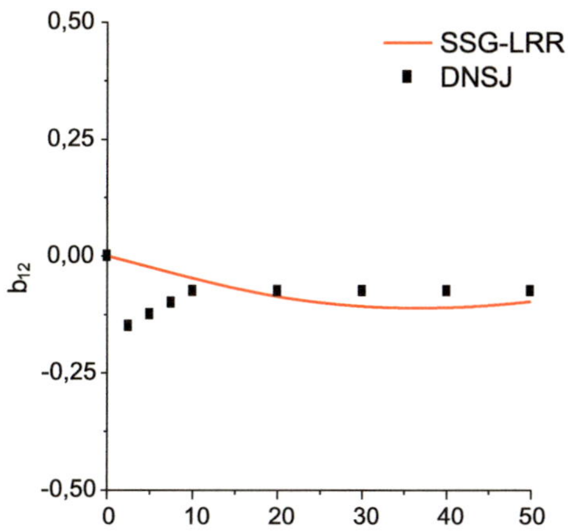

Figure 1.
Time evolution of the principal component of anisotropy b_{12} for low stratification with Fr = 1.11, $\theta = \pi/4$, SK/ε = 2.

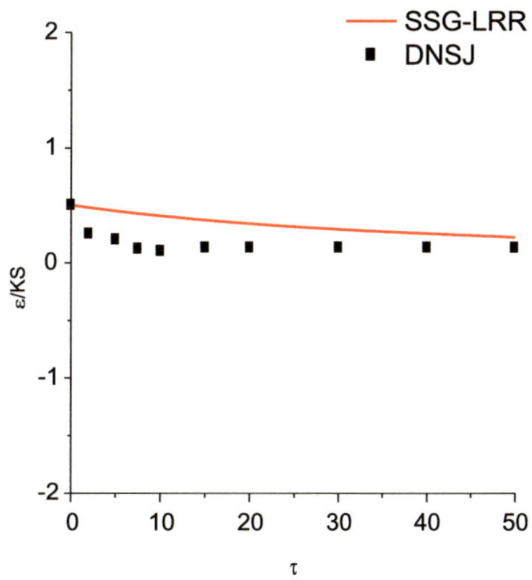

Figure 2.
Time evolution of the normalized turbulence dissipation ε/KS for low stratification with Fr = 1.11, $\theta = \pi/4$, SK/ε = 2.

$$\left(b_{ij}\right)_0 = 0,0, \quad \left(\frac{\varepsilon}{KS}\right)_0 = 0,5, \quad (F_i)_0 = 0,0, \quad (\eta)_0 = 0,0$$

An inclined θ = π/4 is studied for different values of Shear number SK/ε ranging from 2 to 20 and with a constant Froude number Fr equal to 0.35 for high stratification and equal to 1.11 for low stratification. Here θ is the inclination angle between vertical stratification and shear.

A primary focus of research has been the parameterization of the Froude number Fr associated with stationary, where Fr is defined by: $Fr = \sqrt{\frac{0.5^2}{Ri}}$.

The gradient Richardson number Ri is defined by: $R_i = \frac{g}{\rho_0}\frac{S_\rho}{S^2} = \frac{N^2}{S^2}$.

N^2 is the frequency scale relevant to density stratification, and S is the mean vertical shear.

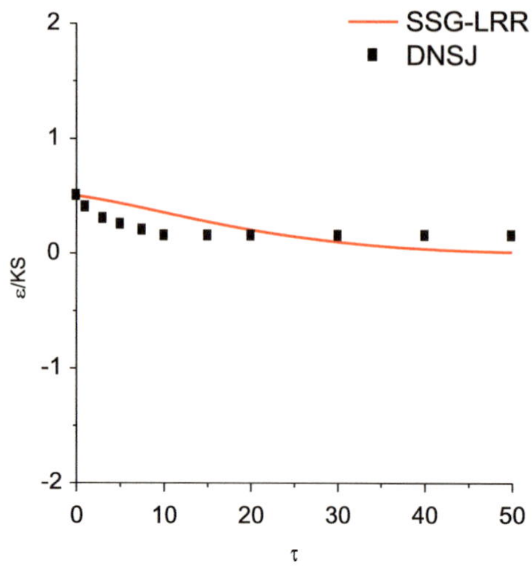

Figure 3.
Time evolution of the normalized turbulence dissipation ε/KS for high stratification with Fr = 0.35, θ = π/4, SK/ε = 2.

The Shear number SK/ε is the ratio of the turbulence time scale K/ε to the shear time scale 1/S.

However, to our knowledge, no previous work is interested in the prediction of equilibrium states by the coupled between SSG and LRR second-order models. This creates the aim of this part of our work. The numerical study is to carried out a comparison between results related to the case of an homogeneous and a stratified (Fr = 0.35 for high stratification and Fr = 1.11 for low stratification) turbulence under inclined shear θ = π/4 for various Shear number ($2 \leq SK/\varepsilon \leq 20$) via SSG-LRR model and findings of those of direct numerical simulation of Jacobitz (DNSJ) [21, 26, 27]. Obtained results via SSG-LRR model and DNSJ [21, 26, 27] are presented in **Figures 1–3**. **Figure 1** shows time evolution of principal component of anisotropy b_{12} according to the SSG-LRR model for a low stratification related to Fr = 1.11 for an intermediate band of τ, in the range [0, 50]. The SSG-LRR model shows a good agreement with the asymptotic value of the component of anisotropy $(b_{12})_\infty$ of DNSJ [21, 26, 27] when a non-dimensional time τ is greater than 20. Using the same model, results of the problem of comparative analysis of coupled second-order models on shear and Richardson numbers effects on homogeneous and stratified turbulence [31] are compared to find numerical approaches with RANS [6]. It has shown a good agreement between different predictions of components of anisotropy b_{12}. Consequently, it can be asserted that these favorable comparisons corroborate the coupled model of SSG-LRR, which can produce reliable numerical findings of turbulent flows.

Figures 2 and **3** show time evolution of the normalized turbulence dissipation ε/KS according to SSG-LRR model compared with values of DNSJ for initial value of shear number SK/ε = 2 and for a constant Froude number equal to 1.11 (low stratification) and Fr = 0.35 (high stratification), respectively. The normalized dissipation ε/KS is affected only slightly by the direction of shear. For high and low stratification, the normalized dissipation ε/KS has a tendency to predict the asymptotic state at long-time evolution for non-dimensional time $\tau \geq 20$. For the case of inclined shear (θ = π/4), an excellent agreement between predictions of the SSG-LRR model and values of DNSJ [21, 26, 27] is detected for higher time ($\tau \geq 30$).

3.2 Effect of shear number

In this section, the influence of shear number (SK/ε) on the evolution of homogeneous and stratified turbulence is addressed. Two cases of stratified turbulence have been performed; a low stratification Fr = 1.29 and a high stratification Fr = 0.35.

3.2.1 Case of low stratification (Fr = 1.29)

Numerical study is presented for various values of shear number SK/ε equal to 2, 6, 14, and 20 in order to simulate the problem of homogeneous and stratified turbulence (Fr = 1.29). A numerical code is taped with Fortran language using a fourth-order Range-Kutta method. All simulations of the present study are approved for initial conditions of Jacobitz [21, 26, 27].

$$\left(b_{ij}\right)_0 = 0,0, \quad \left(\frac{\varepsilon}{KS}\right)_0 = 0,5, \quad (F_i)_0 = 0,0, \quad (\eta)_0 = 0,0$$

In **Figure 4**, the evolution of the principal component of anisotropy b_{12} is shown for a non-dimensional time $0 \leq \tau \leq 40$ with different values of shear number (SK/ε = 2 to 20) for a fixed Froude number at 1.29. The simulations of b_{12} as initial spectra show a large initial decay of b_{12} during the transient phase, and finally, the coupled model SSG-LRR confirms the existence of an asymptotic equilibrium states beyond a non-dimensional time τ = 30 for various values of shear number.

The evolution of the normalized turbulence dissipation ε/KS is shown in **Figure 5**. It appears that the magnitude of ε/KS decreases with increasing shear number (from 2 to 20). Again, an asymptotically constant value of ε/KS is observed. SSG-LRR model confirms the existence of an asymptotic equilibrium states for a long time τ. The time evolution of the thermal parameter (density flux) $\overline{\rho u_1}$as a function of the shear number is illustrated in **Figure 6**. It is found that the $\overline{\rho u_1}$ is affected by the variation of shear number. The SSG-LRR model confirms the existence of asymptotic equilibrium

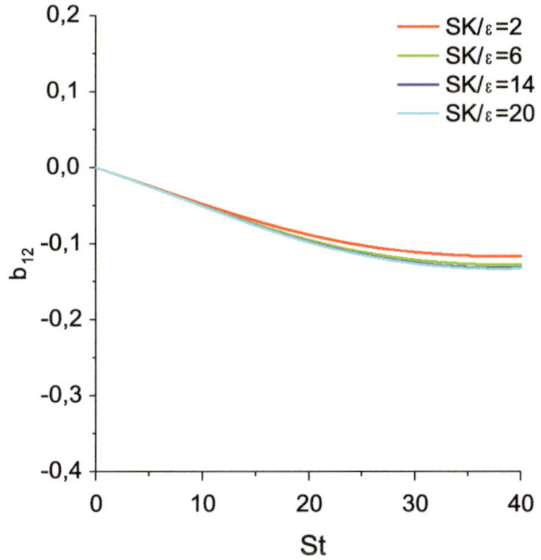

Figure 4.
Time evolution of the principal component of anisotropy b_{12} for several values of SK/ε with $Fr = 1.29$ and $\theta = \pi/4$.

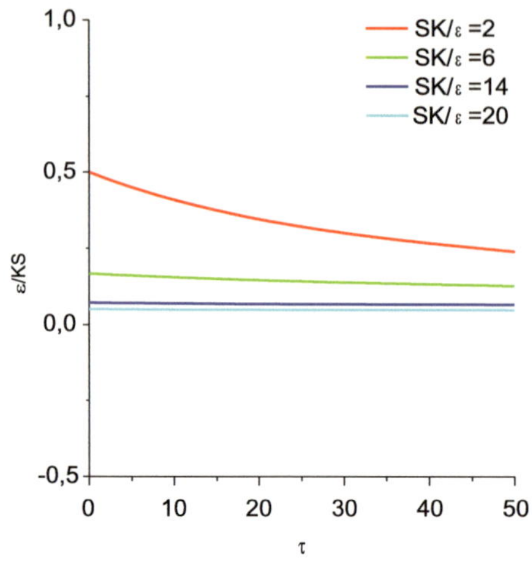

Figure 5.
Time evolution of the normalized turbulence dissipation ε/KS for several values of SK/ε with Fr = 1.29 and θ = π/4.

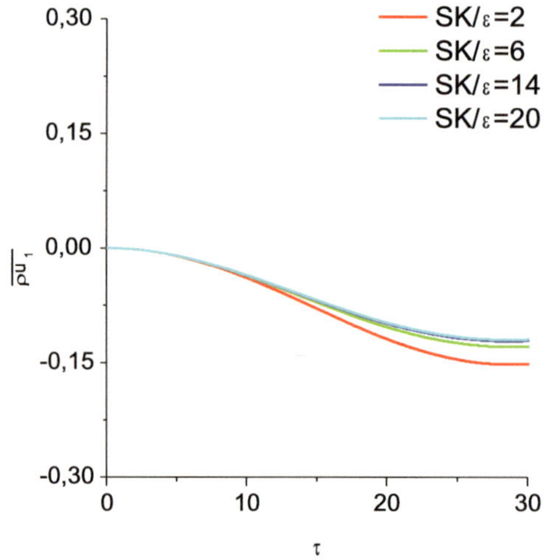

Figure 6.
Time evolution of the density flux $\overline{\rho u_1}$ for several values of SK/ε with Fr = 1.29 and θ = π/4.

states for the component $\overline{\rho u_1}$ beyond non-dimensional time τ = 25. Furthermore, the figure shows that the parameter $(\overline{\rho u_1})_\infty$ increases as increasing $_{SK/\varepsilon}$ from 2 to 20.

3.2.2 Case of high stratification (Fr = 0.35)

In this section, the results of the principal component of anisotropy b_{12}, the normalized turbulence dissipation ε/KS, and the density flux $\overline{\rho u_1}$with different Shear number SK/ε equal to 2, 6, 14, and 20 and for a constant Froude number equal to 0.35 (high stratification) are presented in **Figures** 7, 8, and 9, respectively. All simulation demonstrates that the b_{12} and ε/KS show an initial decay due to the isotropic initial conditions. After this initial phase, the principal component of anisotropy and the

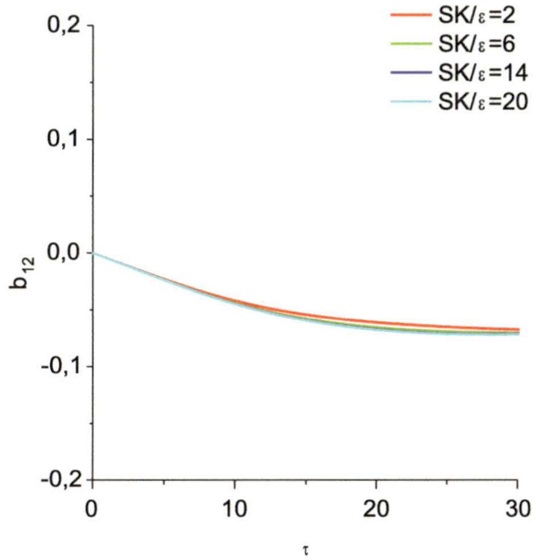

Figure 7.
Time evolution of the principal component of anisotropy b_{12} for several values of SK/ε with Fr = 0.35 and θ = π/4.

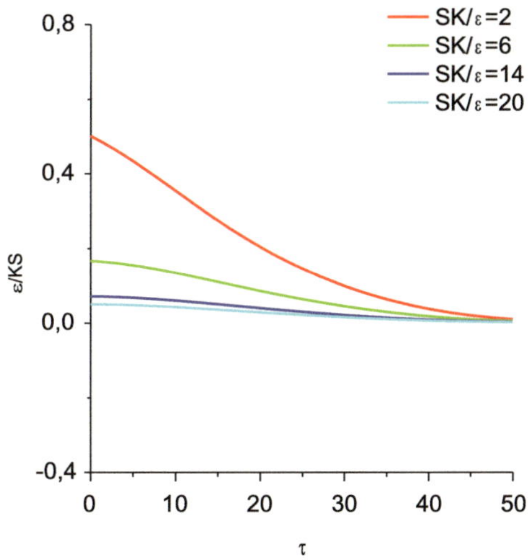

Figure 8.
Time evolution of the normalized turbulence dissipation ε/KS for several values of SK/ε with Fr = 0.35 and θ = π/4.

normalized turbulence dissipation grow for small and decay for large values of the shear number. Contrary, the density flux decays for small and grows for large values of the shear number. **Figures 7** and **8** depict the evolution of the principal component of anisotropy b_{12} and the normalized turbulence dissipation ε/KS, respectively. Obtained by SSG-LRR as a function of the normalized time τ for inclined shear associated to θ = π/4. SSG-LRR confirms the existence of an asymptotic equilibrium states for b_{12} and for ε/KS beyond a non-dimensional time τ = 25 for diverse values of shear number. Also, SSG-LRR indicates that asymptotic equilibrium state $(b_{12})_\infty$ and $(\varepsilon/KS)_\infty$ increase as decreasing shear number from 20 to 2. In **Figure 9** is shown evolution of the density flux $\overline{u_1\rho}$ versus the shear number using SSG-LRR model with Fr = 0.35. For SSG-LRR model, the density flux $\overline{u_1\rho}$ presents a slight decrease with the decrease of shear number from 20 to 2.

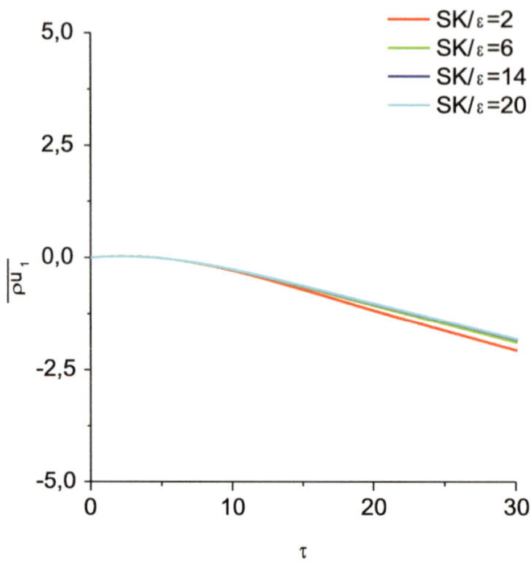

Figure 9.
Time evolution of the density flux $\overline{\rho u_1}$ for several values of SK/ε with Fr = 0.35 and θ = π/4.

4. Conclusion

The effect of shear number on dynamic and thermal components of an homogeneous and stratified turbulence under inclined shear is investigated using coupled second-order model (SSG-LRR). Firstly, a comparison between results obtained with SSG-LRR model and those of DNSJ [21, 26, 27] is performed. A good agreement between predictions of second-order model of SSG-LRR and those of DNSJ is detected for a higher dimensionless time τ. Beyond $\tau \geq 25$, an excellent agreement between results of the SSG-CL model and those of DNSJ is observed for the principal component of anisotropy b_{12} and the normalized turbulence dissipation ε/KS for high and low stratification. Secondly, for high and low stratification Fr = 0.35 and 1.29, respectively, it is found that coupled model of SSG-LRR confirms the existence of an asymptotic equilibrium states of different parameters (b_{12}, ε/KS, and $\overline{\rho u_1}$) for various shear number (2, 6, 14, and 20). Hence, the coupling between the SSG model for the kinematic field and the LRR model for the scalars field proves a significant contribution to predict asymptotic equilibrium states of turbulent thermal and dynamic characteristics. This comparative investigation proves that the coupling between second-order models of SSG and LRR model can present a great solution to modelize homogeneous and stratified turbulence flows under inclined shear. Finally, it can be finished by that coupled second-order models of SSG-LRR present an important contribution for the modeling of turbulence for stratified shear flows and can help to better understand and realized a complex turbulent phenomena of geophysical flows, which propagate in the atmosphere.

Conflict of interest

The authors declare that they have no conflict of interest.

Nomenclature

b	Anisotropic tensor of Reynolds
C_p	Specific heat at constant pressure

g	Acceleration due to gravity
P	Pressure
p	Fluctuation of the pressure
Fr	Froude number
Ri	Richardson number
SK/ε	Shear number
S	Uniform mean shear
S_ρ	Uniform stable stratification
K	Turbulent kinetic energy
$\overline{T_{,i}}$	Gradient of the scalar
t	Time
τ	Non-dimensional time
u_i	i-th component of the fluctuating velocity
U_i	i-th component of mean velocity
$\overline{u_i u_j}$	Reynolds stress tensor
$\overline{u_i \rho}$	Turbulent flux of the scalar
$\overline{U_{p,q}}$	Gradient of mean speed
x_i	Component of an orthonormal Cartesian coordinate system
μ	Dynamic viscosity
λ	Viscosity ratio
ν	Kinematic viscosity
α	Thermal diffusivity
τ	Non-dimensional time
δ_{ij}	Kronecker symbol
ρ	Fluctuation of the scalar
ρ_0	Density of reference
$\overline{\rho^2}$	Density variance
ε	Terms of dissipation of turbulent kinetic energy
$\varepsilon_{\rho\rho}$	Terms of dissipation of variance of the scalar
ε_{ij}	Terms of dissipation of tensor of Reynolds
$\varepsilon_{i\rho}$	Terms of dissipation of the scalar flux turbulent
φ_{ij}	Terms of pressure-strain correlation
$\varphi_{i\rho}$	Terms of pressure- scalar gradient correlation
θ	Angle between the shear and the vertical gradient of stratification
SSG	Speziale, Sarkar, and Gatski model
LRR	Launder, Reece, and Rodi model
DNSJ	Direct numerical simulation of Jacobitz

Author details

Lamia Thamri[1,2*] and Taoufik Naffouti[1,2,3,4]

1 Faculté des Sciences de Tunis, Département de Physique, Université Tunis El-Manar, Tunis, Tunisia

2 Laboratoire d'Energétique et des Transferts Thermique et Massique, Tunis, Tunisia

3 Institut Préparatoire aux Etudes d'Ingénieurs El-Manar, Tunis, Tunisia

4 College of Sciences and Humanities of Dawadmi, Shaqra University, Shaqra, Kingdom of Saudi Arabia

*Address all correspondence to: thamrilamia@yahoo.fr

References

[1] Mauritsen T, Svensson G. Observations of stably stratified shear driven atmospheric turbulence at low and high Richardson numbers. Journal of the Atmospheric Sciences. 2007;**64**: 645-655

[2] Mauritsen T, Svensson G, Zilitinkevich SS, Esau I, Enger L, Grisogono B. A total turbulent energy closure model for neutrally and stably stratified atmospheric boundary layers. Journal of the Atmospheric Sciences. 2007;**64**:4113-4126

[3] Zilitinkevich SS, Elperin T, Kleeorin N, Rogachevskii I, Esau I, Mauritsenec T, et al. Turbulence energetic in stably stratified geophysical flows: Strong and weak mixing regimes. Quarterly Journal of the Royal Meteorological Society. 2008;**134**:793-799

[4] Roy S, Sentchev A, Fourmentin M, Augustin P. Turbulence of landward and seaward wind during sea-breeze days within the lower atmospheric boundary layer. Atmosphere. 2021;**12**:1-13

[5] Jacobitz FG, Sarkar S. The effect of non vertical shear on turbulence in a stably stratified medium. Physics of Fluids. 1998;**10**(5):1158-1168

[6] Pereira JCF, Rocha JMP. Simulation of shear orientation effects on stably stratified homogeneous turbulence with RANS second-order modeling. Journal of Turbulence. 2009;**10**(43):1-35

[7] Craft TJ, Launder BE. A new model for the pressure scalar gradient correlation and its application to Homogeneous and inhomogeneous shear flow. Symposium on Turbulent Shear Flows. 7th Stanford, CA, Aug. 21-23, Proceedings, vol. 2 (A90-35176 15-34). 1989

[8] Speziale CG, Sarkar S, Gatski TB. Modeling the pressure-strain correlation of turbulence: An invariant dynamical systems approach. Journal of Fluid Mechanics. 1991;**227**:245-272

[9] Launder BE, Reece G, Rodi W. Progress in the development of a Reynolds stress turbulence closure. Journal of Fluid Mechanics. 1975;**68**: 537-566

[10] Blumen W. Stability of non-planar shear flow of a stratified fluid. Journal of Fluid Mechanics. 1975;**68**:177-189

[11] Chomaz JM, Bonneton P, Hopfinger EJ. The structure of the near wake of a sphere moving horizontally in a stratified fluid. Journal of Fluid Mechanics. 1993;**254**:1-21

[12] Spedding GR, Browand FK, Fincham AM. Turbulence, similarity scaling and vortex geometry in the wake of a towed sphere in a stably stratified fluid. Journal of Fluid Mechanics. 1996; **314**:53-103

[13] Voropayev SI, Zhang X, Boyer DL, Fernando HJS. Horizontal jets in a rotating stratified fluid. Physics of Fluids. 1997;**9**:115

[14] Komori S, Ueda H, Mizushina T. Turbulence structure in stably stratified open-channel flow. Journal of Fluid Mechanics. 1983;**130**:13-26

[15] Caldwell DR. Small-scale physics of the ocean. Reviews of Geophysics. 1987; **25**:183-192

[16] Chebbi B, Bouzaiane M. On the effects of rotation on the passive scalar and kinematic fields of homogeneous sheared turbulence. Journal of Applied Fluid Mechanics. 2012;**5**(2):55-65

[17] Rohr JJ, Itsweire EC, Helland KN, Van Atta CW. Growth and decay of turbulence in a stably stratified shear flow. Journal of Fluid Mechanics. 1988; **195**:65-77

[18] Piccirillo PS, Van Atta CW. The evolution of a uniformly sheared thermally stratified turbulent flow. Journal of Fluid Mechanics. 1997;**334**: 61-86

[19] Launder BE, Reece G, Rodi W. Progress in the developmenty of a Reynolds stress turbulence closure. Journal of Fluid Mechanics. 1975;**68**: 537-576

[20] Launders BE, Reece G, Rodi W. Advanced turbulence models for industrial application. Turbulence and Transition Modelling. 1996;**2**:193-231

[21] Jacobitz F, Sarkar S, Van Atta C. Direct numerical simulations of the turbulence evolution in a uniformly sheared and stably stratified flow. Journal of Fluid Mechanics. 1997;**342**: 231-261

[22] Sarkar S. The effect of stable stratification on turbulence anisotropy in uniformly sheared flow. Computers and Mathematics with Applications. 2003; **46**:639-646

[23] Bouzaiane M, Ben AH, Lili T. Asecond order modeling of a stably stratified sheared turbulence submitted to a non vertical shear. Journal of Turbulence. 2004;**5**:1-17

[24] Thamri L, Naffouti T, Bouzaiane M. The effect of non-vertical shear on stratified turbulence using the coupling between second-order models. Journal of Turbulence. 2019;**20**(7):439-455

[25] Speziale CG, Sarkar S, Gatski TB. Modeling the Pressure Strain Correlation of Turbulence an Invariant Dynamical Systems Approach. Hampton (VA): Nasa Langley Research Center; 1990. pp. 3665-5225

[26] Jacobitz F, Sarkar S. A direct numerical study of transport and anisotropy in a stably stratified turbulent flow with uniform horizontal shear. Flow Turbulence and combustion. 1999; **63**:343-360

[27] Jacobitz F. A comparison of the turbulence evolution in a stratified fluid with vertical or horizontal shear. Journal of Turbulence. 2002;**3**:1-18

[28] Fatima MA. Numerical simulation with a Reynolds stress turbulence model of flow and heat transfer in rectangular cavities with different aspect ratios. Journal of Thermal Science and Technology. 2016;**11**(1):1-13

[29] Lumley L. Zeman, Siens, the influence of buoyancy on turbulent transport. Journal of Fluid Mechanics. 1978;**84**:581-597

[30] Zeman LL. Modeling buoyancy-driven turbulence mixed layers. Journal of the Atmospheric Sciences. 1976;**33**: 1974-1988

[31] Thamri L, Naffouti T. Comparative analysis of coupled second-order models on shear and Richardson numbers effects on homogeneous and stratified turbulence. European Physical Journal Plus. 2022;**137**:1-17

Chapter 3

Bifurcation and Instabilities in Fluid Flow

Carlos Gavilan

Abstract

In some situations, pipe layout and incoherent behavior of the flow is detected. An instability is generated, and the flow jumps between the values in flow rate or pressure, with no explanation. Usually, the industry solves the problem by creating an exclusion area. It is very important to understand those instabilities, named bifurcation, and why the Navier-Stokes equation has two equiprobable solutions. With this knowledge, the situation can be corrected, and the problem solved. The result is an increase in performance, reliability, etc. and then in the economy of the process. The bistable flow in boiling water reactor (BWR) nuclear power plant is a clear example.

Keywords: bifurcation, instability, Navier-Stokes, symmetry, bistable

1. Introduction

There are many references in the fluid dynamics literature to typical fluid flow instability problems. Some of the existing examples show the existence of pulsating and oscillating anomalies when they should not be present [1–5]. These problems led to suspicions of anomalous solutions to the Navier-Stokes equation [5, 6].

The aim of this work is the determination of instabilities in fluid flow in piping systems. These instabilities will be classified as bifurcations since they constitute situations of two valid solutions of the Navier-Stokes equation. For the specific case, the pipe system is a symmetrical system composed of a riser, a manifold, and five symmetrically arranged rising branches.

These real events in fluid flow in pipes, normally produce a reaction in the operators of the systems, since their existence results in problems of production [7], reliability [8] of components that lead to decisions, which redound in the economy of the productive processes.

Thus, we are faced with systems that, from a certain value of one of their parameters, oscillate between two states. For example, from a given flow rate value (position of a control valve) the system oscillates between two pressures, or for a pressure value, the system has two flow rate values.

Normally, hydraulic studies are carried out on the basis of dimensionless numbers, the Reynolds number being the most commonly used. In this study, because the geometry is fixed, the fluid is in isothermal conditions, and the flow rate, flow velocity, or Reynolds number is equivalent. Likewise, since both flow rate and fluid

IntechOpen

velocity depend only on the position of the control valve, we can also make the equivalence between Re and valve position.

Finally, it must be said that this study is the end of a line of work, starting with the search for a solution to the problem of pulsations in the recirculation system of a boiling water reactor. The initial studies, which are empirical [9–11], have already determined the solution. Years later, CFD studies corroborated the empirical solutions and provided new ones [12–14].

Once all this modeling and analysis work has been completed, it is proposed to determine the true root cause, in order to improve the possibility of other solutions or even the explanation of other effects. This is the aim of the present work.

2. What a bifurcation is

Suppose in a dynamic system, a bifurcation occurs when a small smooth change made to the parameter values of the system, causes a sudden qualitative or topological change in its behavior. The name of bifurcation was first introduced by Henry Poincare [15] in 1885, in the first paper in mathematics showing such behavior. Bifurcation occurs in both continuous systems described by Ordinary Derivative Equations (ODE's), Differential Derivative Equation (DDE's), and Partial Derivative Equations (PDE's) and discrete systems described by maps.

The previous paragraph is a qualitative description of what a bifurcation is, and the next will be a more detailed and mathematical description of it.

The explanation or justification of a bifurcation will start with the concept of the asymptotic solution to an evolution problem. An evolution problem has the form shown in Eq. (1).

$$\frac{dU}{dt} = F(t, \mu, U) \tag{1}$$

where $t \geq 0$ is the time and μ is a parameter that lies on the real line, $-\infty \leq \mu \leq \infty$. The unknown is U(t) and F(.,.,.) is a given function, i.e.: Partial derivative Equation, ordinary derivative equation. When F() is non-time-dependent, $F=F(\mu,U)$, the evolution of U(t) is governed by its initial value $U(0) = U_0$.

An asymptotic solution is defined as the solution to which U(t) evolves after the transient effects associated with the initial value. Asymptotic has two main types: steady solutions and T-periodic solutions for non-autonomous problems.

Once the problem is enunciated, then the bifurcation definition is possible. Bifurcation solutions are asymptotic solutions that form intersecting branches in suitable space of functions. For example, when U lies in R, the bifurcating steady solution form intersecting branches of the curve $F(\mu,U) = 0$ in the μ,U plane.

So, an asymptotic solution bifurcates from another at $\mu = \mu_0$ if there are two distinct asymptotic solutions $U^{(1)}(\mu,t)$ and $U^{(2)}(\mu,t)$ of the evolution problem continuous in μ and such that $U^{(1)}(\mu_0,t) = U^{(2)}(\mu_0,t)$.

Note that not all asymptotic solutions arise from bifurcation, there are other solutions such as isolated solutions or disjoint solutions.

From this point, a one-dimensional problem will be the real problem Eq. (1), could be rewritten as:

$$\frac{du}{dt} = F(\mu, u) \tag{2}$$

where F(.,.) has two continuous derivatives with respect μ and u. Another assumption is that:

$$F(\mu, 0) = 0. \forall \mu \in R \tag{3}$$

The equilibrium solution of Eq. (1) in u = ε satisfied that:

$$F(\mu, \varepsilon) = 0 \tag{4}$$

The study of bifurcation is the study of the equilibrium evolution given by Eq. (4) in the plane (μ,ε).

Theorem:

Let $F(\mu_0,\varepsilon_0) = 0$ and let F be continuously differentiable in some open region containing the point (μ_0,ε_0) of the (μ,ε) plane. Then, if $F_\varepsilon(\mu_0,\varepsilon_0) \neq 0$ there exists $\alpha,\beta > 0$ such that:

i. *The Eq. $F(\mu,\varepsilon) = 0$ has a unique solution $\varepsilon = \varepsilon(\mu)$ when $\mu_0-\alpha < \mu < \mu_0+\alpha$ such that $\varepsilon_0-\beta < \varepsilon < \varepsilon_0+\beta$.*

ii. *The function ε(.) is continuously differentiable when $\mu_0-\alpha < \mu < \mu_0+\alpha$*

iii. $\varepsilon_\mu(\mu) = -F_u(\mu,\varepsilon(\mu))/ F_\varepsilon(\mu,\varepsilon(\mu))$

The solutions of the equilibrium could be the following points:

- Regular point
- Regular turning point
- Singular point
- Double point
- Singular turning (double) point
- Cusp point
- Conjugate point
- High order singular point

In this chapter, the focus will be on the double point. A double point is a point of the curve F(μ,ε) = 0 through which pass two and only two branches of F(μ,ε) = 0 possessing distinct tangents. It is assumed that all second derivatives of F(.,.) do not simultaneously vanish at a double point.

Then, equilibrium curves passing through the singular points satisfy:

$$2F(\mu, \varepsilon) = F_{\mu\mu}\delta\mu^2 + 2F_{\varepsilon\mu}\delta\varepsilon\delta\mu + F_{\varepsilon\varepsilon}\delta\varepsilon^2 + o\left[(|\delta\mu| + |\delta\varepsilon|)^2\right] = 0 \tag{5}$$

In the limit $(\mu,\varepsilon) \longrightarrow (\mu_0,\varepsilon_0)$ for the curves F(.,.) = 0 is reduced to:

$$F_{\mu\mu}d\mu^2 + 2F_{\mu\varepsilon}d\varepsilon d\mu + F_{\varepsilon\varepsilon}d\varepsilon^2 = 0 \tag{6}$$

For the tangents to the curve. Solving Eq. (6) the Eq. (7) is obtained.

$$\begin{bmatrix} \mu_\varepsilon^1(\varepsilon_0) \\ \mu_\varepsilon^2(\varepsilon_0) \end{bmatrix} = -\frac{F_{\varepsilon\mu}}{F_{\mu\mu}}\begin{bmatrix} 1 \\ 1 \end{bmatrix} + \sqrt[2]{\frac{D}{F_{\mu\mu}^2}}\begin{bmatrix} 1 \\ -1 \end{bmatrix} \tag{7}$$

where

$$D = F_{\varepsilon\mu}^2 - F_{\mu\mu}F_{\varepsilon\varepsilon} \tag{8}$$

Then (μ_0,ε_0) is a double point if $D > 0$ additionally to have two tangents with slopes different from zero, the condition has to be $F_{\mu\mu} \neq 0$.

3. Bifurcation in fluid flow

The Navier-Stokes´ equation is a PDE, so it is potentially affected by bifurcation as explained in previous paragraphs. Additionally, the Navier-Stokes´ equation is nonlinear so the occurrence of bifurcation is much more probable.

Now the concept of bifurcation, as described before, will be developed mathematically for fluid flow problems. First, some assumptions have been considered:

- Fluid is in an incompressible state.
- Fluid is at constant temperature.
- The part of the system under study has constant section (pipe).
- There's no fluid sumps or sources.

In this situation, the equation could be written as shown in Eq. (9).

$$\frac{D\vec{u}}{Dt} = \vec{F} - \nabla P + \frac{\mu}{\rho}\left(\frac{1}{3}\nabla\left(\nabla\vec{u}\right) + \nabla^2\vec{u}\right) \tag{9}$$

The equation has the following vectors: u (speed), P (pressure), μ (viscosity), and (fluid density). Since the system has a constant section and there are no fluid sumps or sources, integrated fluid speed is like the scalar flow value. Other supposition is that the system status will depend on a given parameter "A," for example pump speed, control valve position, Pressure regulation vale position, etc. In this situation, the Navier-Stokes´ equation can be reformulated as:

$$\frac{dq}{dt} = F(q, A) \tag{10}$$

where q = flow and A = system parameter.

Primary derivatives of Eq. (10) are determined by Eq. (11), with notation and secondary derivatives being simplified in Eq. (12):

$$\frac{\partial F(q,A)}{\partial q} = F_q y \frac{\partial F(q,A)}{\partial A} = F_A \tag{11}$$

$$\frac{\partial F^2(q,A)}{\partial q^2} = F_{qq}; \frac{\partial F^2(q,A)}{\partial A^2} = F_{AA} y \frac{\partial F^2(q,A)}{\partial qA} = F_{qA} \tag{12}$$

The stability analysis results from establishing a null time derivative. In other words, if the system does not change and A do not vary, flow remains the same. Mathematically, this is seen in Eq. (13).

$$\frac{dq}{dt} = 0 = F(q,A) \tag{13}$$

That means analyzing bifurcations in Eq. (11) is the same as analyzing specific F curve points (q, A) on the A parameter-flow plane. Considering these parameters, the points that confirm the Eq. (13) are considered unique and can be classified as follows: regular points, regular inflection points, unique points, double curve point, double inflection points, and peak points.

The double curve point generates two solutions and a curve running through the unique point that has two slopes. Although there are multiple potential shapes, the one selected provides two stable results as it is coherent with bistable conditions. Once shape is selected, it can have three main sub-types, as seen in **Figure 1**: supercritical, subcritical, and transcritical.

In the case of double point bifurcation, there is a balance point with two curves and two different slopes [16]. Curve tangents conform to Eq. (14)

$$\begin{bmatrix} \frac{dq^{1)}}{dA} \\ \frac{dq^{2)}}{dA} \end{bmatrix} = \begin{bmatrix} q_A^{1)} \\ q_A^{2)} \end{bmatrix} = -\frac{F_{qA}}{Fqq}\begin{bmatrix} 1 \\ 1 \end{bmatrix} - \sqrt[2]{\frac{D}{F_{qq}}} \cdot \begin{bmatrix} 1 \\ -1 \end{bmatrix} \tag{14}$$

where "D" is determined by Eq. (15):

$$D = F_{qA}^2 - F_{AA} \cdot F_{qq} \tag{15}$$

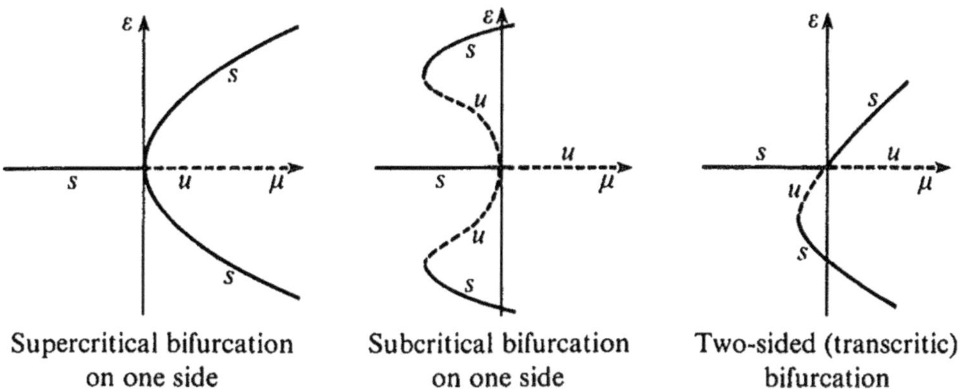

Figure 1.
Sub-types of double point bifurcation (pitchfork type).

Slope analysis results from analyzing the value of parameter D. If $D < 0$, there are no real tangential lines on the point, which means the existence of a double point and two slopes can only be justified when $D > 0$. In this case, if specific concepts are renamed, flow curve slopes in relation to A value are determined by Eq. (16):

$$\frac{dq}{dA}(A_0) = q_A(A_0) \tag{16}$$

For this equation to be true and conformed to, the condition of F_{qq} not being null must be satisfied, which is exactly what happens as this is a process related to fluid flow. The other condition for conformance to $D > 0$ is the verification of Eq. (17):

$$F_{qA} > \sqrt[2]{F_{AA} \cdot F_{qq}} \tag{17}$$

To conclude this theoretical development, the requirements for existence of a double point bifurcation are mathematically conformed because F_{qq} is not null, and D can under specific conditions be higher than zero. And as the solution existence theorem says, if F(q, A) is continuous, then at least one solution of the Eq. (9) exists.

4. A real case of bifurcation in fluid flow. Bistable flow

In 1985, an electrical power fluctuation was detected at the Leibstadt Nuclear Power Plant during startup and load tests at 100%. Event analyses revealed additional fluctuations in steam flow, thermal power, core flow, and recirculation loop flowrates. Loop "A" fluctuation ranged between 2.5% and 3%, whereas in loop "B" it was between 3% and 3.5%. It was concluded that recirculation flow fluctuations were caused by a bistable flow pattern in the pump discharge header.

In 1986, the first regulatory reference on abnormal performance of recirculation loop flows was written [17]. This document recaps events, like that of Leibstadt, at the USA stations of Pilgrim (1985) and Vermont Yankee (1986), and also determines that fluctuations vary from station to station and even within the loops of one single unit. In accordance with the information available at the time, the NRC establishes that magnitude and duration are not predictable using the analytical methods applied to the piping systems under analysis.

In 1988, General Electric, the responsible (OEM) for the design of the recirculation loop, issued a letter [9] establishing that the problem affects Boiling Water Reactors (BWR) from generation 3 to 6. More specifically, the problem is detected at the collectors of feed pipes supplying jet pumps (**Figure 2**). The letter was later revised in 2006 to include operational experiences and operating recommendations.

Around that time, the same phenomenon occurred in Japan, a country with many boiling water reactors. Since 1986 to 1989, a group of researchers applied various hydraulic models to replicate [10], characterize [11], and propose compensatory measures [18]. Eventually, the feed pipe header of jet pumps was successfully modified.

As this issue was considered non-safety related, bistable flow was rendered acceptable at the plants and unit operation took it into account. At times, operational strategies were modified to minimize this phenomenon, although its analysis did not go any further.

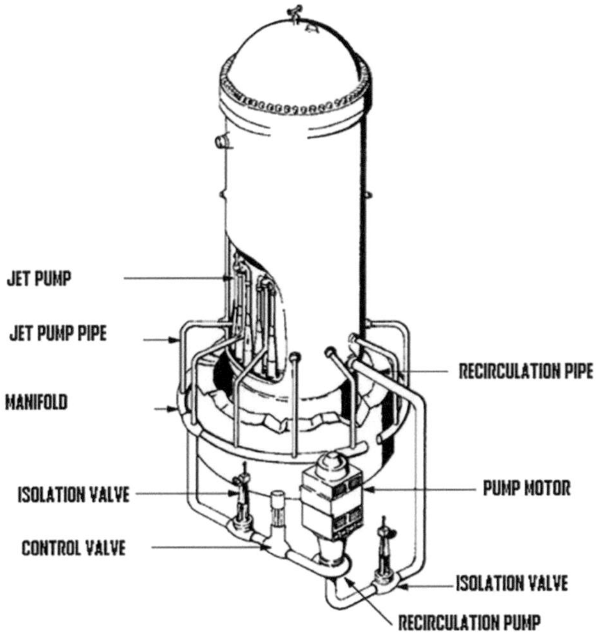

Figure 2.
Recirculation loop schematics in a generation 6 BWR.

In 2006, a bistable flow phenomenon analysis was presented at Laguna Verde nuclear power plant in México [19].

After 2008, new research trends focused on providing a more detailed bistable flow profile [12], concluded that it can be described as a noise-induced transition mechanism [20]. In this theory, noise is identified as flow turbulence under high Reynolds values.

In 2009, bistable flow analyses continued using mathematical techniques such as wavelets [21] and codes for fluid mechanics. These were used to replicate the results of hydraulic models from the 80s [13].

Ever since they were used in nuclear power plants, bistable flow analyses have identified jet pumps as the most critical hydraulic element. In the year 2011, an analysis was undertaken to determine potential bistable flow impact on jet pumps. This analysis [14] presented the spectral situation of jet pumps, which were influenced by resonance frequencies and induced frequencies resulting from bistable flow. Also, in 2011, a new 3D CFD model confirmed the existence of two states and the non-convergence of the stable condition at specific Reynolds values. It was verified that current lines derived from 3D model results are coherent with hydraulic model data and that geometry changes contribute to eliminating this phenomenon as the stable state is quickly reached and calculation converge is ensured. The fact that for similar environment conditions, there were two solutions for the Navier-Stokes equation suggests the existence of a bifurcation.

Operationally speaking, a few works were undertaken to map out recirculation loop flows in relation to reactor power. The aim was to identify the area most prone to developing bistable flow for several reasons, including flowrate [7].

In line with the above mentioned in this section, the state of the art in bistable flow can be summarized as follows:

- Unpredictable, fluctuating hydraulic phenomenon.
- Hydraulic phenomenon only removable at this point through physical hydraulic circuit modifications.
- Phenomenon characterized by high turbulence which at a certain level leads to a bistable state.
- Phenomenon not impacting plant safety.

The methodology used in this study has two main phases.

Phase 1. Analyze recorded recirculation flow and FCV position data for a full 24-month cycle at an operating station. From this point on, the value of the FCV position will be assimilated to the value of "A" in Eqs. (4) and (7)-(11).

- S1. Determines the curve for recirculation flow (%) vs. FCV position (%).
- S2. Establishes the calculated curve adjustment error based on available data. Analyze error behavior in relation to FCV position (%). The analysis of this error, starting point of the bifurcation study, provides a clear picture of bistable flow.

Phase 2. Determine, find, and characterize the mathematical bifurcation physically supporting bistable flow. Apply theoretical results from the previous section to the time series. Considering the noise nature of the series and the difficulty to determine analytical expressions, an empirical approach is developed. The global interpretation includes the definition of bifurcation maps and their characteristics.

4.1 Analysis (phase 1)

FCV position (%) and flow (%) are interrelated time series [22, 23]. It is also important that time series, especially those for flow, are highly noise signals and will require the use of some mathematical techniques [24].

The analysis of recirculation loop flow signal has several phases aimed at determining flow signal features. Once this information is obtained, the entire system is analyzed, with a special focus on the relationship between recirculation loop flow and FCV opening. The limits of bistable occurrence and ideas suggesting bifurcation existence, mostly as a result of error analyses, will appear at this point.

4.1.1 Opening-flow curve analysis

A Boiling Water Reactor (BWR) 6 reactors have two recirculation loops, each with its own flow control valve (FCV) (**Figure 3**). **Figure 3** shows the point cloud generated from loop flow (%) and y FCV position (%) records during reference plant operation over a 24-month operating cycle.

A morphological analysis of the point cloud reveals that after a 60% opening approximately, the cloud begins to widen and after 62% it has a new width which remains constant until the end. This is seen in **Figure 4**.

The system designer and technologist determined, both mathematically and theoretically, a polynomial of degree 3 for the flow equation based on FCV opening.

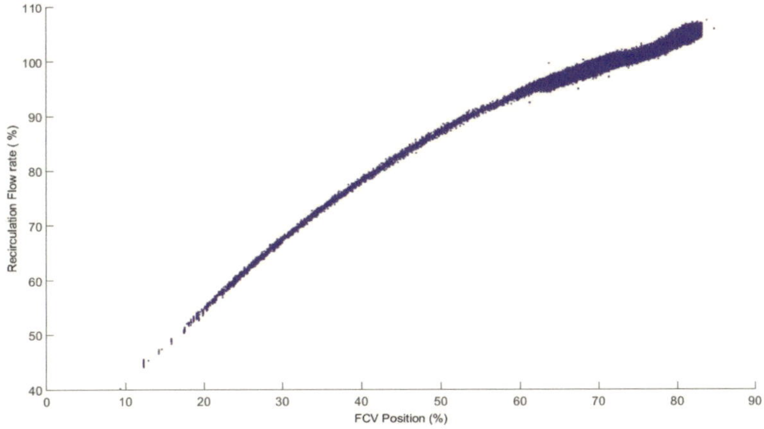

Figure 3.
Plot of operational points (flow % vs. FCV opening %).

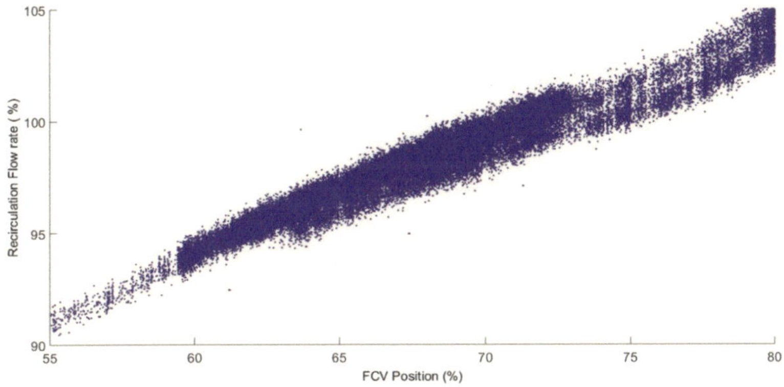

Figure 4.
Details of operational points. (FCV opening >55%).

This adjustment, with a confidence level of 99.999%, provides the coefficients shown in **Table 1**.

Table 1 reveals the coherence between the theoretical equation and that adjusted with actual data.

4.1.2 Analysis of error adjustment (S2)

Adjustment error (Eq. (18)) is established as the difference between the measured flow value (%) and the adjusted curve value, as seen in **Table 2** parameters. Globally speaking, throughout the entire FCV position range there is an error, as seen in **Figure 5**.

$$\mathrm{Error(FCVOpening)} = \mathrm{MeasuredFlow}(\%) - \mathrm{CalculatedFlow}(\%) \tag{18}$$

Figure 5 confirms the idea that there is phenomenon changing flow patterns when FCV opening is 60% or above. This is indicated by the error value, which changes in shape and values. Error varies from about 0% with a + −0.5% band to a + −1.5% band.

	$Flow(\%) = a * (FCV_{POS})^3 + b * (FCV_{POS})^2 + c * (FCV_{POS}) + d$			
	a	**b**	**c**	**d**
Design adjustment	$4{\cdot}10^{-5}$	−0.0149	1.9502	21.191
Loop B adjustment	$6.5801{\cdot}10^{-5}$	−0.0188	2.1469	18.5307

Table 1.
Adjustment polynomial coefficients for the theoretical curve adjusted with actual data.

Lambda	Tau (−)	Tau (+)	Reference figure
−32	−0.577	−0.577	
0	0	0	**Figure 5**
1.45	−0.674	0.293	
1.76	−0.675	0.294	**Figure 6**
3.75	−0.728	0.306	
5.95	−0.785	0.313	
9.55	−0.863	0.256	**Figure 7**
15.25	−0.955	0.287	
19.75	0	1.3	**Figure 8**

Table 2.
Maximum tau and lambda values.

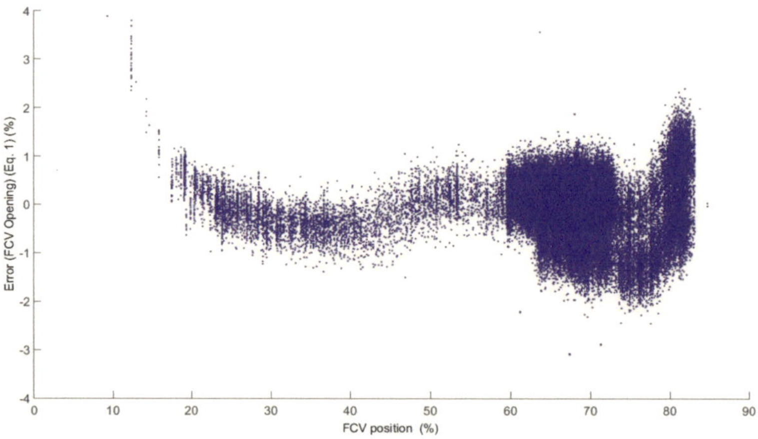

Figure 5.
Adjustment error.

Even for values exceeding 80%, there is a clearly strong, differentiated bias with regards to flow behavior at this FCV opening value.

4.2 Bifurcation determination (phase 2)

The existence of a bifurcation structure will be demonstrated in this phase through the selection of FCV opening as the bifurcation parameter as it is the only one of a variable nature in the system.

This section focuses on characterizing the pitchfork-type bifurcation for recirculation flow under bistable conditions. For clarity purposes and without losing accuracy, the unique point of double slope will be (0,0) on the plane (FCV opening, flow).

The coordinate parameter, known as tau (τ), will be the error included in Eq. (12), whereas the abscissa, known as lambda (λ), will be the difference between FCV position and the initial bistable point. Thus, considering an opening value of 62.3% as the initial bifurcation point, variable λ will be the result of subtracting from FCV opening (%) the previous value of 62.3%. In other words, initial bifurcation will be $\lambda = 0$ and $\tau = 0$.

Instead of interpreting the signal as a whole (FCV opening-flow), recirculation loop flow analysis will be based on various FCV opening levels. The values selected for this task are those in which FCV opening (5) is 29, 62, 62.3, 63, 64, 68, 72, 77, and 82%. These values will be reduced to parameter λ, as see on **Table 2**. The average value for reduced tau flow was selected when there was only a single peak in the normal distribution or two maximum values in two normal distributions due to bistable flow. These tau values can be seen in **Table 2**.

The graphical representation of values in **Table 2** is seen in **Figures 6–9**. These values are used to create a graph considered the bifurcation map (**Figure 10**).

The data in **Table 2** are used to create a set of points on plane λ-τ, meaning points (λ,τ+) and (λ,τ-) are graphically represented for some of the analyzed cases (**Table 3**). The result is seen in the graph of **Figure 10**. An analysis of the graph in **Figure 10** reveals the evident existence of bifurcation on point (0,0), which corresponds to a double point bifurcation of the pitchfork-type (supercritical according to **Figure 1**).

More thorough analysis shows that there are two consecutive bifurcations and two bistable states. The first bistable state is limited to lambda values ranging between 0 and 15 (pitchfork bifurcation), whereas the second bistable state occurs when values are larger than 15 (double point bifurcation).

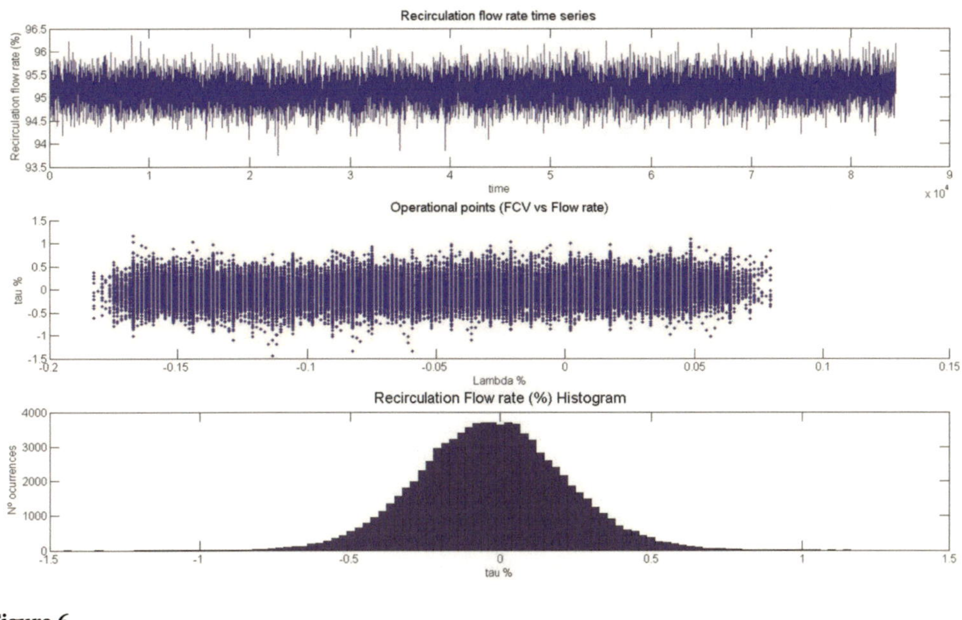

Figure 6.
Recirculation flow evolution with lambda value at 0.00.

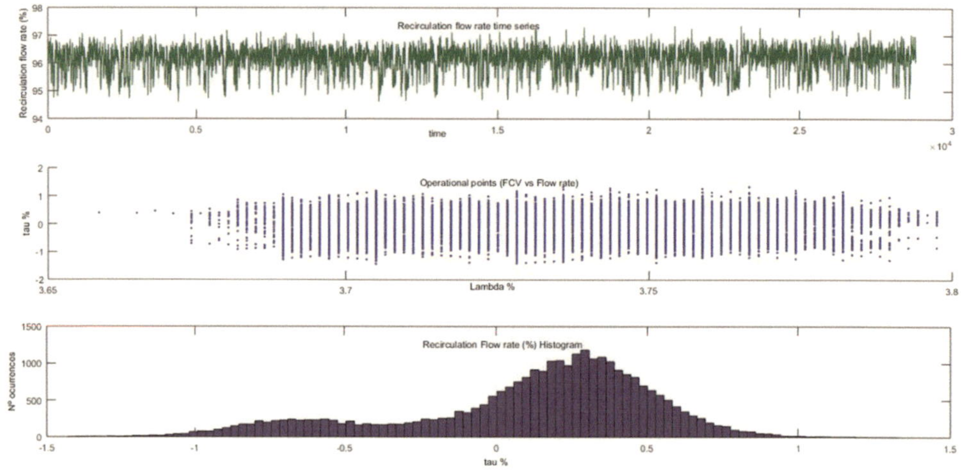

Figure 7.
Recirculation flow evolution with lambda at 3.8.

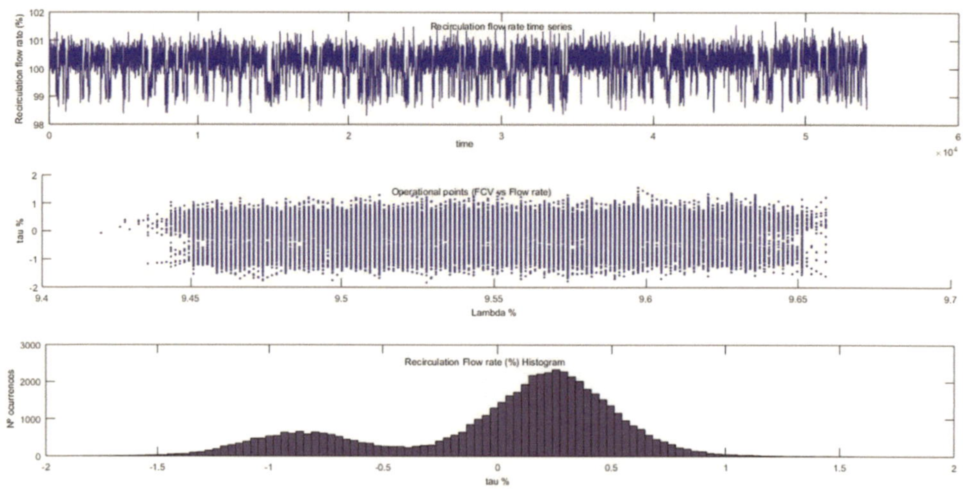

Figure 8.
Recirculation flow evolution with lambda value at 9.55.

4.3 Analysis of results (phase 3)

In **Figure 9** bifurcation diagram is created to justify the behavior of recirculation loop B in a BWR6 unit, which is the system under analysis. As shown in Eq. (8), bifurcation point slopes on stability curves are determined by Eqs. (19) and (20).

$$\frac{dq^{1)}}{dA} = q_A^1 = \frac{\tau^+(\lambda_2) - \tau^+(\lambda_1)}{\lambda_2 - \lambda_1} = \frac{\tau^+(1.45) - \tau^+(0)}{1.45 - 0} = \frac{0.293}{1.45} = 0.2020 \quad (19)$$

$$\frac{dq^{2)}}{dA} = q_A^2 = \frac{\tau^-(\lambda_2) - \tau^-(\lambda_1)}{\lambda_2 - \lambda_1} = \frac{\tau^-(1.45) - \tau^-(0)}{1.45 - 0} = \frac{-0.674}{1.45} = -0.4648 \quad (20)$$

DOI: http://dx.doi.org/10.5772/intechopen.105051

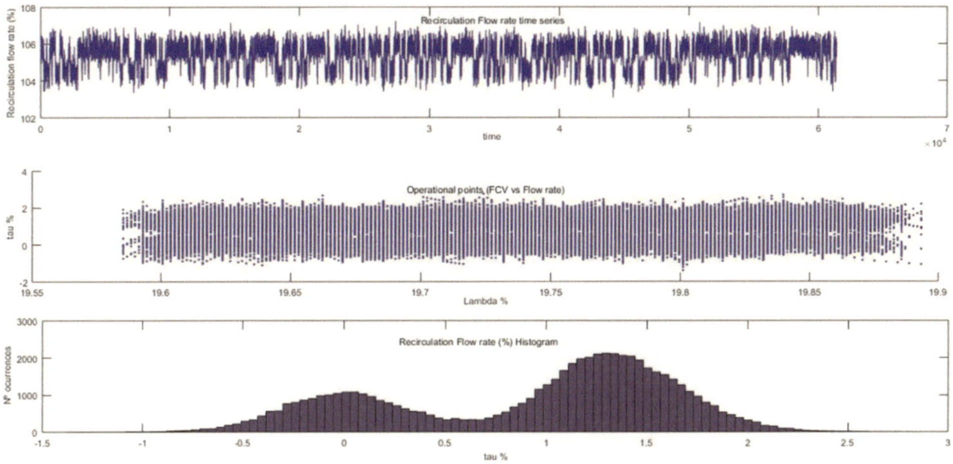

Figure 9.
Recirculation flow evolution with lambda value at 19.75.

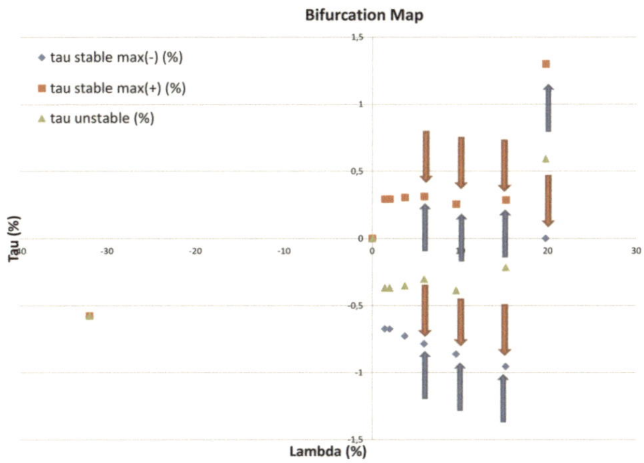

Figure 10.
Recirculation flow bifurcation diagram. Continuous lines represent the values of stable solutions (feasible), and discontinuous lines represent unstable solutions (non-feasible).

Calculations are made using the data in **Table 2**.

The bifurcation diagram curve leads to believe that there is another possible bifurcation (double point), which is not of the pitchfork type. The bifurcation observed when the FCV point equals 77% has two slopes, feature of a double point:

The slope of the upper leg is determined by Eqs. (21) and (22).

$$\frac{dq^{1)}}{dA} = q_A^1 = \frac{\tau^+(\lambda_2) - \tau^+(\lambda_1)}{\lambda_2 - \lambda_1} = \frac{\tau^+(19.75) - \tau^+(15.25)}{19.75 - 15.25} = \frac{1.013}{4.5} = 0.2251 \tag{21}$$

$$\frac{dq^{2)}}{dA} = q_A^2 = \frac{\tau^+(\lambda_2) - \tau^+(\lambda_1)}{\lambda_2 - \lambda_1} = \frac{\tau^+(15.25) - \tau^+(9.55)}{15.25 - 9.55} = 0.0068\ (\text{unstable}) \tag{22}$$

The slope of the lower leg is determined by Eqs. (23) and (24).

$$\frac{dq^{1)}}{dA} = q_A^1 = \frac{\tau^-(\lambda_2) - \tau^-(\lambda_1)}{\lambda_2 - \lambda_1} = \frac{\tau^-(19.75) - \tau^-(15.25)}{19.75 - 15.25} = \frac{0.955}{4.5} = 0.2122 \tag{23}$$

$$\frac{dq^{2)}}{dA} = q_A^2 = \frac{\tau^+(\lambda_2) - \tau^+(\lambda_1)}{\lambda_2 - \lambda_1} = \frac{\tau^+(15.25) - \tau^+(9.55)}{15.25 - 9.55} = 0.0204 \tag{24}$$

The two new curved stability slopes are nearly parallel, meaning stable states are permanent and the bistable state is not eliminated until 82% values (recorded at the plant) are reached. Thus, bifurcation is permanent and stable.

The existence of bistable flow is proven as a physical reality of the mathematical bifurcation concept. However, it is necessary to relate flow states with mathematical states and their physical meaning. This interpretation is based on the results of this work and those of the hydraulic model [10, 11] and computational fluid flow models (CFD) [13].

The recirculation flow rate can have three stable states, grouped in twos. The first two states, corresponding to FCV opening values between 62.3% and 77%, are defined by the vortex dynamics of flow adaption in the so-called cross-piece. These states cause the following effects:

- Vortex formation on the side legs of the manifold, in normal operation.
- The newly formed vortexes cause flow to take two forms: one for direct inlet at high flow and the other for helical current at low flow (**Figure 11**).

In the first flow shape, pressure loss is as designed, and flow is quite similar to its theoretical value. In the second flow shape, pressure loss is higher since helical flow has more internal fluid friction and pipe wall friction, causing flow to be lower.

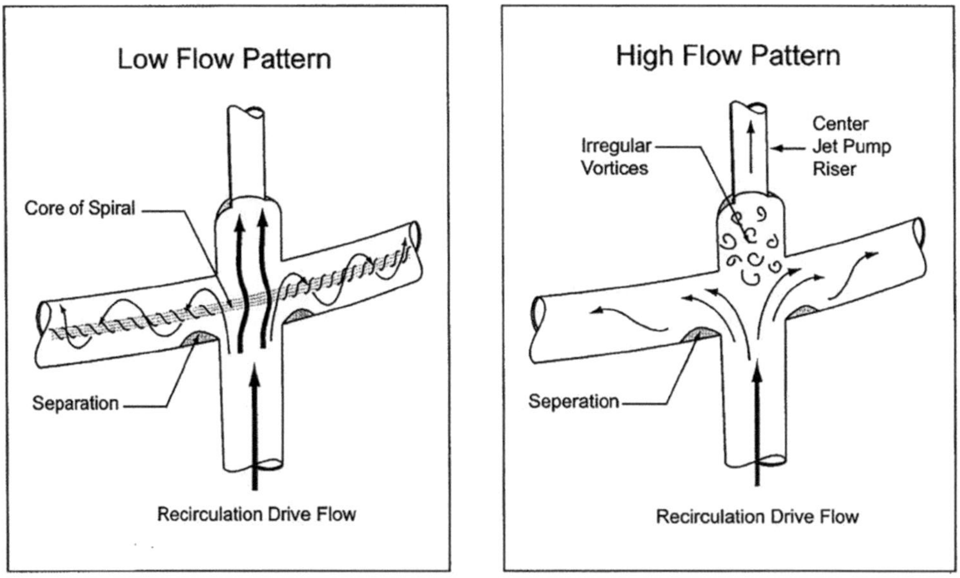

Figure 11.
Recirculation loop cross-piece flow conditions for first bifurcation.

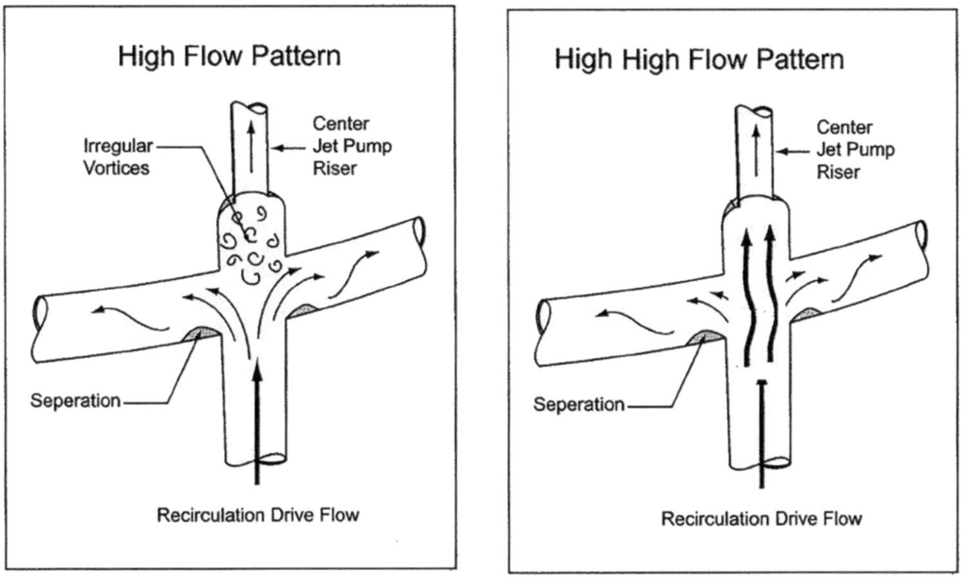

Figure 12.
Recirculation loop cross-piece flow conditions for second bifurcation.

The next bistable scenario occurs when FVC opening is 77%. In this case, flow equals rated values and some are slightly higher (1%). In this state, flow takes two shapes: one of direct inlet at high flow (coinciding with the former high flow value) and the other free of central connection vortexes (which are dragged by the high flow); hence reducing internal friction and enhancing flow (**Figure 12**).

5. Conclusions

The bistable flow state is the result of an evolved dynamic system which is based on a specific FCV opening value, offers two possible states. Transitioning from one to the other is determined by system evolution in all possible values and their flow rate variance.

Graphically, in each moment, the system has a balance point determined by a minimal energy state represented by an attraction basin, causing the system to evolve in it and be attracted to the minimum point, depending on the noise (turbulence). As FCV opening value increases, flow and turbulence also increase, causing the attraction basin to be increasingly larger although the minimum point is preserved. There comes a moment when the attraction basin begins splitting in two, growing in the middle, between both parts, a peak area that separates them. When the bistable flow is fully formed, both attraction basins and their separation are obvious. The system moves between these two "balance" values (**Figure 13**).

The point where the separation appears and both attraction basins begin to take shape is unique and marks the start of bifurcation. This separation coincides with the unstable system solution, meaning that if the system has a solution for such separation, it will evolve towards one of the two values provided by the attraction basin.

With regards to value changes and fluctuations, having two energy pits lead to some changes between them due to the intermediate peak value being exceeded. This

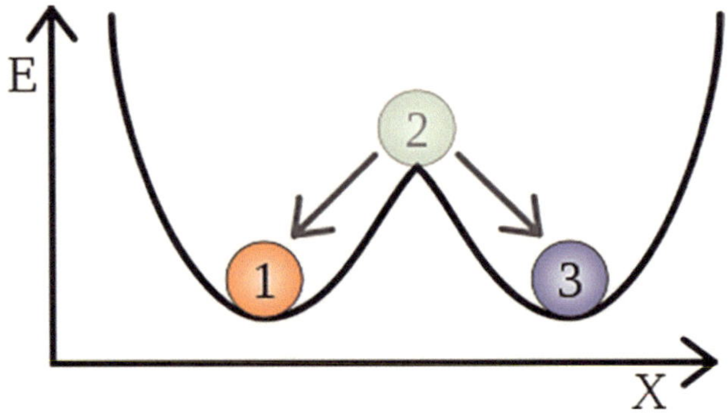

Figure 13.
Energy or attraction basins and separation peak.

is possible because the turbulence generated at high numbers of Reynolds increases the fluid energy locally, which exceeds the intermediate peak and causes the system to enter the other energy state. This theory is coherent with the characterization of bistable flow as a noise-induced transition phenomenon, considering that in this case, noise is the actual flow turbulence (**Figure 14**).

To conclude, the Navier-Stokes equation bifurcation theory for the recirculation system explains all events reported, all simulations made, and all effects observed. Likewise, this theory provides the grounds to justify that only actions intended to

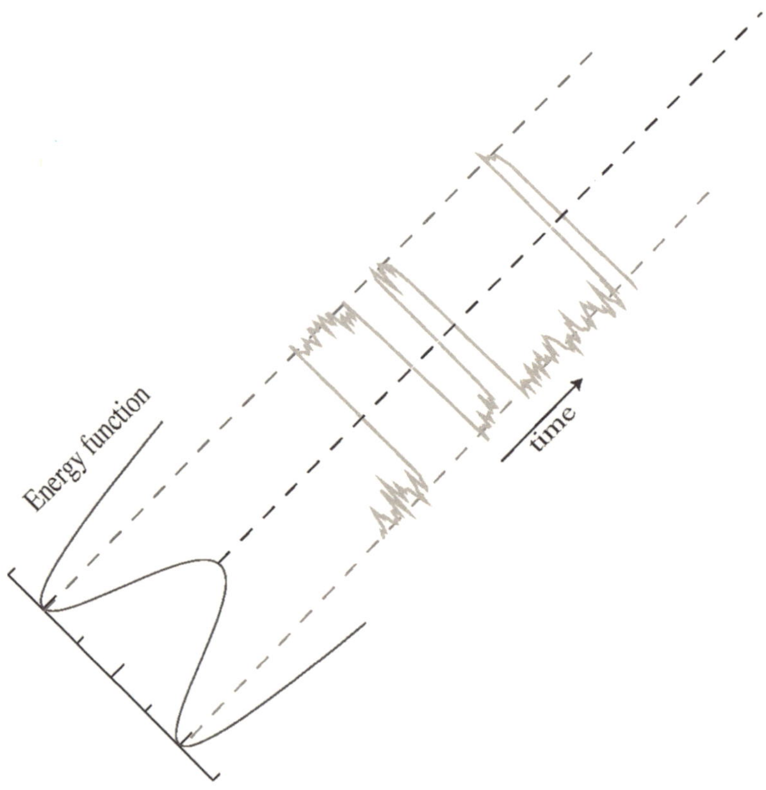

Figure 14.
System energy function and correlation with bistable flow.

eliminate turbulence formation and its detachment from the cross-piece are adequate to remove bistable flow. Thus, cross-piece modification alters the system curve (flow-opening) causing the unique point of double tangent to move to another location and turn currently bifurcated points into regular points. This modification from unique point to regular point causes flow to stop being bistable.

Lastly, it is worth mentioning that this study focuses on a fixed, complex geometry system, proves that turbulence and vorticity evolution are not a continuous function in relation to flow or the number of Reynolds. On the contrary, it has a bifurcation area turbulence that is discontinuous while drifting and has a continuous, non-derivable function.

Author details

Carlos Gavilan
Iberdrola Generacion Nuclear, Spain

*Address all correspondence to: cgavilan@iberdrola.es

References

[1] Fedele F, Hint DL, Prabhu RD. Revisiting the stability of pulsatile pipe flow. European Journal of Mechanics B/Fluids. 2005;**24**:237-254

[2] Olinto RO, Indrusiak ML, Möller SV. Experimental study of the bistable flow in tube arrays. Journal of the Brazilian Society of Mechanical Science and Engineering. 2006;**28**(2):233-241

[3] Mullin T et al. Bifurcation phenomena in the flow through a sudden expansion in a circular pipe. Physics of Fluids. 2009;**21**:014110

[4] Trip R, Kuik DJ, Westerweel J, Poelma C. An experimental study of transitional pulsatile pipe flow. Physics of Fluids. 2012;**24**:014103

[5] Shemer L, Wygnanski I, Kit E. Pulsating flow in a pipe. Journal of Fluid Mechanic. 1985;**153**(-1):313-337

[6] Rummler B. Some strange numerical solutions of the non-stationary Navier-Stokes equation in pipes. Mathematis and Mathematics Education. 2001: 264-280

[7] Baraza A. Determining high bistable flow areas at full power. Review in Nuclear España. 2010;**310**: 62-65

[8] Gavilán Moreno CJ. Bistable flow spectral analyses. Repercussions and jet pumps. Nuclear Engineering and Design. 2011;**241**:2437-2447

[9] Poincare H. Revisiting the stability of pulsatile pipe flow. European Journal of Mechanics B/Fluids. 2005;**24**: 237-254

[10] Miura S et al. Unstable phenomenon in flow through a pipe system with a cross pipe. (1st report, Generation of flow instability and its condition). Journal of Japan Society of Mechanical Engineers. 1986;**86**:35-39

[11] Miura S et al. Unstable Phenomenon in Flow through a Pipe system with a cross pipe. (2nd report, The influence of the branching discharge ratio and structural factor of a pipe upon the unstable discharge phenomenon). Journal of Japan Society of Mechanical Engineers. 1987;**87**:1067-1610

[12] Ohki A et al. Unstable Phenomenon in Flow through a Pipe system with a cross pipe. (3rd report, Investigation of flow stabilizing structure of a pipe for the unstable discharge phenomenon). Journal of Japan Society of Mechanical Engineers. 1989;**88**:3040-3043

[13] Horsthemke, Lefever. Noise Induced Transitions. Springer Series in Synergetics; 2005

[14] Gavilán Moreno CJ. The Bistable flow in a nuclear Power Plant. A bifurcation in the fluid flow in the recirculation loop, modeling and simulation. Proceedings of the Bifurcation in Fluid Dynamics, Barcelona. 2011

[15] Bagarinao E et al. Generalized once-parameter bifurcation diagram reconstruction using time series. Physica D. 1998;**124**:258-270

[16] Ioos G, Joseph DD. Elementary Stability and Bifurcation Theory. Springer; 1997

[17] United Stated Nuclear Regulatory Commission. Anomalous behavior of recirculation loop flow in jet pump BWR plants, Information notice No. 86–110, Washington. 1986

[18] General Electric. Recirculation system bi-stable flow in jet pump BWRs. SIL. 467, Rev. 1. 2006

[19] Carrera AN, Martinez-Mendez E, Paredes GE. Analyses of the bistable flow phenomenon at Laguna Verde Nuclear Power Plant. In: Proceedings of the 2006 International LAS/ANS Congress of Buenos Aires (in Spanish)

[20] Gavilán Moreno CJ. Analysis and simulation of the flow signal in the recirculation loop of a nuclear power station during a bistable flow event. Nuclear Engineering and Design. 2008; **238**(10):2754-2760

[21] Nuñez-Carrera A, Prieto-Guerrero A, Espinosa-Martínez E-G, Espinosa-Paredes G. Analysis of a signal during bistable flow events in Laguna Verde Nuclear Power Station with wavelets techniques. Nuclear Engineering and Design. 2009;**12**:2942-2951

[22] Gavilán Moreno CJ. Hydraulic Study on recirculation loops using computational fluid dynamics (CFD). Design optimization and turbulence reduction. Nuclear Engineering and Design. 2009;**239**:434-441

[23] Bagariano E et al. Reconstructing bifurcation diagrams of dynamical systems using measured time series. Method Information in Medicine. 2000; **39**:146-149

[24] Boudjema G, Cazelles B. Extraction of nonlinear dynamics from short and noisy time series. Chaos Solitons and Fractals. 2001;**12**:2051-2069

Chapter 4

General Drag Correlations for Particle-Fluid System

Zheng Qi, Shibo Kuang, Liangwan Rong, Kejun Dong and Aibing Yu

Abstract

Particle-fluid flows are commonly encountered in industrial applications. It is of great importance to understand the fundamentals governing the behavior of such a flow system for better process design, control, and optimization. Generally, the particle-fluid flow behavior is strongly influenced by the interaction forces between fluid and particles. Among the various kinds of particle-fluid interaction forces, the drag force is the most essential. This chapter reviews the modeling of drag force for particle-fluid systems: from single particle to multiple particles, monosize to multisize, spherical to nonspherical, and Newtonian fluid to non-Newtonian fluid. Typical drag correlations in the literature are compared and assessed in terms of physical meaning, consistency, and generality.

Keywords: drag force, particle-fluid flow, computational fluid dynamics, Lattice-Boltzmann method

1. Introduction

Particle-fluid flows are commonly encountered in industrial applications. It is of great importance to understand the fundamentals governing the behavior of such a flow system for better process design, control, and optimization. The flow behaviors of particles and fluid are strongly influenced by their interaction forces. Thus, it is critical to model the particle-fluid interaction forces accurately when simulating particle-fluid flows. The particle-fluid interaction forces include the pressure gradient force (or buoyancy force), drag force, virtual mass force, Basset force, and lift forces. The drag force, which is usually the dominant force in many particle-fluid flow systems, is undoubtedly the most critical and most studied.

Numerous efforts have been made to quantify the drag force using experimental or numerical methods in the past decades. Early studies in this area were mainly conducted by experiments [1–4]. Due to the limitations of techniques, the experimental conditions are difficult to control. Therefore, the drag force models proposed in these studies somewhat lack generality and consistency. Nevertheless, these pioneer studies [1–4] provide a solid foundation for subsequent research. With the rapid development of computational technology, various numerical methods have become attractive for studying the fluid-particle interaction on a subparticle scale, such as direct numerical simulation (DNS) and Lattice-Boltzmann (LB) model. With these numerical approaches, various fluid-particle systems can be studied under well-controlled conditions, considering more important and complicated factors that affect fluid-particle drag force. Nowadays, the studies on the fluid-particle interaction associated with the development of the drag force model have been extended extensively: from single particle to multiple particles, monosize to multisize, spherical particle to nonspherical particle system, and Newtonian fluid to non-Newtonian fluid. This chapter will review the modeling of particle-fluid drag force from these perspectives.

2. Mathematical model

The flow of a fluid is governed by the momentum and mass conservation equations when ignoring the compressible and viscous heat dissipation effect:

$$\frac{\partial \rho}{\partial t} + \nabla \cdot (\rho \mathbf{u}) = 0 \tag{1}$$

$$\frac{\partial (\rho \mathbf{u})}{\partial t} + \nabla \cdot (\rho \mathbf{u}\mathbf{u}) = \nabla p + \nabla \cdot (\rho \nu \nabla \mathbf{u}) \tag{2}$$

where ρ, p, $\mathbf{u}$, and ν_f are the fluid density, pressure, velocity, and fluid kinematic viscosity, respectively.

In a traditional computational fluid dynamics (CFD) method, Eqs. (1) and (2) with initial and boundary conditions are often solved by the finite volume method (FVM) or finite element method (FEM). However, due to the difficulty of grid generation and boundary treatment in the DNS, the FVM and FEM are not widely used in studying the interaction between fluid and particles on a subparticle scale. Instead of FVM and FEM, the LB model is more widely used to study the drag force in particle-fluid systems. Its detailed description can be found elsewhere [5]. For brevity, the following only briefly introduces the LB model used.

The equations of the LB model are solved in two steps. First is the collision step:

$$f_i^+(\mathbf{x}, t) = f_i(\mathbf{x}, t) - \left[\mathbf{M}^{-1}\mathbf{S}_f \cdot (\mathbf{m} - \mathbf{m}^{eq})(\mathbf{x}, t)\right]_i \tag{3}$$

which is followed by a propagation step:

$$f_i(\boldsymbol{x} + \boldsymbol{c}_i\delta t, t + \delta t) = f_i^+(\boldsymbol{x}, t) \tag{4}$$

where $\mathbf{m} = \mathbf{Mf}$ and $\mathbf{f}$ is the column vector of f_i is the velocity distribution function at the lattice node $\mathbf{x}$ and the time t with the discrete velocity vector $\mathbf{c}_i$ in the ith direction. The number of directions of discrete velocity vectors $\mathbf{c}_i$ depends on the velocity model used. For example, provided that the fluid particle can move in the 19 directions in a three-dimensional case, as shown in **Figure 1**, the velocity model is referred to as a *D3Q19* scheme. For the *D3Q19* scheme, $\mathbf{c}_i$ is given as

$$\mathbf{c}_i = c\begin{bmatrix} 0 & 1 & -1 & 0 & 0 & 0 & 0 & 1 & -1 & 1 & -1 & 1 & -1 & 1 & -1 & 0 & 0 & 0 & 0 \\ 0 & 0 & 0 & 1 & -1 & 0 & 0 & 1 & 1 & -1 & -1 & 0 & 0 & 0 & 0 & 1 & -1 & 1 & -1 \\ 0 & 0 & 0 & 0 & 0 & 1 & -1 & 0 & 0 & 0 & 0 & 1 & 1 & -1 & -1 & 1 & 1 & -1 & -1 \end{bmatrix} \tag{5}$$

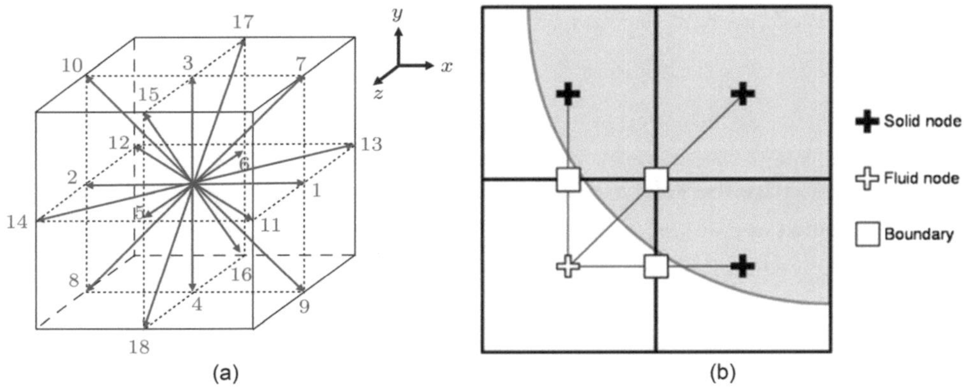

Figure 1.
Schematic illustrations of (a) D3Q19 LB method, and (b) bounce-back boundary treatment.

where $c = \frac{\delta x}{\delta t}$ is the lattice speed, δx is the lattice length, and δt is the time step. Other velocity models are also available, such as D3Q15 and D3Q27, and their details can be found elsewhere [5, 6].

The transfer matrix $\mathbf{M}$ in Eq. (3) defines the transformation of the distribution function to the moment space, which can be chosen to be the same as that of D'Humieres et al. [6], as given in Eq. (6). Note that $\mathbf{S}_f$ in Eq. (7) is a diagonal matrix. Provided that all the elements of $\mathbf{S}_f$ equal the same value as τ, Eq. (3) can be reduced to the well-known LBGK scheme, $f_i^+(\mathbf{x}, t) = f_i(\mathbf{x}, t) - \frac{f_i(\mathbf{x}, t) - f_i^{(eq)}(\mathbf{x}, t)}{\tau}$. D'Humieres et al. [6] gives an optimized value of $\mathbf{S}_f$, where $s_0^f = s_3^f = s_5^f = s_7^f = 0$, $s_1^f = 1.19$, $s_2^f = s_{10}^f = s_{12}^f = 1.4$, $s_{16}^f = s_{17}^f = s_{18}^f = 1.98$, $s_9^f = s_{11}^f = s_{13}^f = s_{14}^f = s_{15}^f = s_\nu$, $s_4^f = s_6^f = s_8^f = s_q$, and $s_q = \frac{8(2-s_\nu)}{(8-s_\nu)}$. s_q are set to $s_q = \frac{8(2-s_\nu)}{(8-s_\nu)}$ to satisfy the nonslip boundary condition. It should be noted that s_ν has a relationship with the kinematic viscosity ν, as given by Eq. (8).

$$\mathbf{M} = \begin{bmatrix} 1 & 1 & 1 & 1 & 1 & 1 & 1 & 1 & 1 & 1 & 1 & 1 & 1 & 1 & 1 & 1 & 1 & 1 & 1 \\ -30 & -11 & -11 & -11 & -11 & -11 & -11 & 8 & 8 & 8 & 8 & 8 & 8 & 8 & 8 & 8 & 8 & 8 & 8 \\ 12 & -4 & -4 & -4 & -4 & -4 & -4 & 1 & 1 & 1 & 1 & 1 & 1 & 1 & 1 & 1 & 1 & 1 & 1 \\ 0 & 1 & -1 & 0 & 0 & 0 & 0 & 1 & -1 & 1 & -1 & 1 & -1 & 1 & -1 & 0 & 0 & 0 & 0 \\ 0 & -4 & 4 & 0 & 0 & 0 & 0 & 1 & -1 & 1 & -1 & 1 & -1 & 1 & -1 & 0 & 0 & 0 & 0 \\ 0 & 0 & 0 & 1 & -1 & 0 & 0 & 1 & 1 & -1 & -1 & 0 & 0 & 0 & 0 & 1 & -1 & 1 & -1 \\ 0 & 0 & 0 & -4 & 4 & 0 & 0 & 1 & 1 & -1 & -1 & 0 & 0 & 0 & 0 & 1 & -1 & 1 & -1 \\ 0 & 0 & 0 & 0 & 0 & 1 & -1 & 0 & 0 & 0 & 0 & 1 & 1 & -1 & -1 & 1 & 1 & -1 & -1 \\ 0 & 0 & 0 & 0 & 0 & -4 & 4 & 0 & 0 & 0 & 0 & 1 & 1 & -1 & -1 & 1 & 1 & -1 & -1 \\ 0 & 2 & 2 & -1 & -1 & -1 & -1 & 1 & 1 & 1 & 1 & 1 & 1 & 1 & 1 & -2 & -2 & -2 & -2 \\ 0 & -4 & -4 & 2 & 2 & 2 & 2 & 1 & 1 & 1 & 1 & 1 & 1 & 1 & 1 & -2 & -2 & -2 & -2 \\ 0 & 0 & 0 & 1 & 1 & -1 & -1 & 1 & 1 & 1 & 1 & -1 & -1 & -1 & -1 & 0 & 0 & 0 & 0 \\ 0 & 0 & 0 & -2 & -2 & 2 & 2 & 1 & 1 & 1 & 1 & -1 & -1 & -1 & -1 & 0 & 0 & 0 & 0 \\ 0 & 0 & 0 & 0 & 0 & 0 & 0 & 1 & -1 & -1 & 1 & 0 & 0 & 0 & 0 & 0 & 0 & 0 & 0 \\ 0 & 0 & 0 & 0 & 0 & 0 & 0 & 0 & 0 & 0 & 0 & 0 & 0 & 0 & 0 & 1 & -1 & -1 & 1 \\ 0 & 0 & 0 & 0 & 0 & 0 & 0 & 0 & 0 & 0 & 0 & 1 & -1 & -1 & 1 & 0 & 0 & 0 & 0 \\ 0 & 0 & 0 & 0 & 0 & 0 & 0 & 1 & -1 & 1 & -1 & -1 & 1 & -1 & 1 & 0 & 0 & 0 & 0 \\ 0 & 0 & 0 & 0 & 0 & 0 & 0 & -1 & -1 & 1 & 1 & 0 & 0 & 0 & 0 & 1 & -1 & 1 & -1 \\ 0 & 0 & 0 & 0 & 0 & 0 & 0 & 0 & 0 & 0 & 0 & 1 & 1 & -1 & -1 & -1 & -1 & 1 & 1 \end{bmatrix} \quad (6)$$

$$\mathbf{S}_f = diag\left(s_0^f, s_1^f, s_2^f, s_3^f, s_4^f, s_5^f, s_6^f, s_7^f, s_8^f, s_9^f, s_{10}^f, s_{11}^f, s_{12}^f, s_{13}^f, s_{14}^f, s_{15}^f, s_{16}^f, s_{17}^f, s_{18}^f\right) \quad (7)$$

$$\nu = \frac{1}{3}\left(\frac{1}{s_\nu} - 0.5\right) \quad (8)$$

$\mathbf{m}^{eq}$ in Eq. (3) are the velocity equilibrium moments defined by Eq. (9). The j_x, j_y, and j_z are defined as $j_x = \rho u_x$, $j_y = \rho u_y$, and $j_z = \rho u_z$. The constant ρ_0 is the mean fluid density of the system, which is set to unity for an incompressible fluid.

$$\mathbf{m}_f^{eq} = \begin{pmatrix} \rho, \ -11\rho + \frac{19\left(j_x^2 + j_y^2 + j_z^2\right)}{\rho_0}, \ 3\rho - \frac{11\left(j_x^2 + j_y^2 + j_z^2\right)}{2\rho_0}, j_x, \ -\frac{2j_x}{3}, j_y, \ -\frac{2j_y}{3}, j_z, \ -\frac{2j_z}{3}, \\ \frac{\left(2j_x^2 - j_y^2 - j_z^2\right)}{\rho_0}, \ -\frac{\left(j_x^2 - j_y^2 - j_z^2\right)}{2\rho_0}, \frac{\left(j_y^2 - j_z^2\right)}{\rho_0}, \frac{\left(j_y^2 - j_z^2\right)}{2\rho_0}, \frac{j_x j_y}{\rho_0}, \frac{j_y j_z}{\rho_0}, \frac{x j_z}{\rho_0}, 0, 0, 0 \end{pmatrix}^T \quad (9)$$

When the effect of fluid rheology is considered, ν is dependent on the shear rate for a non-Newtonian fluid other than a constant for a Newtonian fluid. It is described here by the power-law model, which is suitable for a wide range of non-Newtonian fluids [7]:

$$\nu = \nu_0 \dot{e}^{n-1} = \nu_0\left(2e_{\alpha\beta}e_{\alpha\beta}\right)^{(n-1)/2} \quad (10)$$

where ν_0 is the flow consistency index, $\dot{e}$ denotes the shear rate, and n is the power-law index. The fluid shows shear-thickening behavior when $n > 1$, shear-thinning behavior when $n < 1$, and Newtonian behavior when $n = 1$. In Eq. (10), Einstein's summation convention is applied and $e_{\alpha\beta}$ represents an element of the tensor $\dot{e}$ at the position $\alpha\beta$, calculated by [8]

$$e_{\alpha\beta} = \frac{3}{2\rho\tau}\sum_{i=0} f_i^{(1)} c_{i\alpha} c_{i\beta} \quad (11)$$

where α and β denote spatial indices, $c_{i\alpha}$ and $c_{i\beta}$ are the elements of the vector $\mathbf{c}_i$ at positions α and β, and $f_i^{(1)} = f_i - f_i^{\mathrm{eq}}$ is the non-equilibrium part of the distribution function.

Through the Chapman-Enskog multiple-scale expansion, Eqs. (3) and (4) can be recovered to the Navier-Stokes equations in a low Mach number limit. The macroscopic variables such as density ρ, momentum density $\rho\mathbf{u}$, and stress $\boldsymbol{\sigma}$, can be obtained through the moments of the velocity and temperature distributions:

$$\rho = \sum_i f_i, \ \rho\boldsymbol{u} = \sum_i f_i \boldsymbol{c}_i, \ \boldsymbol{\sigma} = \sum_i f_i \boldsymbol{c}_i \boldsymbol{c}_i \quad (12)$$

The force exerted on a solid particle is calculated by the momentum exchange method proposed by Ladd [9], given by

$$f_i(\mathbf{x}_b, t + 0.5) = 2\left(f_i^+(\mathbf{x}_f, t) - f_{\bar{i}}^+(\mathbf{x}_s, t)\right)\boldsymbol{c}_i \quad (13)$$

where $\mathbf{x}_f$, $\mathbf{x}_s$, and $\mathbf{x}_b$ denote the nodes in the fluid region and solid region and at the boundary for a stationary boundary on a uniform lattice, and the subscript $\bar{i}$ indicates the opposite direction of i. The curved wall of spherical solid particles is treated by the halfway boundary scheme (**Figure 1b**), which sets the wall node at the mid-point between fluid and solid nodes. The selection of the halfway boundary treatment has various advantages. First, it is the simplest and most efficient boundary treatment. Secondly, compared with interpolation schemes, the halfway boundary scheme is very space-saving in modeling particles that are placed closely (e.g., in a packed bed) [10]. Thirdly, the high numerical stability of the halfway boundary scheme ensures simulation reliability under various complicated flow conditions. Lastly, the halfway boundary scheme is a second-order scheme; its accuracy in calculating forces can be as good as other interpolation schemes and better than immersed boundary schemes once a calibration step is applied [11].

3. Model validation

It is necessary to verify the validity of the mathematical model before its application for numerical experiments. The LB model has proven to be valid for studying the fluid-particle interactions under different conditions involving a single particle, two interactive particles, and packed beds of spherical and nonspherical particles over a wide range of conditions [12–15]. For example, the LB prediction of drag coefficient $C_{D0} = \frac{8f_d}{\pi\rho d_e^2 U_0^2}$ for an isolated particle in the Newtonian fluid has been compared with the experimental measurements of Schlichting and Gersten [16], and with those calculated by three correlations given by Stokes [17], Dallavalle [4], and Clift et al. [18], respectively. **Figure 2a** shows good agreement between the simulations and experiments, including the three correlations. Specifically, in the intermediate flow region of Re = 100–1000, the drag coefficient of the sphere for the velocity inlet and outflow conditions under open BCs falls closer to the experimental data than the one under periodic boundary conditions. This may be due to the nonuniformity in the flow far away from the sphere for high values of Re, the state of which can be different from that of the experiments. Drag coefficient in the flow of power-law fluid has also been compared against the measurements by Chhabra [19] and Peden and

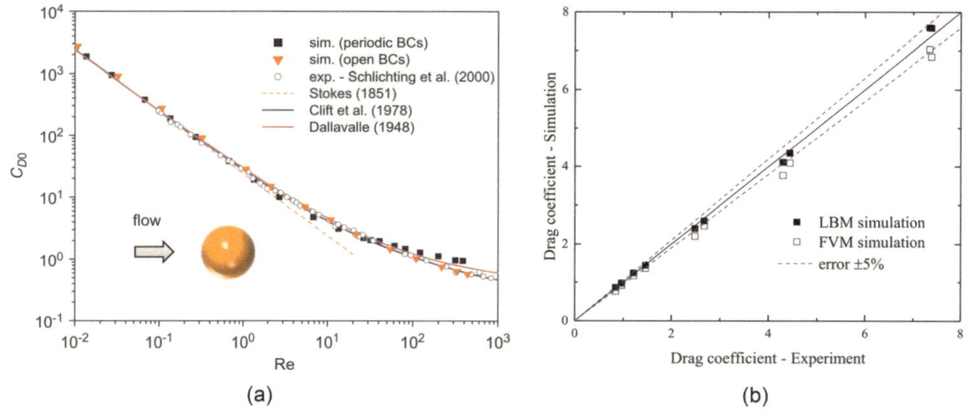

Figure 2.
Comparison between simulated and measured drag coefficients in (a) Newtonian, and (b) non-Newtonian fluids.

Luo [20] over $5 \leq Re \leq 100$ and $0.6 \leq n \leq 1$, as shown in **Figure 2b**. In the comparison, the numerical data obtained by Dhole et al. [21] using the FVM are also considered. **Figure 2** shows that both the LB model and FVM can well predict the drag coefficient on particles in the power-law fluid flows at different Re and n. The prediction errors are less than 4.2% for the present LB model with a calibration step and 7.6% for the FVM.

4. Drag force for single-particle system

The fluid flow past an isolated particle is one of the most basic flow phenomena. It can be seen as a special case of a particle-fluid system when the particle volume fraction is close to zero. Also, it is very natural to start from a single-particle system to investigate the drag force in a complicated particle-fluid system. In the past decades, extensive efforts have been made to study various physical characteristics (e.g., drag, wake, shear, heat transfer, and vortex shedding) of the flow past an isolated particle. Tiwari et al. [22, 23] have given a comprehensive review on this topic, and interested readers can refer to the review for details.

For an isolated particle in fluid flows, the particle-fluid drag force has been well established, which is expressed as [4]

$$\mathbf{f}_d = \frac{1}{8} C_{D0} \pi \rho_f d_p^2 |\mathbf{u}_f - \mathbf{u}_p| (\mathbf{u}_f - \mathbf{u}_p) \tag{14}$$

where $\mathbf{f}_d$ is the drag force, C_{D0} is the drag coefficient for a single particle in fluid flows, which is determined by the Reynolds number $Re = \frac{\rho_f d_p |\mathbf{u}_f - \mathbf{u}_p|}{\mu_f}$, ρ_f is the fluid density, d_p is the particle diameter, μ_f is the fluid viscosity, and $\mathbf{u}_f$ and $\mathbf{u}_p$ are the fluid and particle velocities, respectively.

Figure 3 shows the relationship between C_{D0} and Re in different flow regimes established based on experimental and numerical data. It can be seen that the effect of

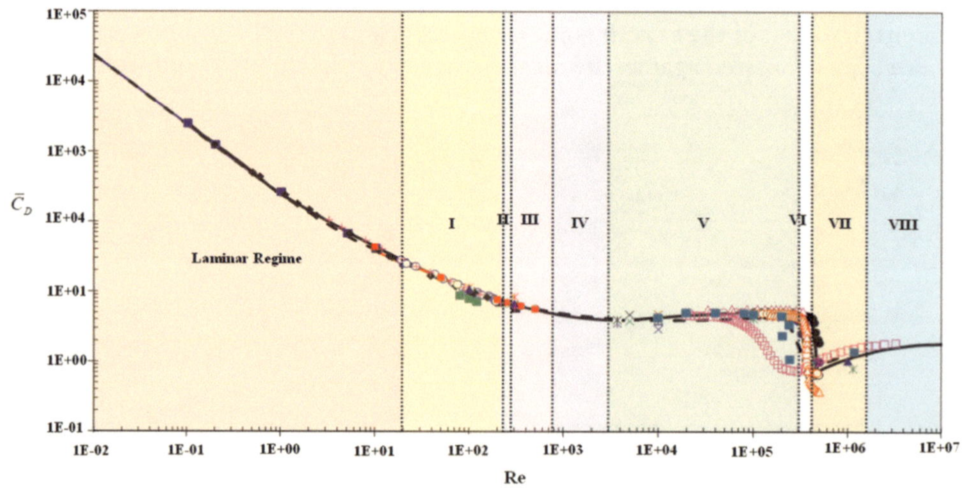

Figure 3.
Drag coefficient for the flows past a stationary sphere in different flow regimes based on experimental and numerical data [23].

References	Drag correlation C_{D0}
Schiller and Nauman (1935) [24]	$C_{D0} = \begin{cases} \frac{24}{Re}\left(1 + 0.15 Re^{0.687}\right) & Re < 1000 \\ 0.44 & Re > 1000 \end{cases}$
Dallavalle (1948) [4]	$C_{D0} = \left(0.63 + \frac{4.8}{\sqrt{Re}}\right)^2$
Turton and Levenspiel (1986) [25]	$C_{D0} = \frac{24}{Re}\left(1 + 0.173 Re^{0.657}\right) + \frac{0.413}{1+16300 Re^{-1.09}}$

Table 1.
Drag correlations for an isolated particle in fluid flows.

Re on C_{D0} varies in different flow regimes. Generally, the flow regimes can be classified into nine groups: laminar regime or axisymmetric wake regime (regime I), planar symmetric wake regime (regime II), vortex shedding regime (regime III), separating vortex tubes regime (regime IV), subcritical regime (regime V), critical regime (regime VI), supercritical regime (regime VII), and transcritical regime (regime VIII). It should be noted that the regimes I–IV may also be called the transition regime. In the laminar regime, the drag coefficient C_{D0} decreases significantly as Re increases. In contrast, in the transition regime (regimes I--IV), the drop of C_{D0} decreases as Re increases and C_{D0} approaches a constant in the subcritical regime. Then, C_{D0} experiences a rapid drop in the critical regime but increases again and reaches a near constant in the subsequent supercritical and transcritical regimes.

Due to the complicated effect of Re on C_{D0}, it is still challenging to formulate a drag coefficient correlation that can accurately describe the complicated variation of C_{D0} in different flow regimes. However, a general drag coefficient can be formulated to meet the needs of engineering applications if the following requirements are satisfied:

1. C_{D0} decreases in the laminar region.
2. The drop of C_{D0} decreases in the transition regime.
3. C_{D0} approaches a positive constant value when Re is relatively large.

Table 1 summarizes some of the well-known drag coefficient correlations. Notably, all these drag coefficient correlations can meet the above requirements.

5. From single-particle system to multiparticle system

Extension from an isolated particle system to a multi-particle system is a big challenge due to the complexity of the particulate system. The presence of other particles reduces the space for fluid and generates a sharp fluid velocity gradient, yielding increased shear stress on particle surfaces. The drag force enhancement is closely associated with particle configuration, particle-fluid slip velocity, and particle and fluid properties.

Generally, the methods to formulate drag correlations for multi-particle systems can be grouped into Ergun and Wen-Yu types. The Ergun-type model, also called the capillary model, focuses on the flow resistance of the whole system, exemplified by the Ergun Eq. (1). This kind of model originates from the idea of treating the void

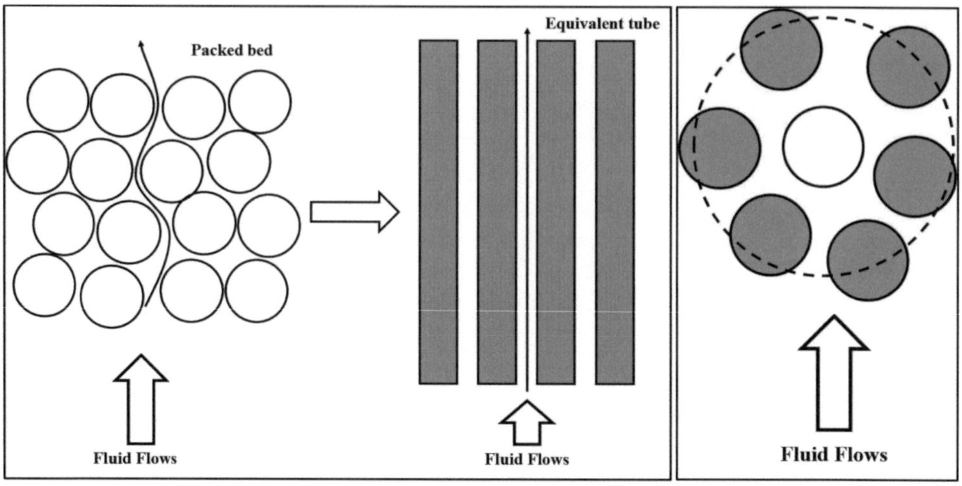

Figure 4.
Schematic representation of idealization of Ergun-type model (a), and Wen-Yu-type model (b).

space in a porous medium as a bunch of tortuous conduits, as illustrated in **Figure 4**. Using equivalent tubes instead of the void space of a packed bed, one can model the total pressure drop across the whole system by considering the contributions of viscous force and inertial force. Correspondingly, its friction factor $f = \frac{d(\Delta P)}{L\rho|\mathbf{U_s}|^2}\left(\frac{\varepsilon^3}{1-\varepsilon}\right)$ falls into the form of $f = \frac{A}{Re} + B$. The Ergun-type model is an easier way to formulate correlations by fitting the data in the creeping flow regime or the extremely high Reynolds flow regime. However, as pointed out previously [26], because of the original idea of treating a multiparticle system as tortuous conduits, the Ergun-type correlations are more suitable for densely packed beds but usually have a discontinuity at high porosity. For example, when particles are immersed in a highly dilute system, the drag force determined by the Ergun-type model is usually inconsistent with the value given by the drag correlation for a single particle, which is apparently against the real physical world. This defect may cause problems in some situations, such as CFD-DEM simulations, and limit the application range of the Ergun-type model. The Wen-Yu-type model, also called the submerge object model, focuses on the enhanced drag force on a particle due to neighboring particles, exemplified by the Wen-Yu Eq. (2). Such a model is more straightforward as it directly considers the effect of neighboring particles on the drag force, as illustrated in **Figure 4**. In the model, the enhancement of drag force on particles in a multi-particle system can be expressed as a ratio to the drag force on a single particle in an unhindered environment. Correspondingly, the drag force in the Wen-Yu type can be written in $\mathbf{f}_d = \frac{1}{8}C_D\pi\rho_f d_p^2|\mathbf{u}_f - \mathbf{u}_p|(\mathbf{u}_f - \mathbf{u}_p)$, where $C_D = C_{D0}{\varepsilon_f}^{2-\chi}$. Unlike Ergun-type models, Wen-Yu-type models naturally ensure the consistency of drag force between the predictions by the correlation for a multi-particle system under dilute conditions and the correlation for a single-particle system, leading to its popularity in CFD-DEM simulations.

Irrespective of the different origins of Ergun-type and Wen-Yu-type models, the drag correlations proposed by either way can be written in the following equation:

$$\mathbf{f}_d = \frac{1}{8}C_D\pi\rho_f d_p^2|\mathbf{u}_f - \mathbf{u}_p|(\mathbf{u}_f - \mathbf{u}_p) \tag{15}$$

References	Drag coefficient C_D	C_D when $\varepsilon_f \to 1$	Req (2-1)	Req (2-2)	Req (2-3)
Ergun (1952) [1]	$C_D = \frac{4}{3}\left(150\frac{1-\varepsilon_f}{Re} + 1.75\right)$	$C_{D,\varepsilon_f \to 1} = 2.33$	×	√	×
Wen and Yu (1966) [2]	$C_D = C_{D0}{\varepsilon_f}^{2-\chi}$, $C_{D0} = \begin{cases} \frac{24}{Re}\left(1+0.15Re^{0.687}\right) & Re < 1000 \\ 0.44 & Re > 1000 \end{cases}$, $\chi = 3.65$	$C_{D,\varepsilon_f \to 1} = \begin{cases} \frac{24}{Re}\left(1+0.15Re^{0.687}\right) & Re < 1000 \\ 0.44 & Re > 1000 \end{cases}$	√	√	√
Gibilaro et al. (1985) [27]	$C_D = C_{D0}{\varepsilon_f}^{2-\chi}$, $C_{D0} = \frac{23}{Re} + 0.448$, $\chi = 3.8$	$C_{D,\varepsilon_f \to 1} = \frac{23}{Re} + 0.448$	√	√	√
Gidaspow (1994) [28]	$C_D = \begin{cases} C_{D0}{\varepsilon_f}^{2-\chi} & \varepsilon_f \geq 0.8 \\ \frac{4}{3}\left(150\frac{1-\varepsilon_f}{Re} + 1.75\right) & \varepsilon_f < 0.8 \end{cases}$, $C_{D0} = \begin{cases} \frac{24}{Re}\left(1+0.15Re^{0.687}\right) & Re < 1000 \\ 0.44 & Re > 1000 \end{cases}$, $\chi = 3.65$	$C_{D,\varepsilon_f \to 1} = \begin{cases} \frac{24}{Re}\left(1+0.15Re^{0.687}\right) & Re < 1000 \\ 0.44 & Re > 1000 \end{cases}$	√	×	√
Di Felice (1994) [29]	$C_D = C_{D0}{\varepsilon_f}^{2-\chi}$, $C_{D0} = \left(0.63 + \frac{4.8}{\sqrt{Re}}\right)^2$, $\chi = 3.7 - 0.65exp\left[-\frac{(1.5-\log Re)^2}{2}\right]$	$C_{D,\varepsilon_f \to 1} = \left(0.63 + \frac{4.8}{\sqrt{Re}}\right)^2$	√	√	√
Van der Hoef et al. (2005) [30]	$C_D = \frac{4}{3}\left(180\frac{1-\varepsilon_f}{Re} + 18\frac{\varepsilon_f^4\left(1+1.5\sqrt{1-\varepsilon_f}\right)}{Re}\right)$	$C_{D,\varepsilon_f \to 1} = \frac{24}{Re}$	×	√	×
Benyahia et al. (2006) [31]	$C_D = \frac{12\varepsilon_f^2 F}{Re_*}$, where $F = \begin{cases} 1 + \frac{3}{8}Re_* & \varepsilon_f \geq 0.99 \; and \; Re_* \leq \frac{F_2 - 1}{3/8 - F_3} \\ F_0 + F_1 {Re_*}^2 & \varepsilon_f < 0.99 \; and \; Re_* \leq \frac{F_3 + \sqrt{F_3^2 - 4F_1(F_0 - F_2)}}{2F_1}, \\ F_2 + F_3 Re_* & others \end{cases}$ $F_0 = \begin{cases} (1-w)\left[\frac{1 + 3\sqrt{(1-\varepsilon_f)/2} + (135/64)(1-\varepsilon_f)\ln(1-\varepsilon_f) + 17.4(1-\varepsilon_f)}{1 + 0.681(1-\varepsilon_f) - 8.48(1-\varepsilon_f)^2 + 8.16(1-\varepsilon_f)^3}\right] + w\left[10\frac{1-\varepsilon_f}{\varepsilon_f^3}\right] & 0.6 < \varepsilon_f < 0.99 \\ 10\frac{1-\varepsilon_f}{\varepsilon_f^3} & \varepsilon_f \leq 0.6 \end{cases}$, $F_1 = \begin{cases} \sqrt{2/(1-\varepsilon_f)}/40 & 0.9 \leq \varepsilon_f < 0.99 \\ 0.11 + 0.00051\exp\left[11.6(1-\varepsilon_f)\right] & \varepsilon_f < 0.9 \end{cases}$, $F_2 = \begin{cases} (1-w)\left[\frac{1 + 3\sqrt{(1-\varepsilon_f)/2} + (135/64)(1-\varepsilon_f)\ln(1-\varepsilon_f) + 17.89(1-\varepsilon_f)}{1 + 0.681(1-\varepsilon_f) - 11.03(1-\varepsilon_f)^2 + 15.41(1-\varepsilon_f)^3}\right] + w\left[10\frac{1-\varepsilon_f}{\varepsilon_f^3}\right] & \varepsilon_f > 0.6 \\ 10\frac{1-\varepsilon_f}{\varepsilon_f^3} & \varepsilon_f \leq 0.6 \end{cases}$,	$C_{C_{D,\varepsilon_f \to 1}} = \begin{cases} \frac{12}{Re_*}\left(1 + 0.375Re_*\right) & Re_* \leq \frac{F_2 - 1}{3/8 - F_3} \\ \frac{12}{Re_*}\left(1 + 0.03667Re_*\right) & Re > \frac{F_2 - 1}{3/8 - F_3} \end{cases}$	×	×	√

References	Drag coefficient C_D	C_D when $\varepsilon_f \to 1$	Req (2-1)	Req (2-2)	Req (2-3)
	$F_3 = \begin{cases} 0.9351(1-\varepsilon_f) + 0.03667\ \varepsilon_f > 0.9047 \\ 0.0673 + 0.212(1-\varepsilon_f) + 0.0232\varepsilon_f^{-5}\varepsilon_f \le 0.9047 \end{cases}$, $Re_* = \frac{\rho_f \varepsilon_f d_p \lvert \mathbf{u}_f - \mathbf{u}_p \rvert}{2\mu_f}$, $w = e^{(-10(\varepsilon_f - 0.6)/(1-\varepsilon_f))}$				
Mazzei and Lettieri (2007) [32]	$C_D = C_{D0}{\varepsilon_f}^{2-\chi}, C_{D0} = \left(0.63 + \frac{4.8}{\sqrt{Re}}\right)^2, \chi = 1 - 2\frac{4.8+0.42 Re_*^{3/4}}{1+0.175 Re_*^{3/4}}$ and $Re_* = \frac{\rho_f d_p \lvert \mathbf{u}_f - \mathbf{u}_p \rvert}{\mu_f \varepsilon_f^{n-1}}$	$C_{D,\varepsilon_f \to 1} = \left(0.63 + \frac{4.8}{\sqrt{Re}}\right)^2$	√	√	√
Beetstra et al. (2007) [26]	$C_D = \frac{4}{3}\left(180\frac{1-\varepsilon_f}{Re} + 18\frac{\varepsilon_f^4(1+1.5\sqrt{1-\varepsilon_f})}{Re} + 0.31\frac{\varepsilon_f^{-1}+3(1-\varepsilon_f)\varepsilon_f+8.4 Re^{-0.343}}{1+10^{3(1-\varepsilon_f)} Re^{-0.5-2(1-\varepsilon_f)}}\right)$	$C_{C_{D\varepsilon_f \to 1}} = \frac{24}{Re} + 0.413\frac{1+8.4 Re^{-0.343}}{1+Re^{-0.5}}$	√	√	√
Cello et al. (2010) [33]	$C_D = \frac{24{\varepsilon_f}^2}{Re}\left[K_1 + K_2\varepsilon_f^4 + K_3\left(1-\varepsilon_f^4\right)\right], K_0 = \frac{1-\varepsilon_f}{1+3\varepsilon_f}, K_1 = \frac{1+128K_0+715K_0^2}{\varepsilon_f(1+49.5K_0)}, K_2 = \frac{1+0.13 Re+0.000666 Re^2}{1+0.0342 Re+0.00000692 Re^2} - 1,$ $K_3 = \left(\frac{2 Re^2}{1+Re}\right)\left(\frac{-410\varepsilon_f+9.2\times10^7 K_0^{20} Re+1900\varepsilon_f^2-0.066 Re}{6600\varepsilon_f+0.000429 Re-43000\varepsilon_f^2-0.000131 Re^2+73800\varepsilon_f^3}\right)$	$C_{D,\varepsilon_f \to 1} = \frac{24}{Re}(1+K_2)$	×	√	×
Tenneti et al. (2011) [34]	$C_D = C_{D0} + \frac{4}{3}\left(104.58\frac{1-\varepsilon_f}{Re} + 8.64\frac{(1-\varepsilon_f)^{1/3}}{Re\,\varepsilon_f} + 18\varepsilon_f^3(1-\varepsilon_f)^3\left(0.95 + \frac{0.61(1-\varepsilon_f)^3}{\varepsilon_f^2}\right)\right),$ $C_{D0} = \begin{cases} \frac{24}{Re}(1+0.15 Re^{0.687})\ Re < 1000 \\ 0.44\ Re > 1000 \end{cases}$	$C_{D,\varepsilon_f \to 1} = \begin{cases} \frac{24}{Re}(1+0.15 Re^{0.687})\ Re < 1000 \\ 0.44\ Re > 1000 \end{cases}$	√	×	√
Rong et al. (2013) [14]	$C_D = C_{D0}{\varepsilon_f}^{2-\chi}, C_{D0} = \left(0.63 + \frac{4.8}{\sqrt{Re}}\right)^2, \chi = 2.65(1+\varepsilon_f) - (5.3 - 3.5\varepsilon_f)\varepsilon_f^2 \exp\left[-\frac{(1.5-\log Re)^2}{2}\right]$	$C_{D,\varepsilon_f \to 1} = \left(0.63 + \frac{4.8}{\sqrt{Re}}\right)^2$	√	√	√
Zaidi et al. (2014) [35]	$C_D = \begin{cases} \frac{4}{3}\left(180\frac{1-\varepsilon_f}{Re\,\varepsilon_f} + 18\frac{\varepsilon_f^4\left(1+1.5\sqrt{1-\varepsilon_f}\right)}{Re} + \frac{0.612}{\varepsilon_f^{0.7}}\right)\ Re \le 200 \\ \frac{4}{3}\left(\frac{196.2(1-\varepsilon_f)^{0.4}}{Re\,\varepsilon_f^{-0.3}} + \frac{0.432}{\varepsilon_f^{0.86}}\right)\ Re \ge 200 \end{cases}$	$C_{D,\varepsilon_f \to 1} = \begin{cases} \frac{24}{Re} + 0.816\ Re \le 200 \\ 0.576\ Re \ge 200 \end{cases}$	×	×	×
Bogner et al. (2015) [36]	$C_D = \frac{24}{Re\,\varepsilon_f^{2.726}}\left[1.751 + 0.151 Re^{0.684} - 0.445(1+Re)^{1.04(1-\varepsilon_f)} - 0.16(1+Re)^{0.0003(1-\varepsilon_f)}\right]$	$C_{D,\varepsilon_f \to 1} = \frac{24}{Re}(1.146 + 0.151 Re^{0.684})$	×	√	√
Tang et al. (2015) [37]	$C_D = \frac{4}{3}\left(180\frac{1-\varepsilon_f}{Re} + 18\frac{\varepsilon_f^4(1+1.5\sqrt{1-\varepsilon_f})}{Re} + 18\varepsilon_f^2\left[0.11(1-\varepsilon_f)(2-\varepsilon_f) - \frac{0.00456}{\varepsilon_f^4} + \left(0.169\varepsilon_f + \frac{0.0644}{\varepsilon_f^4}\right) Re^{-0.343}\right]\right)$	$C_{D,\varepsilon_f \to 1} = \frac{24}{Re}(1 - 0.0046 Re + 0.233 Re^{0.657})$	×	√	×

DOI: http://dx.doi.org/10.5772/intechopen.106427

References	Drag coefficient C_D	C_D when $\varepsilon_f \to 1$	Req (2-1)	Req (2-2)	Req (2-3)
Zhou and Fan (2015) [38]	$C_D = F_0 \frac{24\varepsilon_f^2}{Re} + \left(0.256 + 1.41(1-\varepsilon_f) - 5.61(1-\varepsilon_f)^2 + 6.04(1-\varepsilon_f)^3\right) \frac{\left[\varepsilon_f^{-1} + 3\varepsilon_f(1-\varepsilon_f) + 8.4 Re^{-0.343}\right]}{\varepsilon_f \left[1 + 10^{3(1-\varepsilon_f)} Re^{-2.5+2\varepsilon_f}\right]}$ where $F_0 = \begin{cases} \frac{9.9(1-\varepsilon_f)}{\varepsilon_f^2} + \varepsilon_f^3\left[1 + 3(1-\varepsilon_f)^{0.6}\right] & \varepsilon_f > 0.45 \\ \frac{5.87 \sin\left[\frac{\pi}{2}\left(\frac{1-\varepsilon_f}{0.637}\right)^{1.75}\right]}{\varepsilon_f^2} & 0.363 \le \varepsilon_f \le 0.45 \\ \frac{5.87(1-\varepsilon_f)}{0.637\varepsilon_f^2} & \varepsilon_f < 0.363 \end{cases}$	$C_{D,\varepsilon_f \to 1} = \frac{24}{Re} + 0.256 \frac{1 + 8.4 Re^{-0.343}}{1 + Re^{-0.5}}$	√	×	√
Sheikh and Qiu (2018) [39]	$C_D = \frac{24}{Re\,\varepsilon_f^{0.5}}\left[3^{1.25\left(1-\varepsilon_f^3\right)} + Re\left(0.2 Re^{0.5\varepsilon_f - 1} - 0.053\varepsilon_f + 0.073\right)\right]$	$C_{D,\varepsilon_f \to 1} = \frac{24}{Re}\left[1 + Re\left(0.2 Re^{-0.5} + 0.02\right)\right]$	√	√	√
Kravets et al. (2019) [40]	$C_D = 240 \frac{1-\varepsilon_f}{Re} + 24 \frac{\varepsilon_f^4\left(1 + 1.5\sqrt{1-\varepsilon_f}\right)}{Re} + 24\varepsilon_f^2\left[0.1695(1-\varepsilon_f)(2-\varepsilon_f) - \frac{0.004321}{\varepsilon_f^4} + \left(0.0719\varepsilon_f + \frac{0.02169}{\varepsilon_f^4}\right) Re^{-0.2017}\right]$	$C_{D,\varepsilon_f \to 1} = \frac{24}{Re}\left[1 - 0.00432 Re + 0.0936 Re^{0.7983}\right]$	×	√	×

Table 2.
Summary of the drag correlations for multiparticle systems.

where C_D is a coefficient relying on $Re = \frac{\rho_f d_p \varepsilon_f |\mathbf{u}_f - \mathbf{u}_p|}{\mu_f}$ and ε_f is the porosity. C_D is difficult to derive theoretically for a multiparticle system. Formulating an empirical correlation of C_D should meet the following requirements for physical consistency and generality:

1. C_D should be able to reduce to C_{D0} when ε_f is approaching 1;

2. C_D should be continuous over the whole application range.

3. C_D should always be larger than C_{D0} when ε_f is smaller than 1;

Note that the requirement (2-1) must be satisfied in a CFD-DEM simulation. Otherwise, inconsistency may occur in the dilute regime. The requirement (2-1)

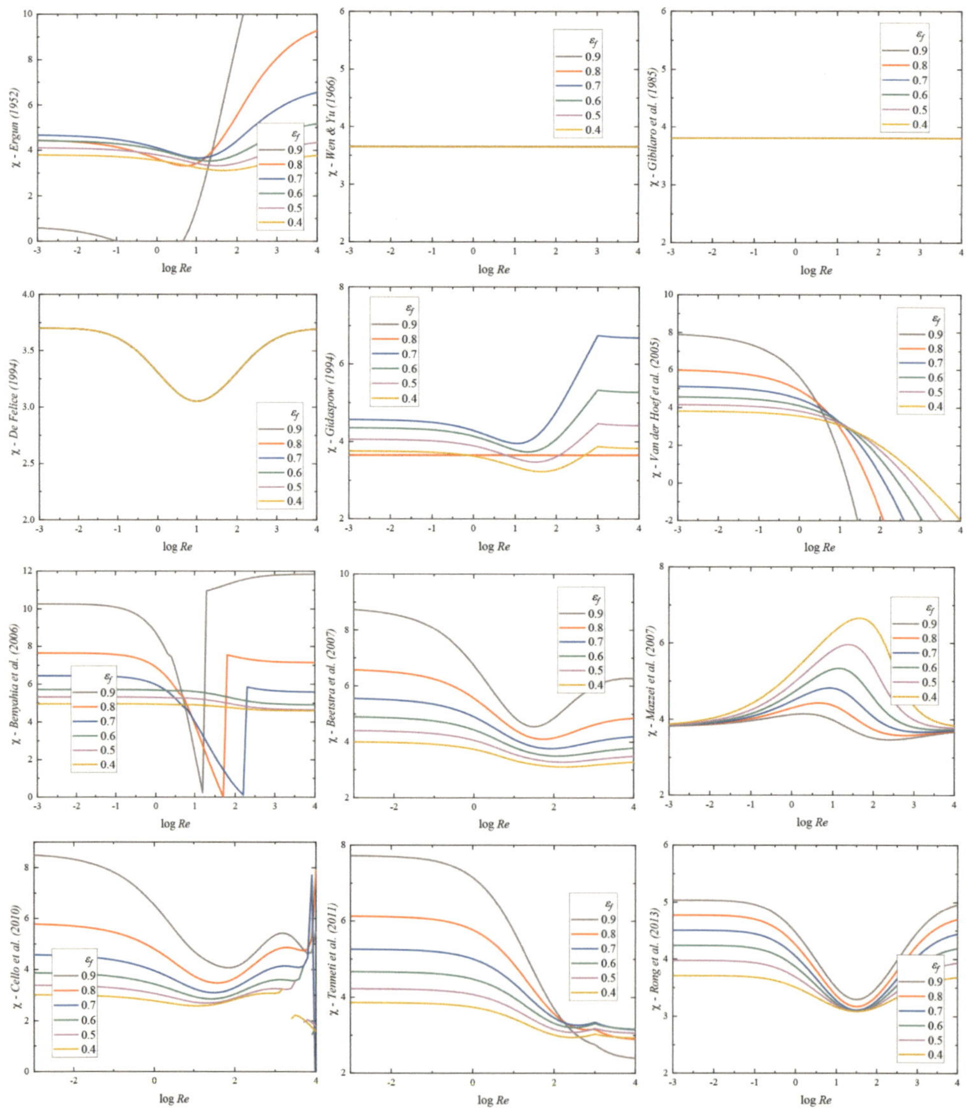

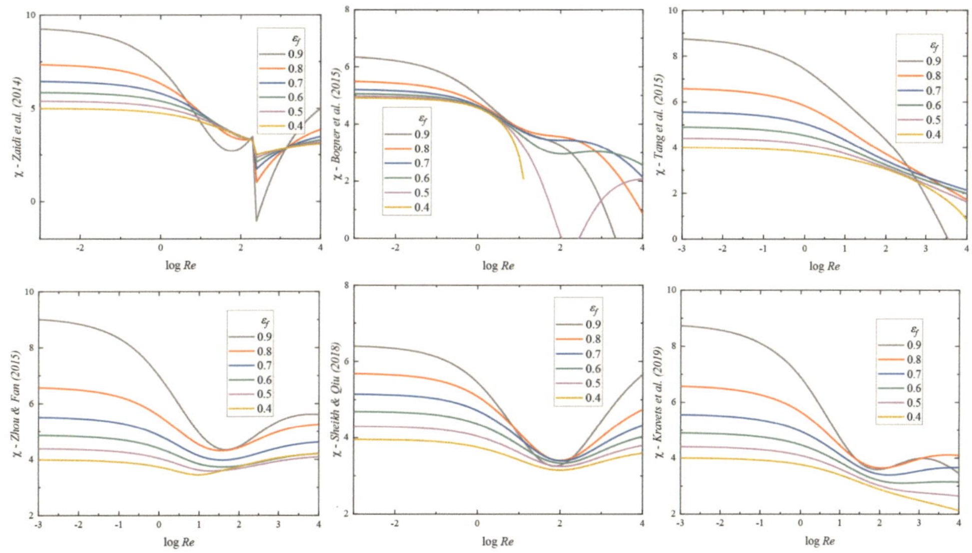

Figure 5.
Exponent χ as a function of porosity ε_f and Reynolds number Re calculated from different drag correlations.

can be written in $C_{D0} = \lim_{\varepsilon \to 1} C_D$, which should satisfy the requirements (1-1) to (1-3). This is important for evaluating different drag correlations when new drag correlations are considered for a single particle. The requirement (2-3) ensures that C_D is physically meaningful because neighboring particles reduce the space for fluid and generate a higher particle-fluid velocity in the surrounding region under the same Re, increasing the drag force.

Table 2 summarizes the correlations of drag coefficient C_D reported for multi-particle systems in recent years. For easy comparison, the value of C_D when $\varepsilon_f = 1$ is given to assess if $\lim_{\varepsilon \to 1} C_D$ satisfies the requirements from (1-1) to (1-3). **Table 2** also indicates whether the correlations meet the requirements from (2-1) to (2-3). Note that the requirement (2-3) is not easy to be assessed directly. The values of void function $\chi = 2 - \log_{\varepsilon_f} \frac{C_D}{C_{D0}}$ (**Figure 5**) are calculated from different drag correlations under different conditions to solve this problem. The value of χ in **Figure 5** must be positive to satisfy the requirement (2-3). Otherwise, C_D would be smaller than C_{D0}.

Table 2 shows that most of the drag correlations in Ergun-type cannot satisfy the requirement (2-1), thereby generating nonphysical results in a dilute regime. Combining different correlations was employed in the past to overcome this problem. For example, the Gidaspow correlation [28], adopted in many commercial CFD software such as ANSYS Fluent, combined the Ergun correlation under dense conditions and the Wen-Yu correlation under dilute conditions. A similar treatment is also taken by Benyahia et al. [31] and Zhou and Fan [38]. However, this treatment may cause apparent discontinuity at the switching point, as shown in **Figure 5**, which is against the requirement (2-2). Some other investigators attempted to formulate an Ergun-type correlation by adding additional items to guarantee continuity based on DNS or LB simulation data for randomly distributed particle systems generated with the Monte Carlo method. Typical work can be exemplified by Van der Hoef et al. [30],

Beetstra et al. [26], Cello et al. [33], Tang et al. [37], and Kravets et al. [40]. Among all these Ergun-type correlations, only the correlations proposed by Beetstra et al. [26] and Sheikh and Qiu [39] can meet all the requirements. The correlations of Tenneti et al. [34] and Zhou and Fan [38] can fulfill the requirements (2-1) and (2-3). However, with $\lim_{\varepsilon \to 1} C_D$, the correlation of Zhou and Fan [38] cannot be reduced to a single particle. Specifically, the value of the $\lim_{\varepsilon \to 1} C_D$ is much smaller than the experimental value at a high Re. The void function value χ predicted by the correlation of Tenneti et al. [34] has an opposite trend of other correlations. Also, a discontinuous point exists at a relatively high Re, as shown in **Figure 5**.

Compared with Ergun-type correlations, Wen-Yu-type drag correlations inherently have the advantage of satisfying both requirements (2-1) and (2-2) due to the introduction of the voidage function χ. In fact, all the Wen-Yu-type correlations in **Table 2** can satisfy the requirements from (2-1) to (2-3). One of the most widely used Wen-Yu-type correlations is the one proposed by Di Felice [29], which has good performance in both dilute and dense regimes. However, its voidage function is too simple and ignores the effect of Re and ε_f. Later, Rong et al. [14] used the LB method to study the drag force in different packed beds with a wide range of porosity generated by the DEM simulations, and on this basis, modified the voidage function of Di Felice [29] to incorporate the effect of porosity. Compared with other formulations, for example, the correlation of Beetstra et al. [26], that of Rong et al. [14] has a simpler form and better performance in a broader range of applications. Note that the existing drag correlations were often established on experimental or numerical data obtained in low-to-intermediate flow regimes ($Re < 1000$). The same expression has been adopted to deal with the high Re regime in applications. Nevertheless, drag force correlations for mono-size particle systems have been firmly established with the extensive efforts from different investigators in the last decade.

6. From monosize to multisize particle systems

The correlations in **Table 2** are all formulated for systems with monosize spheres. However, particle systems composed of multisize spheres are encountered in most applications. The drag correlation formulated for monosize spheres cannot be directly applied to multisize particle systems. Therefore, some investigators have studied the drag forces using LB or DNS simulations for different components in multisize particle systems generated with DEM simulations or Monte Carlo methods, formulating new correlations based on numerical data.

Generally, two main treatments are used to estimate the fluid-particle drag force in a mixture of particles. The first one is to estimate the drag force on each particle directly, calculate the mean drag forces on different components, and finally obtain the total fluid-particle drag force by summing the forces of all particles. Due to the lack of reliable drag correlations for mixtures, different investigators used this treatment in early studies, such as Feng and Yu [41] and Bokkers et al. [42]. It has proven to have a poor performance by Rong et al. [13]. Thus, investigators turned their sight to the second approach, which uses an opposite calculation path. It first estimates the total fluid-particle drag force and then distributes the total force among different components to obtain their mean drag forces according to a specific rule. In this approach, the total drag force is estimated by treating the particle mixture as a monosize particle system with a representative average diameter, that is, the Sauter mean diameter $\langle d_p \rangle = \left[\sum_i x_i/d_i\right]^{-1}$, where x_i and d_i are the volume fraction and diameter of component i. The distributing function can be formulated in different ways. However, it should meet some basic requirements for physical consistency and generality, which have been well-established by Rong et al. [13], given by:

1. $\frac{\tilde{f}_{d,i}}{\tilde{f}_d} = 1$ when all d_i are equal, where $\tilde{f}_{d,i} = \mathbf{f}_{d,i}/3\pi\mu_f d_i \varepsilon_f (\mathbf{u}_f - \mathbf{u}_p)$ is the drag force for component i normalized by the Stokes drag and $\tilde{f}_d$ is normalized drag force for the equivalent mono-size particle.

2. $\tilde{f}_d = \sum_{i=1}^{N} \frac{x_i \tilde{f}_{d,i}}{y_i^2}$, where $y_i = d_i/\langle d_p \rangle$.

3. $\tilde{f}_{d,i} > 0$.

Note that the requirement (3-1) is required for the drag force model to reproduce the drag law for mono-size particle systems. The requirement (3-2) ensures the sum of the drag forces of all components equals the total fluid-particle drag force. The requirement (3-3) ensures positive mean drag forces.

Table 3 summarizes the distribution functions for multisize particle systems proposed by different investigators. The predictions by different correlations for

References	Treatments	Drag correlation	Req. (3-1)	Req. (3-2)	Req. (3-3)
Sarkar et al. (2009) [43]	$\frac{\tilde{f}_{d,i}}{\tilde{f}_d} = y_i + 0.064y_i^3$	Van der Hoef et al. (2005) [30]	×	×	√
Yin and Sundaresan (2009) [44]	$\tilde{f}_{d,i} = \frac{1}{\varepsilon_f} + \left(\tilde{f}_d - \frac{1}{\varepsilon_f}\right)\left(ay_i + (1-a)y_i^2\right)$ $a = 1 - 2.66(1-\varepsilon_f) + 9.096(1-\varepsilon_f)^2 - 11.338(1-\varepsilon_f)^3$	Van der Hoef et al. (2005) [30]	√	×	×
Cello et al. (2010) [33]	$\frac{\tilde{f}_{d,i}}{\tilde{f}_d} = y_i + \frac{1-\varepsilon_f}{\varepsilon_f}\left(\frac{1-\varepsilon_f-0.27}{1-0.27}\right)\frac{(y_i^2-y_i)}{\sum_{j=1}^{N} x_j y_j}$	Cello et al. (2010) [33]	√	×	×
Rong et al. (2014) [13]	$\frac{\tilde{f}_{d,i}}{\tilde{f}_d} = \frac{0.5\varepsilon_f}{\sum_{i=1}^{N}(x_i/y_i^2)} + 0.5(1-\varepsilon_f)y_i^2 + 0.5y_i$	Rong et al. (2013) [14]	√	√	√
Mehrabadi et al. (2016) [45]	$\frac{\tilde{f}_{d,i}}{\tilde{f}_d} = \varepsilon_f y_i + (1-\varepsilon_f)y_i^2$	Tenneti et al. (2011) [34]	√	√	√
Duan et al. (2020) [46]	$\tilde{f}_{d,i} = \frac{1}{\varepsilon_f} + \left(\tilde{f}_d - \frac{1}{\varepsilon_f}\sum_{i=1}^{N}\left(\frac{x_i}{y_i^2}\right)\right)\left(0.37y_i + 0.63y_i^{2-\eta}\right)$ $\eta = 2.169\ln\left[0.466 + \left(\frac{1-\varepsilon_f}{\varepsilon_{ij,max}}\right)^{0.2025}\right]$ $\varepsilon_{ij,max} = \begin{cases} 0.313\left(1-\chi_{ij}\right)x_i + 0.64, x_i < 0.414 \\ 0.870\left(1-\chi_{ij}\right)x_j + 0.64, x_i > 0.414 \end{cases}$ $\chi_{ij} = \sqrt{d_j/d_i}(d_j > d_i)$	Zhou and Fan (2015) [38]	√	×	×

Table 3.
Summary of the drag coefficient for multisize particle system.

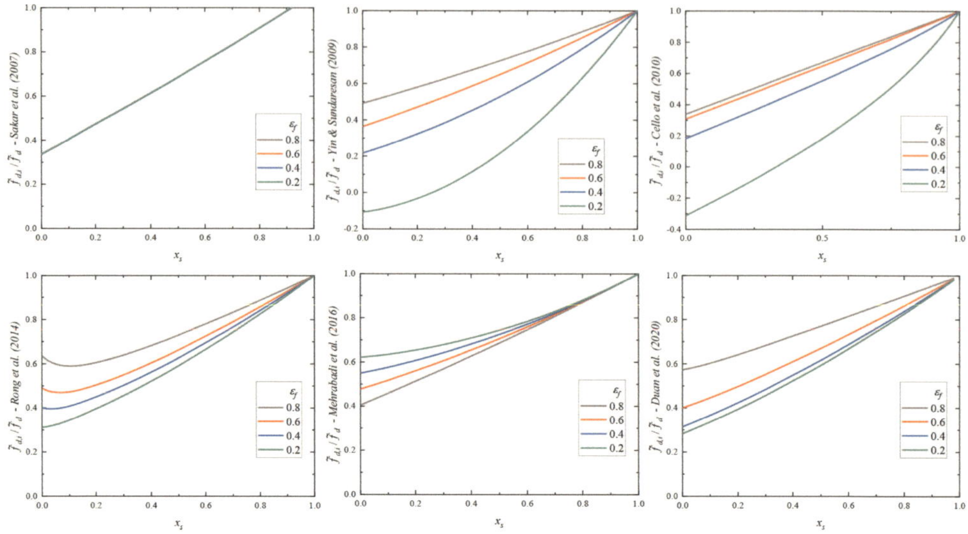

Figure 6.
$\tilde{f}_{d,i}/\tilde{f}_d$ as a function of porosity ε_f and volume fraction of small particle x_s calculated from different drag correlations.

comparison are given in **Figure 6**. These correlations were extended from monosize particle systems through distributing functions. Therefore, the performance of the drag correlations of multisize particle systems needs to be evaluated based on the requirements for both monosize and multi-size particle systems. Only the correlations of Rong et al. [13] and Mehrabadi et al. [45] can satisfy all the requirements. Others suffer various discrepancies to different extents. For example, the correlation of Sarkar et al. [43] fails to meet the criteria (3-1) and (3-2), whereas the correlations by Yin and Sundaresan [44] and Cello et al. [33] may generate negative drag forces, which is obviously against the requirement (3-3). It should also be noted that Rong et al. [13] and Mehrabadi et al. [45] used the correlations proposed in their previous studies to estimate the total fluid-particle drag force. The discussion about these two correlations for monosize systems can be found in Section 5.

7. From spherical to nonspherical particle system

Particle shape could strongly affect drag forces. Rather than perfectly spherical particles, real particles often show diverse morphology. For example, they can be cubes, cylinders, and ellipsoids, or more generally, particles of irregular shapes. Such morphological diversity adds further complexity to the modeling of particle-fluid interaction forces. In earlier years, nonspherical particles are usually treated as volume-equivalent spheres. However, this simple treatment gives inaccurate results, even for an isolated particle. Several studies have been conducted to overcome this problem, particularly based on ellipsoidal particles. Such efforts are discussed in this section.

First, the definition of the geometry of an ellipsoidal particle is given to understand the factors that should be considered in drag correlations. For a standard axis-aligned ellipsoid particle, the Cartesian coordinates are given by $\frac{x^2}{a^2} + \frac{y^2}{b^2} + \frac{z^2}{c^2} = 1$, where a, b, and c are the principal semi-axes. In most works for ellipsoidal particles, the

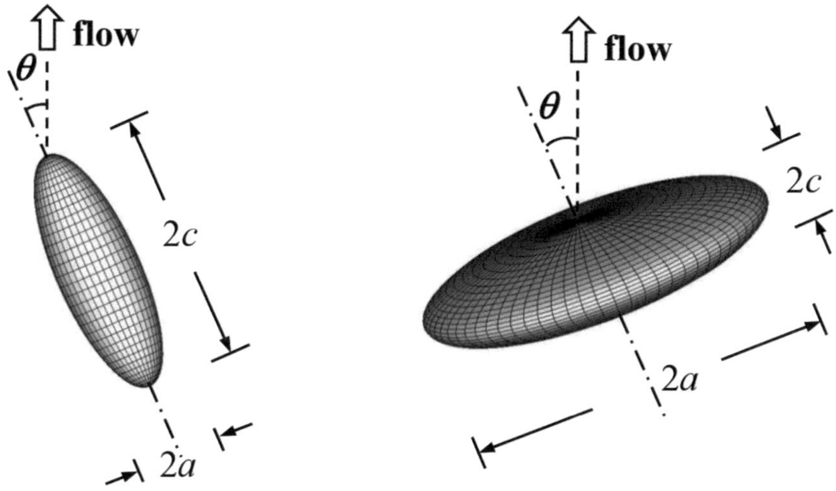

Figure 7.
Characteristics of spheroidal particles prolate spheroid (left) and oblate spheroid (right).

degenerate cases are considered, where the ellipsoid particles have two equal axes, say $a = b$, generally referred to as spheroid or ellipsoid of revolution. Variation of a results in different shapes of particles, which can be represented by aspect ratio, defined as $Ar = \frac{c}{a} = \frac{c}{b}$. Obviously, for an oblate spheroid, $Ar < 1$; for a sphere, $Ar = 1$; and for a prolate spheroid, $Ar > 1$. The shapes of oblate and prolate spheroids are schematically shown in **Figure** 7. Note that the angle between the direction of the main flow and the symmetric axis of the ellipsoid body is defined as incident angle θ.

Based on a large number of experimental data, Holzer and Sommerfeld [47] established a drag correlation for an isolated nonspherical particle, given as:

$$C_D = \frac{8}{Re}\frac{1}{\sqrt{\psi_\perp}} + \frac{16}{Re}\frac{1}{\sqrt{\psi}} + \frac{3}{\sqrt{Re}}\frac{1}{\psi^{3/4}} + 0.4210^{0.4(-\log\psi)^{0.2}}\frac{1}{\psi_\perp} \tag{16}$$

where ψ is the sphericity, which is defined as the ratio between the surface area of the equivalent-volume sphere and that of the considered particle. $\psi_\perp$ is the mean crosswise sphericity, $\psi_\perp = \langle \psi_{\perp,i} \rangle$, when applying to a multi-particle system. Here, the individual crosswise sphericity $\psi_{\perp,i}$ represents the ratio between the cross-sectional area of the equivalent-volume sphere and the projected cross-sectional area of the considered ith particle. For spheroidal particles, these two parameters can be calculated by $\psi = \frac{Ar^{2/9}}{\left(\frac{1+2Ar^{1.61}}{3}\right)^{1/1.61}}$ and $\psi_{\perp,i} = \frac{Ar^{2/9}}{\left(Ar^2\sin^2\theta_i + Ar^2\cos^2\theta_i\right)^{1/2}}$. Later, Zastawny et al. [48] and Ouchene et al. [49] established correlations based on LB and DNS simulation data.

The procedure of extension from single sphere to multi-spheres particle systems is also used in nonspherical particle systems. In this direction, various drag correlations have been established in recent years based on numerical data. Similar to the requirements as discussed for multiparticle systems, the drag correlation $C_D(Re, \varepsilon_f, \psi)$ for nonspherical particles should also meet the following requirements for physical consistency and generality:

References	Drag coefficient C_D	Req (4-1)	Req (4-2)	Req (4-3)	Req (4-4)
Rong et al. (2015) [13]	$C_D = C_{D0}\varepsilon_f{}^{2-\chi}, \chi = \beta + \lambda, \beta = 2.65(1+\varepsilon_f) - (5.3 - 3.5\varepsilon_f)\varepsilon_f^2 \exp\left[-\frac{(1.5-\log Re)^2}{2}\right]$ $\lambda = (1-\psi)\left\{101.8(\psi - 0.81)^2 + 2.4 - (39\psi - 20.6)\exp\left[-\frac{(3.5-\log Re)^2}{2}\right]\right\}, \psi = \frac{Ar^{2/9}}{\left(\frac{1+2Ar^{1.61}}{3}\right)^{1/1.61}}$	√	√	√	√
Li et al. (2019, 2020) [50, 51]	$C_D = C_{D,0^\circ}\cos^2\theta + C_{D,90^\circ}\sin^2\theta$ $C_{D,0^\circ} = C_{D,sphere}F_{D0,0^\circ}\psi_{0^\circ},$ $F_{D0,0^\circ} = \begin{cases} \frac{8}{3}\left[-\frac{2Ar}{Ar^2-1} + \frac{2Ar^2-1}{(Ar^2-1)^{3/2}}\ln\left(\frac{Ar+\sqrt{Ar^2-1}}{Ar-\sqrt{Ar^2-1}}\right)\right]^{-1}, & Ar>1 \\ 1, & Ar=1 \\ \frac{8}{3Ar^{1/3}}\left[\frac{2Ar}{1-Ar^2} + \frac{2(1-2Ar^2)}{(1-Ar^2)^{3/2}}\tan^{-1}\left(\frac{\sqrt{1-Ar^2}}{Ar}\right)\right]^{-1}, & Ar<1 \end{cases}$ $\psi_{0^\circ} = \begin{cases} (1-\varepsilon_f)^{-0.103S+0.167}(Ar-1)\left(0.326 - 0.374Ar^{-0.34}(S+0.5)^{1.05}\right) + 1, & Ar>1 \\ 1, & Ar=1 \\ (1-\varepsilon_f)^{-0.2S+0.498}Ar^{1.65}\left(\frac{1}{Ar}-1\right)^{0.706}\left(-0.550 + 0.302Ar^{-2.3}(S+0.5)^{1.08}\right) + 1, & Ar<1 \end{cases}$ $C_{D,90^\circ} = C_{D,sphere}F_{D0,90^\circ}\psi_{90^\circ}$ $F_{D0,90^\circ} = \begin{cases} \frac{8}{3}\left[\frac{Ar}{Ar^2-1} + \frac{2Ar^2-3}{(Ar^2-1)^{3/2}}\ln\left(Ar+\sqrt{Ar^2-1}\right)\right]^{-1}, & Ar>1 \\ 1, & Ar=1 \\ \frac{8}{3Ar^{1/3}}\left[-\frac{Ar}{1-Ar^2} - \frac{2Ar^2-3}{(1-Ar^2)^{3/2}}\sin^{-1}\left(\sqrt{1-Ar^2}\right)\right]^{-1}, & Ar<1 \end{cases}$ $\psi_{90^\circ} = \begin{cases} (1-\varepsilon_f)^{-0.344S+0.716}(Ar-1)^{0.281}\left(0.232(S+Ar)^{0.851} - 0.401\right) + 1, & Ar>1 \\ 1, & Ar=1 \\ (1-\varepsilon_f)^{-0.160S+0.293}Ar^{2.18}\left(\frac{1}{Ar}-1\right)^{2.65}\left(1.54 - 1.30Ar^{0.266}(S+0.5)^{1.13}\right) + 1, & Ar<1 \end{cases}$ $S = \frac{1}{2}\left(3\overline{\cos^2\phi} - 1\right)$	√	√	×	√
Cao et al. (2020) [52]	$C_D = C_{D0}\varepsilon_f{}^{2-\chi}, \chi = \beta + \lambda, \beta = 4.988(0.5139 + \varepsilon_f) - (3.175 - 1.493\varepsilon_f)\varepsilon_f^2 \exp\left[-\frac{(1.5-\log Re)^2}{2}\right] - 0.5884\log Re$ $\lambda = (0.5201Ar^2 + 0.6094Ar - 11.74)\varepsilon_f{}^3 + (-1.713Ar^2 - 3.467Ar + 12.65)\varepsilon_f + (1.056Ar^2 + 0.4316Ar - 4.168)$,	√	√	√	×

Table 4.
Summary of the drag correlations for nonspherical particle system.

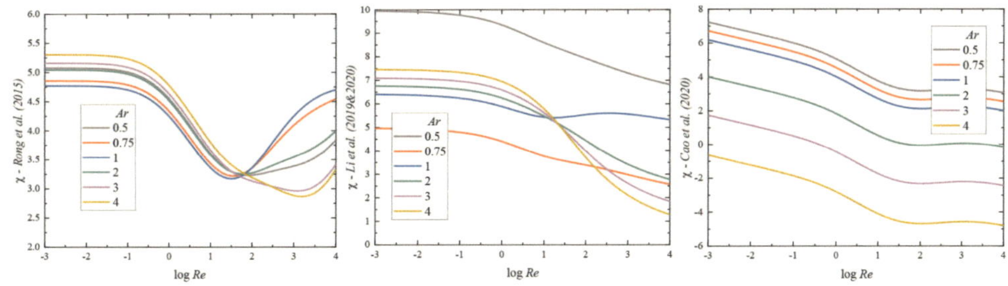

Figure 8.
Exponent χ as a function of porosity ε_f, Reynolds numbers Re and Ar calculated from different drag correlations.

1. $C_D(Re, \varepsilon_f, \psi)$ should be reduced to the drag correlation for spherical particles when $\psi = 1$.

2. $C_D(Re, \varepsilon_f, \psi)$ should be reduced to the drag correlation $C_{D0}(Re, \psi)$ for single nonspherical particle when porosity ε_f approaches unity.

3. $C_D(Re, \varepsilon_f, \psi)$ should be continuous over the entire application range of Re, ε_f, ψ.

4. $C_D(Re, \varepsilon_f, \psi)$ should always be larger than $C_{D0}(Re, \psi)$ when ε_f is smaller than 1;

Table 4 summarizes the drag correlations proposed by different investigators. **Figure 8** shows the values of χ calculated from different drag correlations. All the correlations can meet the requirements (4-1) & (4-2) and can be reduced to the drag correlation for a single spherical particle. These correlations are established based on those for a nonspherical particle and are extensions of earlier studies from different investigators. Specifically, Rong et al. [12] used the drag correlation of Holzer and Sommerfeld [47], and their correlation is the extension of their work for monosize and multisize particle systems [13, 14]. Li et al. [50, 51] extended the work of Zhou and Fan [38]. Cao et al. [52] extended the work of Tenneti et al. [34] and adopted the

Viscosity model	Correlation
Power-law fluid model	$\mu = K\dot{\gamma}^{n-1}$, where K is the flow consistency index and n is the power index.
Spriggs fluid model	$\mu = \begin{cases} \mu_0, & \dot{\gamma} \le \dot{\gamma_0} \\ \mu_0\left(\frac{\dot{\gamma}}{\dot{\gamma_0}}\right)^{n-1}, & \dot{\gamma} > \dot{\gamma_0} \end{cases}$, where $\dot{\gamma_0}$ is the reference shear rate.
Carreau fluid model	$\mu = \mu_\infty + (\mu_0 - \mu_\infty)\left[1 + (\lambda\dot{\gamma})^2\right]^{\frac{n-1}{2}}$
Bingham fluid model	$\mu = \begin{cases} \mu + \frac{\tau}{\dot{\gamma}}, & \tau_{max} \ge \tau \\ \infty, & \tau_{max} < \tau \end{cases}$, where τ_{max} is the maximum shear stress.
Casson fluid model	$\mu = \begin{cases} \left(\sqrt{\mu} + \sqrt{\frac{\tau}{\dot{\gamma}}}\right)^2, & \tau_{max} \ge \tau \\ \infty, & \tau_{max} < \tau \end{cases}$

Table 5.
Summary of the viscosity models for non-Newtonian fluids.

References	Drag coefficient C_D	Req (2-1)	Req (2-2)	Req (2-3)	Req (5-1)
Srinivas and Chhabra (1995) [53]	$C_D = \frac{4}{3}\left(\frac{150}{Re_p} + 1.75\right),\ Re_p = \frac{\left(15\sqrt{2}\right)^{1-n} \rho_f \varepsilon_f{}^n \left\lvert \mathbf{u}_f - \mathbf{u}_p \right\rvert^{2-n} d_p{}^n}{K\left(1-\varepsilon_f\right)^n} \left(\frac{4n}{3n+1}\right)^n$	×	√	×	√
Sabiri and Comiti (2000) [54]	$C_D = 4\tau^3\left(\frac{16}{Re_p} + 0.194\right),\ Re_p = \frac{\rho_f \varepsilon_f{}^n \left\lvert \mathbf{u}_f - \mathbf{u}_p \right\rvert^{2-n} d_p{}^n \tau^{2-n}}{2^{2n-3} K [3(3n+1)/4n]^n \left(1-\varepsilon_f\right)^n},\ \tau = 1 - 0.41 \ln \varepsilon_f\ S$	×	√	×	×
Dhole et al. (2004) [55]	$C_D = \frac{4}{3}\left(\frac{230.4}{Re_k} + 0.33\right),\ Re_k = \frac{\rho_f \varepsilon_f{}^{2-n} \left\lvert \mathbf{u}_f - \mathbf{u}_p \right\rvert^{2-n} d_p}{\varphi\left(1-\varepsilon_f\right)},$ $\varphi = \frac{K}{12}\left(9 + \frac{3}{n}\right)^n \left(150 S \varepsilon_f\right)^{(1-n)/2},\ S = \frac{d_p{}^2 \varepsilon_f{}^3}{150\left(1-\varepsilon_f\right)^2}$	×	√	√	×
Qi et al. [57]	$C_D = C_{D0} \varepsilon_f{}^{2-\chi},\ C_{D0} = \left(\frac{2.654}{n^{-1.9} + 3.213} + \frac{4.8}{n^{0.08}\sqrt{Re}}\right)^2,\ Re = \frac{\rho_f \varepsilon_f{}^{2-n} \left\lvert \mathbf{u}_f - \mathbf{u}_p \right\rvert^{2-n} d_p{}^n}{K}$ $\chi = 2.65n(\varepsilon + 1)^n - (5.3 - 3.5\varepsilon^n)\varepsilon^2 \exp\left[-\frac{(\log Re - 1.5)^2}{2}\right] + \frac{2.65(\varepsilon+1)^{2-n} - 2.65n(\varepsilon+1)^n}{1 + \exp[-2(\log Re - 1.5)]}$	√	√	√	√

Table 6.
Summary of the drag correlations for non-Newtonian fluid-particle system.

drag correlation for ellipsoids proposed by Ouchene et al. [49]. Because Li et al. [50, 51] conducted their works for oblate and prolate particles separately, the drag correlation discontinues at the switching point as Ar approaches 1. Cao et al. [52] only examined the drag correlation for prolate particles and their drag correlation may generate the negative value of the void function at relatively high Ar or Re. So far, only the drag correlation proposed by Rong et al. [12] can satisfy all the requirements (4-1) to (4-4). However, it should be noted that all these studies only focused on particular particle shapes and a unified drag correlation with better generality for particles of irregular shapes is still lacking.

8. From Newtonian fluid to non-Newtonian fluid system

Another important factor affecting the fluid-particle interaction force is fluid rheology. In some applications, the fluid does not necessarily follow Newton's law of viscosity. Fluid may present shear-thinning/shear-thickening behaviors. That is, the viscosity of fluid μ increases/decreases as the shear stress τ or shear rate $\dot{\gamma}$ increases. **Table 5** lists some widely used viscosity models for non-Newtonian fluid. Among these models, the power-law model is the most popular one in engineering applications because it provides a unified and simple way of describing the rheological characteristics of shear-thinning fluid ($n < 1$), Newtonian fluid ($n = 1$), and shear-thickening fluid ($n > 1$). However, even with the simplest power-law fluid model, predicting the behaviors of particles and fluid or quantifying the fluid-particle interaction force in non-Newtonian fluid is extremely challenging due to the variable viscosity.

To date, attention paid to the effect of fluid rheology on drag force is much less than other factors. **Table 6** summarizes the drag correlations proposed for a non-Newtonian fluid-particle system. Note that the rheology of all the considered non-Newtonian fluids obeys the power law. As Newtonian fluid is a special case when $n = 1$ for the power-law viscosity model, the drag correlation for a non-Newtonian fluid-particle system should not only satisfy requirements (2-1) to (2-3) but also the following requirements:

1. $C_D\left(Re, \varepsilon_f, n\right)$ should be reduced to the drag correlation C_D for a Newtonian fluid-particle system in **Table 2** when $n = 1$.

The correlation proposed by Srinivas and Chhabra [53] is established based on the Ergun-drag correlation with a different definition of Reynolds number for non-Newtonian fluids and can satisfy requirements (2-2) and (5-2). However, it also has the same problem faced in the Ergun correlation. Sabiri and Comiti [54] and Dhole et al. [55] further revised the definition of the Reynolds number and introduced the tortuosity of packed beds to consider the effect of fluid rheology. However, their correlations cannot overcome the discrepancies of the Ergun drag correlation. Besides, the correlations of Srinivas and Chhabra [53] and Sabiri and Comiti [54] may generate a negative value of void function χ, as shown in **Figure 9**, which is against the requirement (2-3). To overcome this problem, Qi et al. [15, 56, 57] conducted a series of studies by extending the studies of Rong et al. [13, 14] from Newtonian fluid to non-Newtonian fluid. Qi et al. [56, 57] also considered both monosize and multisize particle systems. Overall, the resulting drag correlation is consistent with the drag correlation proposed for Newtonian fluid-particle systems by Rong et al. [13, 14] and satisfies all the requirements (**Table 6**).

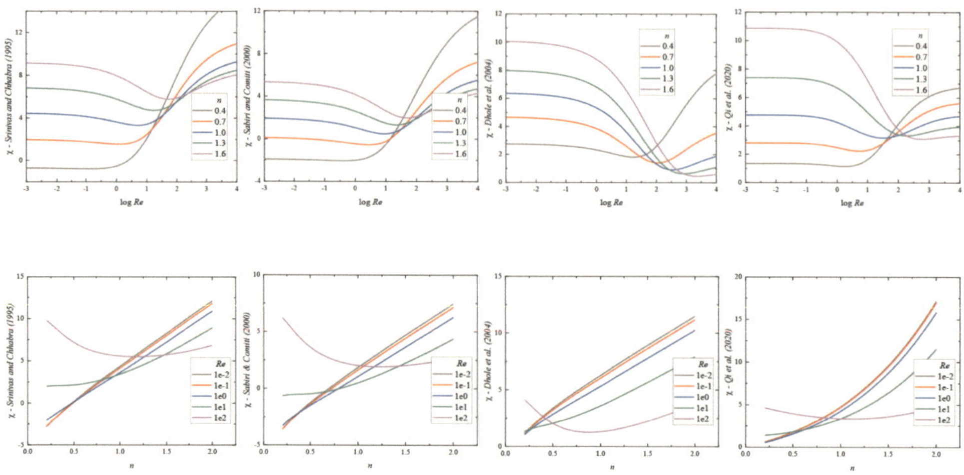

Figure 9.
Exponent χ as a function of porosity ε_f, Reynolds number Re and power-law index n *calculated from different drag correlations.*

9. Conclusions

Drag force correlations for particle-fluid systems are reviewed, covering from simple to complicated systems, including from single particle to multiple particles, monosize to multisize, spherical to nonspherical, and Newtonian fluid to non-Newtonian fluid. The drag correlations for mono-size and multi-size spherical particle systems are more mature. However, the physical consistency and generality of different drag correlations could be more carefully considered, which are discussed in this review. Several drag correlations show superiority in generality. The understanding of fluid-particle interactions in complicated systems involving factors such as particle shape or fluid rheology is still lacking. For example, the studies on the effect of fluid rheology are largely limited to non-Newtonian fluids obeying the power-law fluid model. Similarly, the studies on nonspherical particles are mostly limited to ellipsoids. Further studies should be conducted to generally consider these important factors to meet various engineering needs.

Author details

Zheng Qi[1], Shibo Kuang[2], Liangwan Rong[3], Kejun Dong[4] and Aibing Yu[2,5]*

1 China Iron and Steel Research Institute Group, Beijing, China

2 ARC Research Hub for Computational Particle Technology, Department of Chemical Engineering, Monash University, Melbourne, VIC, Australia

3 School of Civil Engineering and Transportation, South China University of Technology, Guangzhou, Guangdong, China

4 School of Engineering, Design and Built Environment, Western Sydney University, Penrith, NSW, Australia

5 Center for Simulation and Modeling of Particulate Systems, Southeast University-Monash University Joint Research Institute, Suzhou, Jiangsu, China

*Address all correspondence to: aibing.yu@monash.edu

References

[1] Ergun S. Fluid flow through packed columns. Chemical Engineering Progress. 1952;**48**:89-94

[2] Wen CY, Yu YH. Mechanics of fluidisation. Chemical Engineering Progress Symposium Series. 1966;**62**: 100-111

[3] Richardson JF, Jerónimo S. Velocity-voidage relations for sedimentation and fluidisation. Chemical Engineering Science. 1979;**34**:1419-1422

[4] Dallavalle JM. Micromeritics: The Technology of Fine Particles. New York: Pitman Publishing corporation; 1948

[5] Chen S, Doolen GD. Lattice Boltzmann method for fluid flows. Annual Review of Fluid Mechanics. 1998;**30**:329-364

[6] Dhumieres D, Ginzburg I, Krafczyk M, Lallemand P, Luo LS. Multiple-relaxation-time lattice Boltzmann models in three dimensions. Philosophical Transaction A Mathematical Physics, Engineering and Science. 2002;**360**:437-451

[7] Denier JP, Dabrowski PP. On the boundary-layer equations for power-law fluids. Proceedings of the Royal Society Mathematical Physical and Engineering Sciences. 2004;**460**:3143-3158

[8] Artoli AM, Hoekstra AG, Sloot PMA. Optimizing lattice Boltzmann simulations for unsteady flows. Computers & Fluids. 2006;**35**:227-240

[9] Ladd AJC. Numerical simulations of particulate suspensions via a discretized Boltzmann equation. Part 1. Theoretical foundation. Journal of Fluid Mechanics. 1994;**271**:285-309

[10] Stobiac V, Tanguy PA, Bertrand F. Boundary conditions for the lattice Boltzmann method in the case of viscous mixing flows. Computers & Fluids. 2013; **73**:145-161

[11] Chen L, Yu Y, Lu JH, Hou GX. A comparative study of lattice Boltzmann methods using bounce-back schemes and immersed boundary ones for flow acoustic problems. International Journal for Numerical Methods in Fluids. 2014; **74**:439-467

[12] Rong LW, Zhou ZY, Yu AB. Lattice-Boltzmann simulation of fluid flow through packed beds of uniform ellipsoids. Powder Technology. 2015;**285**:146-156

[13] Rong LW, Dong KJ, Yu AB. Lattice-Boltzmann simulation of fluid flow through packed beds of spheres: Effect of particle size distribution. Chemical Engineering Science. 2014;**116**:508-523

[14] Rong LW, Dong KJ, Yu AB. Lattice-Boltzmann simulation of fluid flow through packed beds of uniform spheres: Effect of porosity. Chemical Engineering Science. 2013;**99**:44-58

[15] Qi Z, Kuang S, Rong L, Yu A. Lattice Boltzmann investigation of the wake effect on the interaction between particle and power-law fluid flow. Powder Technology. 2018;**326**:208-221

[16] Schlichting H, Gersten K, Gersten K. Boundary-Layer Theory. Berlin: Spring; 2000

[17] Stokes GG. On the Effect of the Internal Friction of Fluids on the Motion of Pendulums. Cambridge: Pitt Press; 1851

[18] Clift R, Grace JR, Weber ME. Bubbles, Drops, and Particles. New York: Dover Publications; 2005

[19] Chhabra RP. Non-Newtonian Fluid Particle Systems: Sphere Drag. Monash University; 1980

[20] Peden JM, Luo Y. Settling velocity of variously shaped particles in drilling and fracturing fluids. SPE Drilling Engineering. 1987;**2**:337-343

[21] Dhole SD, Chhabra RP, Eswaran V. Flow of power-law fluids past a sphere at intermediate Reynolds numbers. Industrial & Engineering Chemistry Research. 2006;**45**:4773-4781

[22] Tiwari SS, Pal E, Bale S, Minocha N, Patwardhan AW, Nandakumar K, et al. Flow past a single stationary sphere, 1. Experimental and numerical techniques, Powder Technology. 2020; **365**:115-148

[23] Tiwari SS, Pal E, Bale S, Minocha N, Patwardhan AW, Nandakumar K, et al. Flow past a single stationary sphere, 2. Regime mapping and effect of external disturbances, Powder Technology. 2020; **365**:215-243

[24] Schiller L, Naumann Z. A drag coefficienty correlation. Z. Ver. Deutsch. Ing. 1933;**77**:318-320

[25] Turton R, Levenspiel O. A short note on the drag correlation for spheres. Powder Technology. 1986;**47**: 83-86

[26] Beetstra R, Van Der Hoef MA, Kuipers JM. Drag force of intermediate Reynolds number flow past mono- and bidisperse arrays of spheres. AICHE Journal. 2007;**53**:489-501

[27] Gibilaro LG, Di Felice R, Waldram SP, Foscolo PU. Generalized friction factor and drag coefficient correlations for fluid-particle interactions. Chemical Engineering Science. 1985;**40**:1817-1823

[28] Gidaspow D. Multiphase Flow and Fluidization: Continuum and Kinetic Theory Descriptions. New York: Academic Press; 1994

[29] Di Felice R. The voidage function for fluid-particle interaction systems. International Journal of Multiphase Flow. 1994;**20**:153-159

[30] Van Der Hoef MA, Beetstra R, Kuipers JM. Lattice-Boltzmann simulations of low-Reynolds-number flow past mono- and bidisperse arrays of spheres: Results for the permeability and drag force. Journal of Fluid Mechanics. 2005;**528**:233-254

[31] Benyahia S, Syamlal M, O'brien TJ. Extension of Hill–Koch–Ladd drag correlation over all ranges of Reynolds number and solids volume fraction. Powder Technology. 2006;**162**: 166-174

[32] Mazzei L, Lettieri P. A drag force closure for uniformly dispersed fluidized suspensions. Chemical Engineering Science. 2007;**62**:6129-6142

[33] Cello F, Di Renzo A, Di Maio FP. A semi-empirical model for the drag force and fluid–particle interaction in polydisperse suspensions. Chemical Engineering Science. 2010;**65**:3128-3139

[34] Tenneti S, Garg R, Subramaniam S. Drag law for monodisperse gas–solid systems using particle-resolved direct numerical simulation of flow past fixed assemblies of spheres. International Journal of Multiphase Flow. 2011;**37**: 1072-1092

[35] Zaidi AA, Tsuji T, Tanaka T. A new relation of drag force for high Stokes number monodisperse spheres by direct numerical simulation. Advanced Powder Technology. 2014;**25**:1860-1871

[36] Bogner S, Mohanty S, Rüde U. Drag correlation for dilute and moderately dense fluid-particle systems using the lattice Boltzmann method. International Journal of Multiphase Flow. 2015;**68**: 71-79

[37] Tang Y, Peters EJF, Kuipers JAM, Kriebitzsch SHL, Van Der Hoef MA. A new drag correlation from fully resolved simulations of flow past monodisperse static arrays of spheres. AICHE Journal. 2015;**61**:688-698

[38] Zhou Q, Fan L-S. Direct numerical simulation of low-Reynolds-number flow past arrays of rotating spheres. Journal of Fluid Mechanics. 2015;**765**: 396-423

[39] Sheikh B, Qiu T. Pore-scale simulation and statistical investigation of velocity and drag force distribution of flow through randomly-packed porous media under low and intermediate Reynolds numbers. Computers & Fluids. 2018;**171**:15-28

[40] Kravets B, Rosemann T, Reinecke SR, Kruggel-Emden H. A new drag force and heat transfer correlation derived from direct numerical LBM-simulations of flown through particle packings. Powder Technology. 2019;**345**: 438-456

[41] Feng YQ, Yu AB. Assessment of model formulations in the discrete particle simulation of gas–solid flow. Industrial and Engineering Chemistry Research. 2004;**43**:8378-8390

[42] Bokkers GA, Annaland MVS, Kuipers JM. Mixing and segregation in a bidisperse gas-solid fluidised bed: A numerical and experimental study. Powder Technology. 2004;**140**:176-186

[43] Sarkar S, Van Der Hoef MA, Kuipers JAM. Fluid–particle interaction from lattice Boltzmann simulations for flow through polydisperse random arrays of spheres. Chemical Engineering Science. 2009;**64**:2683-2691

[44] Yin X, Sundaresan S. Fluid-particle drag in low-Reynolds-number polydisperse gas-solid suspensions. AICHE Journal. 2009;**55**:1352-1368

[45] Mehrabadi M, Tenneti S, Subramaniam S. Importance of the fluid-particle drag model in predicting segregation in bidisperse gas-solid flow. International Journal of Multiphase Flow. 2016;**86**:99-114

[46] Duan F, Zhao L, Chen X, Zhou Q. Fluid–particle drag and particle–particle drag in low-Reynolds-number bidisperse gas–solid suspensions. Physics of Fluids. 2020;**32**:113311

[47] Hölzer A, Sommerfeld M. New simple correlation formula for the drag coefficient of non-spherical particles. Powder Technology. 2008;**184**:361-365

[48] Zastawny M, Mallouppas G, Zhao F, Van Wachem B. Derivation of drag and lift force and torque coefficients for non-spherical particles in flows. International Journal of Multiphase Flow. 2012;**39**: 227-239

[49] Ouchene R, Khalij M, Arcen B, Tanière A. A new set of correlations of drag, lift and torque coefficients for non-spherical particles and large Reynolds numbers. Powder Technology. 2016;**303**: 33-43

[50] Li X, Jiang M, Huang Z, Zhou Q. Effect of particle orientation on the drag force in random arrays of prolate ellipsoids in low-Reynolds-number flows. AICHE Journal. 2019;**65**:e16621

[51] Li X, Jiang M, Huang Z, Zhou Q. Effect of particle orientation on the drag

force in random arrays of oblate ellipsoids in low-Reynolds-number flows. AICHE Journal. 2020;**67**:e17040

[52] Cao Z, Tafti DK, Shahnam M. Development of drag correlation for suspensions of ellipsoidal particles. Powder Technology. 2020;**369**:298-310

[53] Srinivas BK, Chhabra RP. An experimental-study of non-Newtonian fluid-flow in fluidized-beds - minimum fluidization velocity and bed expansion characteristics. Chemical Engineering and Processing. 1991;**29**:121-131

[54] Sabiri NE, Comiti J. Experimental validation of a model allowing pressure gradient determination for non-Newtonian purely viscous fluid-flow through packed beds. Chemical Engineering Science. 1997;**52**:3589-3592

[55] Dhole SD, Chhabra RP, Eswaran V. Power law fluid flow through beds of spheres at intermediate Reynolds numbers. Chemical Engineering Research and Design. 2004;**82**:642-652

[56] Qi Z, Kuang S, Yu A. Lattice Boltzmann investigation of non-Newtonian fluid flow through a packed bed of uniform spheres. Powder Technology. 2019;**343**:225-236

[57] Qi Z, Kuang SB, Qiu TS, Yu AB. Lattice Boltzmann investigation on fluid flows through packed beds: Interaction between fluid rheology and bed properties. Powder Technology. 2020; **369**:248-260

Section 2

Mathematical Modelling Computer Implementation

Chapter 5

A Study for Coupled Systems of Nonlinear Boundary Value Problem

Noureddine Bouteraa and Habib Djourdem

Abstract

This chapter deals with the existence and uniqueness of solutions for a coupled system of fractional differential equations with coupled nonlocal and integral boundary conditions and for the system of two-point boundary value problem when we take the case of integer derivative. The existence results for the fist problem are obtained by using Leray-Shauder nonlinear alternative and Banach contraction principle and for the second problem, we derive explicit eigenvalue intervals of λ for the existence of at least one positive solution by using Krasnosel'skii fixed point theorem. An illustrative examples is presented at the end for each problem to illustrate the validity of our results.

Keywords: positive solution, uniqueness, Green's function, system of fractional differential equations, system of differential equations, existence, nonlocal boundary value problem, fixed point theorem

1. Introduction

In this chapter, we are interested in the existence of solutions for the nonlinear fractional boundary value problem (BVP)

$$\begin{cases} {}^cD^\alpha u(t) = f(t,u(t),v(t)), & t\in[0,1],\ 2<\alpha\leq 3, \\ {}^cD^\beta v(t) = g(t,u(t),v(t)), & t\in[0,1],\ 2<\beta\leq 3, \\ \lambda u(0)+\gamma u(1) = u(\eta), & \lambda v(0)+\gamma v(1) = v(\eta), \\ u(0) = \int_0^\eta u(s)ds, & v(0) = \int_0^\eta v(s)ds, \\ \lambda^c D^p u(0) + \gamma^c D^p u(1) = {}^cD^p u(\eta), & 1<p\leq 2. \end{cases} \tag{1}$$

We also study the integer case of problem

$$\begin{cases} u^{(4)}(t) = \lambda a(t) f(v(t)), & 0<t<1, \\ v^{(4)}(t) = \lambda b(t) g(u(t)), & 0<t<1, \\ u(0)=0, u'(0)=0, u''(1)=0, & u'''(1)=0, \\ v(0)=0, v'(0)=0, v''(1)=0, & v'''(1)=0. \end{cases} \tag{2}$$

where $D^{\alpha}_{0^+}, D^{\beta}_{0^+}$ are the standard Riemann-Liouville fractional derivative of order α and β, the functions $f, g \in C((0,1) \times \mathbb{R}^2, \mathbb{R})$, the functions $f, g \in C((0,1) \times \mathbb{R}, \mathbb{R})$ in the second problem and $\lambda > 0, a, b \in C([0,1], [0,\infty))$.

The first definition of fractional derivative was introduced at the end of the nineteenth century by Liouville and Riemann, but the concept of non-integer derivative and integral, as a generalization of the traditional integer order differential and integral calculus, was mentioned already in 1695 by Leibniz and L'Hospital. In fact, fractional derivatives provide an excellent tool for the description of memory and hereditary properties of various materials and processes. The mathematical modeling of systems and processes in the fields of physics, chemistry, aerodynamics, electrodynamics of complex medium, polymer rheology, Bode's analysis of feedback amplifiers, capacitor theory, electrical circuits, electro-analytical chemistry, biology, control theory, fitting of experimental data, involves derivatives of fractional order. In consequence, the subject of fractional differential equations is gaining much importance and attention. For more details we refer the reader to [1–6] and the references cited therein.

Boundary value problems for nonlinear differential equations arise in a variety of areas of applied mathematics, physics and variational problems of control theory. A point of central importance in the study of nonlinear boundary value problems is to understand how the properties of nonlinearity in a problem influence the nature of the solutions to the boundary value problems. The multi-point boundary conditions are important in various physical problems of applied science when the controllers at the end points of the interval (under consideration) dissipate or add energy according to the sensors located, at intermediate points, see [7, 8] and the references therein. We quote also that realistic problems arising from economics, optimal control, stochastic analysis can be modelled as differential inclusion. The study of fractional differential inclusions was initiated by EL-Sayad and Ibrahim [9]. Also, recently, several qualitative results for fractional differential inclusion were obtained in [10–13] and the references therein.

The techniques of nonlinear analysis, as the main method to deal with the problems of nonlinear differential equations (DEs), nonlinear fractional differential equations (FDEs), nonlinear partial differential equations (PDEs), nonlinear fractional partial differential equations (FPDEs), nonlinear stochastic fractional partial differential equations (SFPDEs), plays an essential role in the research of this field, such as establishing the existence, uniqueness and multiplicity of solutions (or positive solutions) and mild solutions for nonlinear of different kinds of FPDEs, FPDEs, SFPDEs, inclusion differential equations and inclusion fractional differential equations with various boundary conditions, by using different techniques (approaches). For more details, see [14–37] and the references therein. For example, iterative method is an important tool for solving linear and nonlinear Boundary Value Problems. It has been used in the research areas of mathematics and several branches of science and other fields. However, Many authors showed the existence of positive solutions for a class of boundary value problem at resonance case. Some recent devolopment for resonant case can be found in [38, 39]. Let us cited few papers. Zhang et al. [40] studied the existence of two positive solutions of following singular fractional boundary value problems:

$$\begin{cases} D^{\alpha}_{0+}u(t) + f(t, u(t)) = 0, \quad t \in (0,1) \\ u(0) = 0, \ D^{\beta}_{0+}u(0) = 0, \quad D^{\beta}_{0+}u(1) = \sum_{j=1}^{\infty} D^{\beta}_{0+}u(\eta_j), \end{cases} \tag{3}$$

where D^{α}_{0+}, D^{β}_{0+} are the stantard Riemann-Liouville fractional derivative of order $\alpha \in (2,3], \beta \in [1,2]$, $f \in C([0,1] \times \mathbb{R}, \mathbb{R})$ and $a_j, \eta_j \in (0,1), \alpha - \beta \geq 1$ with $\sum_{\infty}^{i=0} a_j \eta_j^{\alpha-\beta-1} < 1$.

In [41], the authors studied the boundary value problems of the fractional order differential equation:

$$\begin{cases} D^{\alpha}_{0+}u(t) = f(t, u(t)) = 0, \quad t \in (0,1), \\ u(0) = 0, \ \ D^{\beta}_{0+}u(1) = aD^{\beta}_{0+}u(\eta), \end{cases} \tag{4}$$

where $1 < \alpha \leq 2$, $0 < \eta < 1$, $0 < a, \beta < 1, f \in C([0,1] \times \mathbb{R}^2, \mathbb{R})$ and D^{α}_{0+}, D^{β}_{0+} are the stantard Riemann-Liouville fractional derivative of order α. They obtained the multiple positive solutions by the Leray-Schauder nonlinear alternative and the fixed point theorem on cones.

In 2015, Alsulami et al. [42] studied the existence of solutions of the following nonlinear third-order ordinary differential inclusion with multi-strip boundary conditions

$$\begin{cases} u^{(3)}(t) \in F(t, u(t)), \ \ t \in (0,1), \\ u(0) = 0, \ \ u'(0) = 0, \\ u(1) = \sum\limits_{i=1}^{n-2} \alpha_i \int_{\zeta_i}^{\eta_i} u(s)ds, \\ 0 < \zeta_i < \eta_i < 1, i = 1, 2, \ldots, n-2, n \geq 3. \end{cases} \tag{5}$$

In 2017, Resapour et al. [43] investigated a Caputo fractional inclusion with integral boundary condition for the following problem

$$\begin{cases} {}^cD^{\alpha}u(t) \in F\big(t, u(t), {}^cD^{\beta}u(t), u'(t)\big), \\ u(0) + u'(0) + {}^cD^{\beta}u(0) = \int_0^{\eta} u(s)ds, \\ u(1) + u'(1) + {}^cD^{\beta}u(1) = \int_0^{\nu} u(s)ds, \end{cases} \tag{6}$$

where $1 < \alpha \leq 2$, $\eta, \nu, \beta \in (0,1)$, $F : [0,1] \times \mathbb{R} \times \mathbb{R} \times \mathbb{R} \to 2^{\mathbb{R}}$ is a compact valued multifunction and ${}^cD^{\alpha}$ denotes the Caputo fractional derivative of order α.

Inspired and motivated by the works mentioned above, The goal of this chapter is to establish the existence and uniqueness results for the nonlocal boundary value problem system (1) by using some well-known tools of fixed point theory such as Banach contraction principle and Leray-Shauder nonlinear alternative and the existence of at least one positive solution for the system of two-point boundary value problem (2) by using Krasnosel'skii fixed point theorem. The aim of the last results is to establish some simple criteria for the existence of single positive solutions of the BVPs (2) in explicit intervals for λ. The chapter is organized as follows. In Section 2, we recall some preliminary facts that we need in the sequel, for more details; see [44] and we give main results of problem (1). Finally, we give an example to illustrate our result. In Section 3, deals with main results of problem (2) and we give an example to illustrate our results.

2. Existence and uniqueness results for problem (1)

2.1 Preliminaries

In this section, we introduce some definitions and lemmas, see [2, 4, 44–46].

Definition 2.1. Let $\alpha > 0,\ \ n-1 < \alpha < n,\ \ n = [\alpha]+1$ and $u \in C([0,\infty),\mathbb{R})$. The Caputo derivative of fractional order α for the function u is defined by

$$^{c}D^{\alpha}u(t) = \frac{1}{\Gamma(n-\alpha)}\int_0^t (t-s)^{n-\alpha-1}u^{(n)}(s)ds, \tag{7}$$

where $\Gamma(\cdot)$ is the Eleur Gamma function.

Definition 2.2. The Riemann-Liouville fractional integral of order $\alpha > 0$ of a function $u : (0,\infty) \to \mathbb{R}$ is given by

$$I^{\alpha}u(t) = \frac{1}{\Gamma(\alpha)}\int_0^t (t-s)^{\alpha-1}u(s)ds, \quad t > 0, \tag{8}$$

where $\Gamma(\cdot)$ is the Eleur Gamma function, provided that the right side is pointwise defined on $(0,\infty)$.Lemma 2.1. *Let $\alpha > 0,\ \ n-1 < \alpha < n$ and the function $g : [0,T] \to \mathbb{R}$ be continuous for each $T > 0$. Then, the general solution of the fractional differential equation $^{c}D^{\alpha}g(t) = 0$ is given by*

$$g(t) = c_0 + c_1 t + \cdots + c_{n-1}t^{n-1}, \tag{9}$$

where $c_0, c_1, \ldots, c_{n-1}$ are real constants and $n = [\alpha]+1$.

Also, in [19], authors have been proved that for each $T > 0$ and $u \in C([0,T])$ we have

$$I^{\alpha c}D^{\alpha}u(t) = u(t) + c_0 + c_1 t + \cdots + c_{n-1}t^{n-1}, \tag{10}$$

where $c_0, c_1, \ldots, c_{n-1}$ are real constants and $n = [\alpha]+1$.

2.2 Existence results

Let $X = \{u(t) : u(t) \in C([0,1],\mathbb{R})\}$ endowed with the norm $\|u\| = \sup_{t\in[0,1]} |u(t)|$ such that $\|u\| < \infty$. Then $(X, \|.\|)$ is a Banach space and the product space $(X \times X, \|(u,v)\|)$ is also a Banach space equipped with the norm $\|(u,v)\| = \|u\| + \|v\|$.

Throughout the first section, we let

$$M = \frac{\Gamma(3-p)}{|\gamma-\eta^{2-p}|} \neq 0, |\lambda+\gamma-1| \neq 0, |\gamma-\eta^2| \neq 0, \quad Q = |2(1-\eta)(\gamma-\eta) + \eta^2|\lambda+\gamma-1|| \neq 0,$$
$$A(t) = |\Lambda_1(t)| = |\lambda+\gamma-1|(\eta^2 + 2(1-\eta)t),$$
$$B(t) = |\Lambda_2(t)| = (\eta^3|\lambda+\gamma-1| + 3|\gamma-\eta^2|(1-\eta))(\eta^2+2(1-\eta)t) - Q(\eta^3 + 3(1-\eta)t^2), \tag{11}$$

and

$$Q = 2(1-\eta)(\gamma-\eta) + \eta^2(\lambda+\gamma-1) \neq 0. \tag{12}$$

Lemma 2.2. *Let $y \in C([0,1], \mathbb{R})$. Then the solution of the linear differential system*

$$\begin{cases} {}^cD^{\alpha}u(t) = y(t), \ {}^cD^{\beta}v(t) = h(t), \ t \in [0,1], \ 2 < \alpha, \beta \leq 3 \\ \lambda u(0) + \gamma u(1) = v(\eta), \ \lambda v(0) + \gamma v(1) = u(\eta), \\ u(0) = \int_0^{\eta} v(s)ds, \quad v(0) = \int_0^{\eta} u(s)ds, \\ \lambda^c D^p u(0) + \gamma^c D^p u(1) = {}^cD^p v(\eta), \quad 1 < p \leq 2, \\ \lambda^c D^p v(0) + \gamma^c D^p v(1) = {}^cD^p u(\eta), \quad 1 < p \leq 2, \end{cases} \tag{13}$$

is equivalent to the system of integral equations

$$\begin{aligned} u(t) = & \int_0^t \frac{(t-s)^{\alpha-1}}{\Gamma(\alpha)} y(s)ds + \frac{1}{1-\eta_0}^{\eta} \left(\int_0^s \frac{(s-\tau)^{\beta-1}}{\Gamma(\beta)} h(\tau)d\tau \right) ds \\ & - \frac{\Lambda_1(t)}{Q(1-\eta)} \int_0^{\eta} \left(\int_0^s \frac{(s-\tau)^{\beta-1}}{\Gamma(\beta)} h(\tau)d\tau \right) ds \\ & - \frac{\Lambda_2(t)M}{6(1-\eta)Q} \left[\int_0^{\eta} \frac{(\eta-s)^{\beta-p-1}}{\Gamma(\beta-p)} h(s)ds - \gamma \int_0^1 \frac{(1-s)^{\alpha-p-1}}{\Gamma(\alpha-p)} y(s)ds \right] \\ & + \frac{\Lambda_1(t)}{Q(\lambda+\gamma-1)} \left[\int_0^{\eta} \frac{(\eta-s)^{\beta-1}}{\Gamma(\beta)} h(s)ds - \gamma \int_0^1 \frac{(1-s)^{\alpha-1}}{\Gamma(\alpha)} y(s)ds \right], \end{aligned} \tag{14}$$

and

$$\begin{aligned} v(t) = & \int_0^t \frac{(t-s)^{\beta-1}}{\Gamma(\beta)} y(s)ds + \frac{1}{1-\eta} \int_0^{\eta} \left(\int_0^s \frac{(s-\tau)^{\alpha-1}}{\Gamma(\alpha)} h(\tau)d\tau \right) ds \\ & - \frac{\Lambda_1(t)}{Q(1-\eta)} \int_0^{\eta} \left(\int_0^s \frac{(s-\tau)^{\alpha-1}}{\Gamma(\alpha)} h(\tau)d\tau \right) ds \\ & - \frac{\Lambda_2(t)M}{6(1-\eta)Q} \left[\int_0^{\eta} \frac{(\eta-s)^{\alpha-p-1}}{\Gamma(\alpha-p)} h(s)ds - \gamma \int_0^1 \frac{(1-s)^{\beta-p-1}}{\Gamma(\alpha-p)} y(s)ds \right] \\ & + \frac{\Lambda_1(t)}{Q(\lambda+\gamma-1)} \left[\int_0^{\eta} \frac{(\eta-s)^{\alpha-1}}{\Gamma(\alpha)} h(s)ds - \gamma \int_0^1 \frac{(1-s)^{\beta-1}}{\Gamma(\alpha)} y(s)ds \right], \end{aligned} \tag{15}$$

where

$$\Lambda_1(t) = (\lambda + \gamma - 1)\left(\eta^2 + 2(1-\eta)t\right), \tag{16}$$

and

$$\Lambda_2(t) = \left(\eta^3(\lambda+\gamma-1) + 3(\gamma-\eta^2)(1-\eta)\right)\left(\eta^2 + 2(1-\eta)t\right) - Q\left(\eta^3 + 3(1-\eta)t^2\right).$$

Proof. It is well known that the solution of equation ${}^cD^\alpha u(t) = y(t)$ can be written as

$$u(t) = I^\alpha y(t) + c_0 + c_1 t + c_2 t^2, \tag{17}$$

$$v(t) = I^\beta h(t) + d_0 + d_1 t + d_2 t^2, \tag{18}$$

where $c_0, c_1, c_2 \in \mathbb{R}$ and and $d_0, d_1, d_2 \in \mathbb{R}$ are arbitrary constants.
Then, from (68) we have

$$u'(t) = I^{\alpha-1} y(t) + c_1 + 2c_2 t, \tag{19}$$

and

$${}^cD^p u(t) = I^{\alpha-p} y(t) + c_2 \frac{2t^{2-p}}{\Gamma(3-p)}, \quad 1 < p \leq 2. \tag{20}$$

By using the three-point boundary conditions, we obtain.

$$c_2 = \frac{M}{2}\left(I^{\beta-p} y(\eta) - \gamma I^{\alpha-p} y(1)\right),$$

$$\begin{aligned} c_0 = & -\frac{2\eta^2(\lambda+\gamma-1)}{2(1-\eta)Q}\int_0^\eta\left(\int_0^s \frac{(s-\tau)^{\beta-1}}{\Gamma(\beta)} h(\tau)d\tau\right)ds + \frac{1}{1-\eta}\int_0^\eta\left(\int_0^s \frac{(s-\tau)^{\beta-1}}{\Gamma(\beta)} h(\tau)d\tau\right)ds \\ & - \frac{(\eta^2[\eta^3(\lambda+\gamma-1) + 3(\gamma-\eta^2)(1-\eta)] - \eta^3 Q)M}{2(1-\eta)Q}\left[\int_0^\eta \frac{(\eta-s)^{\beta-p-1}}{\Gamma(\beta-p)} h(s)ds\right. \\ & \left. -\gamma\int_0^1 \frac{(1-s)^{\alpha-p-1}}{\Gamma(\alpha-p)} y(s)ds\right] + \frac{\eta^2}{Q}\left[\int_0^\eta \frac{(\eta-s)^{\beta-1}}{\Gamma(\beta)} h(s)ds - \gamma\int_0^1 \frac{(1-s)^{\alpha-1}}{\Gamma(\alpha)} y(s)ds\right], \end{aligned} \tag{21}$$

and

$$\begin{aligned} c_1 = & \frac{-2(\lambda+\gamma-1)}{Q}\int_0^\eta\left(\int_0^s \frac{(s-\tau)^{\beta-1}}{\Gamma(\beta)} h(\tau)d\tau\right)ds \\ & - \frac{(\eta^3(\lambda+\gamma-1) + 3(\gamma-\eta^2)(1-\eta))M}{3Q}\left[\int_0^\eta \frac{(\eta-s)^{\beta-p-1}}{\Gamma(\beta-p)} h(s)ds - \gamma\int_0^1 \frac{(1-s)^{\alpha-p-1}}{\Gamma(\alpha-p)} y(s)ds\right] \\ & + \frac{2(1-\eta)}{Q}\left[\int_0^\eta \frac{(\eta-s)^{\beta-1}}{\Gamma(\beta)} y(s)ds - \gamma\int_0^1 \frac{(1-s)^{\alpha-1}}{\Gamma(\alpha)} y(s)ds\right]. \end{aligned} \tag{22}$$

Substituting the values of constants c_0, c_1 and c_2 in (68), we get solution (64). Similarly, we obtain solution (65). The proof is complete.
The following rolations hold:

$$|A(t)| \leq |\beta + \gamma - 1|(\eta^2 + 2(1-\eta)) = A_1, \tag{23}$$

and

$$|B(t)| \leq \left| \left(\eta^3|\beta+\gamma-1| + 3\left|\gamma-\eta^2\right|(1-\eta)\right)\left(\eta^2+2(1-\eta)\right) - Q\left(\eta^3+3(1-\eta)\right)\right| = B_1, \tag{24}$$

For the sake of brevity, we set

$$\begin{aligned}
\Delta_1 &= \frac{\eta^{\beta+1}}{(1-\eta)\Gamma(\beta+2)} + \frac{A_1\eta^{\beta+1}}{Q(1-\eta)\Gamma(\beta+2)} + \frac{MB_1\eta^{\beta-p}}{(1-\eta)Q\Gamma(\lambda-p+1)} \\
&\quad + \frac{A_1\eta^{\beta}}{Q|\beta+\gamma-1|\Gamma(\beta+1)} \\
\Delta_2 &= \frac{MB_1\gamma}{6(1-\eta)Q\Gamma(\alpha-p+1)} + \frac{A_1\gamma}{Q|\lambda+\gamma-1|\Gamma(\alpha+1)} + \frac{1}{\Gamma(\alpha+1)} \\
\Delta_3 &= \frac{\eta^{\alpha+1}}{(1-\eta)\Gamma(\alpha+2)} + \frac{A_1\eta^{\alpha+1}}{Q(1-\eta)\Gamma(\alpha+2)} + \frac{MB_1\eta^{\alpha-p}}{(1-\eta)Q\Gamma(\alpha-p+1)} \\
&\quad + \frac{A_1\eta^{\alpha}}{Q|\lambda+\gamma-1|\Gamma(\alpha+1)},
\end{aligned} \tag{25}$$

and

$$\Delta_4 = \frac{MB_1\gamma}{6(1-\eta)Q\Gamma(\beta-p+1)} + \frac{A_1\gamma}{Q|\lambda+\gamma-1|\Gamma(\beta+1)} + \frac{1}{\Gamma(\beta+1)}. \tag{26}$$

In view of Lemma 2, we define the operator $T : X \times X \to X \times X$ by

$$T(u,v)(t) = \begin{pmatrix} T_1(u,v)(t) \\ T_2(u,v)(t) \end{pmatrix}, \tag{27}$$

where

$$\begin{aligned}
T_1(u,v)(t) &= \int_0^t \frac{(t-s)^{\alpha-1}}{\Gamma(\alpha)} f(s,u(s),v(s))ds + \frac{1}{1-\eta}\int_0^\eta \left(\int_0^s \frac{(s-\tau)^{\beta-1}}{\Gamma(\beta)} g(\tau,u(\tau),v(\tau))d\tau \right) ds \\
&\quad - \frac{B(t)M}{6(1-\eta)Q}\left[\int_0^\eta \frac{(\eta-s)^{\beta-p-1}}{\Gamma(\beta-p)} g(s,u(s),v(s))ds - \gamma \int_0^1 \frac{(1-s)^{\alpha-p-1}}{\Gamma(\alpha-p)} f(s,u(s),v(s))ds \right] \\
&\quad + \frac{A(t)}{Q|\beta+\gamma-1|}\left[\int_0^\eta \frac{(\eta-s)^{\beta-1}}{\Gamma(\beta)} g(s,u(s),v(s))ds - \gamma \int_0^1 \frac{(1-s)^{\alpha-1}}{\Gamma(\alpha)} f(s,u(s),v(s))ds. \right] \\
&\quad - \frac{A(t)}{Q(1-\eta)} \int_0^\eta \left(\int_0^s \frac{(s-\tau)^{\beta-1}}{\Gamma(\beta)} g(\tau,u(\tau),v(\tau))d\tau \right) ds,
\end{aligned} \tag{28}$$

and

$$T_2(u,v)(t) = \int_0^t \frac{(t-s)^{\beta-1}}{\Gamma(\beta)} g(s,u(s),v(s))ds + \frac{1}{1-\eta}\int_0^\eta \left(\int_0^s \frac{(s-\tau)^{\alpha-1}}{\Gamma(\alpha)} f(\tau,u(\tau),v(\tau))d\tau \right) ds$$

$$-\frac{B(t)M}{6(1-\eta)Q}\left[\int_0^\eta \frac{(\eta-s)^{\alpha-p-1}}{\Gamma(\alpha-p)} f(s,u(s),v(s))ds - \gamma \int_0^1 \frac{(1-s)^{\beta-p-1}}{\Gamma(\beta-p)} g(s,u(s),v(s))ds \right]$$

$$+\frac{A(t)}{Q|\beta+\gamma-1|}\left[\int_0^\eta \frac{(\eta-s)^{\alpha-1}}{\Gamma(\alpha)} f(s,u(s),v(s))ds - \gamma \int_0^1 \frac{(1-s)^{\beta-1}}{\Gamma(\beta)} g(s,u(s),v(s))ds. \right]$$

$$-\frac{A(t)}{Q(1-\eta)}\int_0^\eta \left(\int_0^s \frac{(s-\tau)^{\alpha-1}}{\Gamma(\alpha)} f(\tau,u(\tau),v(\tau))d\tau \right) ds. \tag{29}$$

Observe that the boundary value problem (1) has solutions if the operator equation $(u,v) = T(u,v)$ has fixed points.

Now we are in a position to present the first main results of this paper. The existence results is based on Leray-Shauder nonlinear alternative.

Lemma 2.3. [44] *(Leray-Schaucler alternative). Let E be a Banach space and* $T : E \to E$ *be a completely continuous operator (i.e., a map restricted to any bounded set in E is compact). Let*

$$\varepsilon = \{(u,v) \in X \times X : (u,v) = \lambda T(u,v), for\ some\ 0 < \lambda < 1\}. \tag{30}$$

Then either the $\varepsilon(T)$ is unbounded or T has at least one fixed point.

Theorem 1.1 Assume that $f,g : [0,1] \times \mathbb{R} \times \mathbb{R} \to \mathbb{R}$ are a continuous function and.

(H_1) there exist a function $k_i \geq 0 m_i \geq 0, i = 1,2$ and $k_0 > 0, m_0 > 0$ such that $\forall u \in \mathbb{R}, \forall v \in \mathbb{R}, i = 1,2$, we have

$$|f(t,u,v)| \leq k_0 + k_1|u| + k_2|v|, \tag{31}$$

and

$$|g(t,u,v)| \leq m_0 + m_1|u| + m_2|v|. \tag{32}$$

If $(\Delta_2 + \Delta_3)k_1 + (\Delta_1 + \Delta_4)m_1 < 1$ and $(\Delta_2 + \Delta_3)k_2 + (\Delta_1 + \Delta_4), m_3 < 1$, where $\Delta_i,\ i = 1,2,3,4$ are given above. Then the boundary value problem (1)–(58) has at least one solution on [0,1].

Proof. It is clear that T is a continuous operator where $T : X \times X \to X \times X$ is defined above. Now, we show that T is completely continuous. Let $\Omega \subset X \times X$ be bounded. Then there exist positive constants L_1 and L_2 such that

$$|f(t,u(t),v(t))| \leq L_1, \quad |g(t,u(t),v(t))| \leq L_2, \quad \forall (u,v) \in \Omega. \tag{33}$$

Then for any $(u,v)\in\Omega$, we have

$$\begin{aligned}|T_1(u,v)(t)| &\le \frac{L_2}{1-\eta}\int_0^\eta\left(\int_0^s\frac{(s-\tau)^{\beta-1}}{\Gamma(\beta)}d\tau\right)ds\\
&+\frac{|A(t)|L_2}{Q(1-\eta)}\int_0^\eta\left(\int_0^s\frac{(s-\tau)^{\beta-1}}{\Gamma(\beta)}d\tau\right)ds+L_1\int_0^t\frac{(t-s)^{\alpha-1}}{\Gamma(\alpha)}ds\\
&+\frac{M|B(t)|}{6(1-\eta)Q}\left[L_2\int_0^\eta\frac{(\eta-s)^{\beta-p-1}}{\Gamma(\beta-p)}ds+\gamma L_1\int_0^0\frac{(1-s)^{\alpha-p-1}}{\Gamma(\alpha-p)}ds\right]\\
&+\frac{|A(t)|}{Q|\lambda+\gamma-1|}\left[L_2\int_0^\eta\frac{(\eta-s)^{\beta-1}}{\Gamma(\beta)}ds+\gamma L_1\int_0^1\frac{(1-s)^{\alpha-1}}{\Gamma(\alpha)}ds\right],\\
&\le L_2\left\{\frac{1}{1-\eta}\int_0^\eta\left(\int_0^s\frac{(s-\tau)^{\beta-1}}{\Gamma(\beta)}d\tau\right)ds+\frac{A_1}{Q(1-\eta)}\int_0^\eta\left(\int_0^s\frac{(s-\tau)^{\beta-1}}{\Gamma(\alpha\beta)}d\tau\right)ds\right.\\
&\left.+\frac{MB_1}{6(1-\eta)Q}\int_0^\eta\frac{(\eta-s)^{\beta-p-1}}{\Gamma(\beta-p)}ds+\frac{A_1}{6|\lambda+\gamma-1|}\int_0^\eta\frac{(\eta-s)^{\beta-1}}{\Gamma(\beta)}ds\right\}\\
&+L_1\left\{\frac{M\gamma B_1}{6(1-\eta)Q}\int_0^1\frac{(1-s)^{\alpha-p-1}}{\Gamma(\alpha-p)}ds+\frac{A_1\gamma}{Q|\lambda+\gamma-1|}\int_0^1\frac{(1-s)^{\alpha-1}}{\Gamma(\alpha)}ds+\int_0^t\frac{(t-s)^{\alpha-1}}{\Gamma(\alpha)}ds\right\},\\
&\le L_2\Delta_1+L_1\Delta_2.\end{aligned}\tag{34}$$

Hence

$$\|T_1(u,v)\|\le L_2\Delta_1+L_1\Delta_2.\tag{35}$$

In the same way, we can obtain that

$$\|T_2(u,v)\|\le L_1\Delta_3+L_2\Delta_4.\tag{36}$$

Thus, it follows from (78) and (95) that the operator T is uniformly bounded, since $\|T(u,v)\|\le L_1(\Delta_1+\Delta_3)+L_2(\Delta_2+\Delta_4)$. Now, we show that T is equicontinuous. Let $t_1,t_2\in[0,1]$ with $t_1<t_2$. Then we have

$$\begin{aligned}|T_1(u(t_2),v(t_2))-T_1(u(t_1),v(t_1))| &\le L_1\int_0^{t_1}\frac{(t_2-s)^{\alpha-1}-(t_1-s)^{\alpha-1}}{\Gamma(\alpha)}ds\\
&+L_1\int_{t_1}^{t_2}\frac{(t_2-s)^{\alpha-1}}{\Gamma(\alpha)}ds+\frac{|A(t_2)-A(t_1)|L_2}{Q(1-\eta)}\int_0^\eta\left(\int_0^s\frac{(s-\tau)^{\beta-1}}{\Gamma(\beta)}d\tau\right)ds\\
&+\frac{(B(t_2)-B(t_1))M}{6(1-\eta)Q}\left[L_2\int_0^\eta\frac{(\eta-s)^{\beta-p-1}}{\Gamma(\beta-p)}ds+\gamma L_1\int_0^1\frac{(1-s)^{\alpha-p-1}}{\Gamma(\alpha-p)}ds\right]\\
&+\frac{A(t_2)-A(t_1)}{Q|\lambda+\gamma-1|}\left[L_2\int_0^\eta\frac{(\eta-s)^{\beta-1}}{\Gamma(\beta)}ds-\gamma L_1\int_0^1\frac{(1-s)^{\alpha-1}}{\Gamma(\alpha)}ds\right].\end{aligned}\tag{37}$$

Obviously, the right-hand side of the above inequality tends to zero as $t_2 \to t_1$. Similarly, we have

$$
\begin{aligned}
|T_2(u(t_2),v(t_2)) - T_2(u(t_1),v(t_1))| \le & L_2 \int_0^{t_1} \frac{(t_2-s)^{\beta-1}-(t_1-s)^{\beta-1}}{\Gamma(\beta)}ds \\
& + L_2 \int_{t_1}^{t_2} \frac{(t_2-s)^{\beta-1}}{\Gamma(\beta)}ds + \frac{|A(t_2)-A(t_1)|L_1}{Q(1-\eta)} \int_0^{\eta}\left(\int_0^{s}\frac{(s-\tau)^{\alpha-1}}{\Gamma(\alpha)}d\tau\right)ds \\
& + \frac{(B(t_2)-B(t_1))M}{6(1-\eta)Q}\left[L_1\int_0^{\eta}\frac{(\eta-s)^{\alpha-p-1}}{\Gamma(\alpha-p)}ds + \gamma L_2 \int_0^1 \frac{(1-s)^{\beta-p-1}}{\Gamma(\beta-p)}ds\right] \\
& + \frac{A(t_2)-A(t_1)}{Q|\lambda+\gamma-1|}\left[L_1\int_0^{\eta}\frac{(\eta-s)^{\alpha-1}}{\Gamma(\alpha)}ds - \gamma L_2\int_0^1\frac{(1-s)^{\beta-1}}{\Gamma(\beta)}ds\right].
\end{aligned}
\tag{38}
$$

Again, it is seen that the right-hand side of the above inequality tends to zero as $t_2 \to t_1$. Thus, the operator T is equicontinuous.

Therefore, the operator T is completely continuous.

Finally, it will be verified that the set $\varepsilon = \{(u,v) \in X \times X : (u,v) = \lambda T(u,v), 0 \le \lambda \le 1\}$ is bounded. Let $(u,v) \in \varepsilon$, with $(u,v) = \lambda T(u,v)$ for any $t \in [0,1]$, we have

$$u(t) = \lambda T_1(u,v)(t), \quad v(t) = \lambda T_2(u,v)(t). \tag{39}$$

Then

$$
\begin{aligned}
|u(t)| &\le \Delta_2(k_0 + k_1|u| + k_2|v|) + \Delta_1(m_0 + m_1|u| + m_2|v|), \\
&= \Delta_2 k_0 + \Delta_1 m_0 + (\Delta_2 k_1 + \Delta_1 m_1)|u| + (\Delta_2 k_2 + \Delta_1 m_2)|v|,
\end{aligned}
\tag{40}
$$

and

$$
\begin{aligned}
|v(t)| &\le \Delta_3(k_0 + k_1|u| + k_2|v|) + \Delta_4(m_0 + m_1|u| + m_2|v|), \\
&= \Delta_3 k_0 + \Delta_4 m_0 + (\Delta_3 k_1 + \Delta_4 m_1)|u| + (\Delta_3 k_2 + \Delta_4 m_2)|v|.
\end{aligned}
\tag{41}
$$

Hence we have

$$\|u\| = \Delta_2 k_0 + \Delta_1 m_0 + (\Delta_2 k_1 + \Delta_1 m_1)\|u\| + (\Delta_2 k_2 + \Delta_1 m_2)\|v\|, \tag{42}$$

and

$$\|v\| = \Delta_3 k_0 + \Delta_4 m_0 + (\Delta_3 k_1 + \Delta_4 m_1)|u| + (\Delta_3 k_2 + \Delta_4 m_2)|v|, \tag{43}$$

which imply that

$$
\begin{aligned}
\|u\| + \|v\| = & (\Delta_2 + \Delta_3)k_0 + (\Delta_1 + \Delta_4)m_0 + [(\Delta_2 + \Delta_3)k_1 + (\Delta_1 + \Delta_4)m_1]\|u\| \\
& + [(\Delta_2 + \Delta_3)k_2 + (\Delta_1 + \Delta_4)m_2]\|v\|.
\end{aligned}
\tag{44}
$$

Consequently,

$$\|(u,v)\| = \frac{(\Delta_2+\Delta_3)k_0 + (\Delta_1+\Delta_4)m_0}{\Delta_0}, \tag{45}$$

where

$$\Delta_0 = min\,\{1-[(\Delta_2+\Delta_3)k_1+(\Delta_1+\Delta_4)m_1], 1-[(\Delta_2+\Delta_3)k_2+(\Delta_1+\Delta_4)m_2]\}, \tag{46}$$

which proves that ε is bounded. Thus, by Lemma 15, the operator T has at least one fixed point. Hence boundary value problem (1) has at least one solution. The proof is complete.

Now, we are in a position to present the second main results of this paper.

Theorem 1.2 Assume that $f, g : [0,1]\times\mathbb{R}^2 \to \mathbb{R}$ are continuous functions and there exist positive constants L_1 and L_2 such that for all $t\in[0,1]$ and $u_i, v_i \in \mathbb{R}, i=1,2$, we have.

1. $|f(t,u_1,u_2) - f(t,v_1,v_2)| \le L_1(|u_1-v_1|+|u_2-v_2|)$,

2. $|g(t,u_1,u_2) - g(t,v_1,v_2)| \le L_2(|u_1-v_1|+|u_2-v_2|)$.

Then the boundary value problem (1) has a unique solution on [0,1] provided

$$(\Delta_1+\Delta_3)L_1 + (\Delta_2+\Delta_4)L_2 < 1. \tag{47}$$

Proof. Let us set $\sup_{t\in[0,1]} |f(t,0,0)| = N_1 < \infty$ and $\sup_{t\in[0,1]} |g(t,0,0)| = N_2 < \infty$.
For $u \in X$, we observe that

$$\begin{aligned} |f(t,u(t),v(t))| &\le |f(t,u(t)) - f(t,0,0)| + |f(t,0,0)|,\\ &\le L_1(|u(t)|+|v(t)|) + N_1,\\ &\le L_1(\|u\|+\|v\|) + N_1, \end{aligned} \tag{48}$$

and

$$|g(t,u(t),v(t))| \le |g(t,u(t)) - g(t,0,0)| + |g(t,0,0)| \le L_2\|u\| + N_2. \tag{49}$$

Then for $u \in X$, we have

$$\begin{aligned} |T_1(u,v)(t)| &\le \frac{1}{1-\eta}\int_0^\eta\left(\int_0^s \frac{(s-\tau)^{\beta-1}}{\Gamma(\beta)}[L_2\|(u,v)\|+N_2]d\tau\right)ds\\ &+\frac{|A(t)|}{Q(1-\eta)}\int_0^\eta\left(\int_0^s \frac{(s-\tau)^{\beta-1}}{\Gamma(\beta)}[L_2\|(u,v)\|+N_2]d\tau\right)ds + \int_0^t \frac{(t-s)^{\alpha-1}}{\Gamma(\alpha)}[L_1\|(u,v)\|+N_1]ds\\ &+\frac{M|B(t)|}{6(1-\eta)Q}\left[\int_0^\eta \frac{(\eta-s)^{\beta-p-1}}{\Gamma(\beta-p)}[L_2\|(u,v)\|+N_2]ds + \gamma\int_0^1 \frac{(1-s)^{\alpha-p-1}}{\Gamma(\alpha-p)}[L_1\|(u,v)\|+N_1]ds\right]\\ &+\frac{|A(t)|}{Q|\lambda+\gamma-1|}\left[\int_0^\eta \frac{(\eta-s)^{\beta-1}}{\Gamma(\beta)}[L_2\|(u,v)\|+N_2]ds + \gamma\int_0^1 \frac{(1-s)^{\alpha-1}}{\Gamma(\alpha)}[L_1\|(u,v)\|+N_1]ds,\right] \end{aligned}$$

$$\leq (L_2\|(u,v)\| + N_2)\left\{\frac{1}{1-\eta}\int_0^\eta\left(\int_0^s \frac{(s-\tau)^{\beta-1}}{\Gamma(\beta)}d\tau\right)ds + \frac{A_1}{Q(1-\eta)}\int_0^\eta\left(\int_0^s \frac{(s-\tau)^{\beta-1}}{\Gamma(\beta)}d\tau\right)ds\right.$$
$$+\frac{MB_1}{6(1-\eta)Q}\int_0^\eta \frac{(\eta-s)^{\beta-p-1}}{\Gamma(\beta-p)}ds + \frac{A_1}{6|\lambda+\gamma-1|}\int_0^\eta \frac{(\eta-s)^{\beta-1}}{\Gamma(\beta)}ds\Bigg\}$$
$$+\frac{MB_1}{6(1-\eta)Q}\int_0^\eta \frac{(\eta-s)^{\beta-p-1}}{\Gamma(\beta-p)}ds + \frac{A_1}{6|\lambda+\gamma-1|}\int_0^\eta \frac{(\eta-s)^{\beta-1}}{\Gamma(\beta)}ds\Bigg\} \tag{50}$$
$$+(L_1\|(u,v)\| + N_1)\left\{\int_0^t \frac{(t-s)^{\alpha-1}}{\Gamma(\alpha)}ds\right.$$
$$\left.+\frac{M\gamma B_1}{6(1-\eta)Q}\int_0^1 \frac{(1-s)^{\alpha-p-1}}{\Gamma(\alpha-p)}ds + \frac{A_1\gamma}{Q|\lambda+\gamma-1|}\int_0^1 \frac{(1-s)^{\alpha-1}}{\Gamma(\alpha)}ds\right\},$$
$$\leq (L_2 r + N_2)\Delta_1 + (L_1 r + N_1)\Delta_2$$

Hence

$$\|T_1(u,v)\| \leq (L_2\Delta_1 + L_1\Delta_2)r + N_2\Delta_1 + N_1\Delta_2 \tag{51}$$

In the same way, we can obtain that

$$\|T_2(u,v)\| \leq (L_1\Delta_3 + L_2\Delta_4)r + N_2\Delta_4 + N_1\Delta_3. \tag{52}$$

Consequently,

$$\|T(u,v)\| \leq ((\Delta_2+\Delta_3)L_1 + (\Delta_1+\Delta_4)L_2)r + N_2(\Delta_1+\Delta_4) + N_1(\Delta_2+\Delta_3) \leq r. \tag{53}$$

Now, for $(u_1,v_1), (u_2,v_2) \in X \times X$ and for each $t \in [0,1]$, it follows from assumption (H_3) that

$$|T_1(u_2,v_2)(t) - T_1(u_1,v_1)(t)| \leq L_2(\|u_2-u_1\| + \|v_2-v_1\|)\left\{\frac{1}{1-\eta}\int_0^\eta\left(\int_0^s \frac{(s-\tau)^{\beta-1}}{\Gamma(\beta)}d\tau\right)ds\right.$$
$$+\frac{A_1}{Q(1-\eta)}\int_0^\eta\left(\int_0^s \frac{(s-\tau)^{\beta-1}}{\Gamma(\beta)}d\tau\right)ds$$
$$\left.+\frac{MB_1}{6(1-\eta)Q_0}\,{}^{\eta}\frac{(\eta-s)^{\beta-p-1}}{\Gamma(\beta-p)}ds + \frac{A_1}{6|\lambda+\gamma-1|}\int_0^\eta \frac{(\eta-s)^{\beta-1}}{\Gamma(\beta)}ds\right\}$$
$$+L_1(\|u_2-u_1\| + \|v_2-v_1\|)\left\{\int_0^t \frac{(t-s)^{\alpha-1}}{\Gamma(\alpha)}ds\right.$$
$$\left.+\frac{M\gamma B_1}{6(1-\eta)Q}\int_0^1 \frac{(1-s)^{\alpha-p-1}}{\Gamma(\alpha-p)}ds + \frac{A_1\gamma}{Q|\lambda+\gamma-1|}\int_0^1 \frac{(1-s)^{\alpha-1}}{\Gamma(\alpha)}ds\right\}$$
$$\leq (L_2\Delta_1 + L_1\Delta_2)(\|u_2-u_1\| + \|v_2-v_1\|). \tag{54}$$

Thus

$$\|T_1(u_2, v_2) - T_1(u_1, v_1)\| \leq (L_2\Delta_1 + L_1\Delta_2)(\|u_2 - u_1\| + \|v_2 - v_1\|). \tag{55}$$

Similarly,

$$\|T_2(u_2, v_2) - T_2(u_1, v_1)\| \leq (L_2\Delta_3 + L_1\Delta_4)(\|u_2 - u_1\| + \|v_2 - v_1\|). \tag{56}$$

It follows from (101) and (110) that

$$\|T(u_2, v_2) - T(u_1, v_1)\| \leq (L_2(\Delta_1 + \Delta_3) + L_1(\Delta_2 + \Delta_4))(\|u_2 - u_1\| + \|v_2 - v_1\|). \tag{57}$$

Since $L_2(\Delta_1 + \Delta_3) + L_1(\Delta_2 + \Delta_4) < 1$, thus T is a contraction operator. Hence it follows by Banach's contraction principle that the boundary value problem (1) has a unique solution on $[0, 1]$.

We construct an example to illustrate the applicability of the results presented.

Example 2.1. Consider the following system fractional differential equation

$$\begin{cases} {}^cD^3u(t) = \frac{t}{8}\left((\cos(t))\sin\left(\frac{|u(t)| + |v(t)|}{2}\right)\right) + \frac{e^{-(u(t)+v(t))^2}}{1+t^2}, & t \in [0,1], \\ {}^cD^3v(t) = \frac{1}{32}\sin(2\pi u(t)) + \frac{|v(t)|}{16(1 + |v(t)|)} + \frac{1}{2}, & t \in [0,1], \end{cases} \tag{58}$$

subject to the three-point coupled boundary conditions

$$\begin{cases} \frac{1}{100}u(0) + \frac{1}{10}u(1) = u\left(\frac{1}{2}\right), \\ u(0) = \int_0^{0,5} u(s)ds, \\ \frac{1}{100}{}^cD^{\frac{3}{2}}u(0) + \frac{1}{10}{}^cD^{\frac{3}{2}}u(1) = {}^cD^{\frac{3}{2}}u\left(\frac{1}{2}\right), \end{cases} \tag{59}$$

where $f(t, u, v) = \frac{t}{8}\left((\cos(t))\sin\left(\frac{|u|+|v|}{2}\right)\right) + \frac{e^{-(u+v)^2}}{1+t^2}, t \in [0, 1], \eta = 0, 5, \lambda = 0, 01, \gamma = 0, 1, p = 1, 5$ *and* $g(t, u, v) = \frac{1}{32\pi}\sin(2\pi u(t)) + \frac{|v(t)|}{16(1+|v(t)|)} + \frac{1}{2}$.

It can be easily found that $M = \frac{20}{3}$ and $Q = \frac{9}{400}$.

Furthermore, by simple computation, for every $u_i, v_i \in \mathbb{R}, \ i = 1, 2$, we have

$$|f(t, u_1, u_2) - f(t, v_1, v_2)| \leq L(|u_1 - v_1| + |u_2 - v_2|), \tag{60}$$

and

$$|g(t, u_1, u_2) - g(t, v_1, v_2)| \leq L(|u_1 - v_1| + |u_2 - v_2|), \tag{61}$$

where $L_1 = L_2 = L = \frac{1}{16}$. It can be easily found that $\Delta_1 = \Delta_3 \cong 0, 799562, \ \Delta_2 = \Delta_4 \cong 1, 182808$.

Finally, since $L_1(\Delta_1 + \Delta_3) + L_2(\Delta_2 + \Delta_4) = 2L(\Delta_1 + \Delta_2) \cong 0, 247796 < 1$, thus all assumptions and conditions of Theorem 1.2 are satisfied. Hence, Theorem implies that the three-point boundary value problem (58, 59) has a unique solution.

3. Existence result for second problem (2)

We provide another results about the existence of solutions for the problem (2) by using the assumption.

We shall consider the Banach space $B = C([0,1])$ equipped with usual supermum norm and $B^+ = C^+([0,1])$. In arriving our results, we present some notation and preliminary lemmas. The first is well known.

Lemma 3.1. *Let* $y(t) \in C[0,1]$. *If* $u \in C^4[0,1]$, *then the BVP*

$$\begin{cases} u^{(4)}(t) = y(t), & 0 \le t \le 1, \\ u(0) = u'(0) = u''(1) = u'''(1) = 0, \end{cases} \tag{62}$$

has a unique solution

$$u(t) = \int_0^1 G(t,s)y(s)ds, \tag{63}$$

where

$$G(t,s) = \begin{cases} \frac{1}{6}t^2(3s - t), & 0 \le t \le s \le 1, \\ \frac{1}{6}s^2(3t - s), & 0 \le s \le t \le 1. \end{cases} \tag{64}$$

Lemma 3.2. *For any* $(t,s) \in [0,1] \times [0,1]$, *we have*

$$0 \le G(t,s) \le G(1,s) = \frac{1}{6}s^2(3 - s) = \psi(s). \tag{65}$$

Proof. The derivatives of the function G with respect to t is

$$\frac{\partial}{\partial t}G(t,s) = \begin{cases} \frac{1}{2}s^2 - \frac{1}{2}(s - t)^2, & 0 \le t \le s \le 1 \\ \frac{1}{2}s^2, & 0 \le s \le t \le 1. \end{cases} \tag{66}$$

Since the derivative of the function G with respect to t is nonnegative for all $t \in [0,1]$, G is nondecreasing function of t that attaints its maximum when $t = 1$. Then

$$\max_{0 \le t \le 1} G(t,s) = G(1,s) = \frac{1}{2}s^2 - \frac{1}{6}s^3. \tag{67}$$

Lemma 3.3. *Let* $0 < \theta < 1$. *Then for* $y(t) \in C^+[0,1]$, *the unique solution* $u(t)$ *of BVP* (14) *is nonnegative and satisfies*

$$\min_{t \in [\theta,1]} u(t) \ge \frac{2\theta^3}{3}\|u\|. \tag{68}$$

Proof. Let $y(t) \in C^+[0,1]$, then from $G(t,s) \ge 0$ we know $u \in C^+[0,1]$. Set $u(t_0) = \|u\|$, $t_0 \in (0,1]$. We first prove that

$$\frac{G(t,s)}{G(t_0,s)} \geq \frac{2}{3}t^3, \quad t, t_0, s \in (0,1]. \tag{69}$$

In fact, we can consider four cases:

1. if $0 < t, t_0 \leq s \leq 1$, then

$$\frac{G(t,s)}{G(t_0,s)} = \frac{t^2(3s-t)}{t_0^2(3s-t_0)} \geq \frac{t^2(2s)}{3s-t_0} \geq \frac{t^2(2s)}{3} \geq \frac{t^2(2t)}{3} = \frac{2t^3}{3}, \tag{70}$$

2. if $0 < t \leq s \leq t_0 \leq 1$, then

$$\frac{G(t,s)}{G(t_0,s)} = \frac{t^2(3s-t)}{t_0^2(3s-t_0)} \geq \frac{t^2(2s)}{3s-t_0} \geq \frac{t^2(2s)}{3} \geq \frac{t^2(2t)}{3} = \frac{2t^3}{3}, \tag{71}$$

3. if $0 < s \leq t, t_0 \leq 1$, then

$$\frac{G(t,s)}{G(t_0,s)} = \frac{s^2(3t-s)}{s^2(3t_0-s)} = \frac{3t-s}{3t_0-s} \geq \frac{3t-s}{3t_0} \geq \frac{3t-s}{3} \geq \frac{2t+t-s}{3} \geq \frac{2t}{3} \geq \frac{2t^3}{3}, \tag{72}$$

4. if $0 < t_0 \leq s \leq t \leq 1$, then

$$\frac{G(t,s)}{G(t_0,s)} = \frac{s^2(3t-s)}{t_0^2(3s-t_0)} \geq \frac{t_0^2(3t-s)}{t_0^2(3t-t_0)} \geq \frac{3t-s}{3t} \geq \frac{3t-t}{3t} \geq \frac{2t}{3} \geq \frac{2t^3}{3}, \tag{73}$$

Therefore, for $t \in [\theta, 1]$, we have

$$u(t) = \int_0^1 G(t,s)y(s)ds = \int_0^1 \frac{G(t,s)}{G(t_0,s)} G(t_0,s)y(s)ds \geq \frac{2t^3}{3}u(t_0) \geq \frac{2\theta^3}{3}\|u\|. \tag{74}$$

The proof is complete.
If we let

$$K = \left\{ x \in B : \ x(t) \geq 0 \ \ on \ \ [0,1], \ \ and \ \ \min_{t \in [\theta,1]} x(t) \geq \frac{2\theta^3}{3}\|x\| \right\} \tag{75}$$

then it is easy to see that K a cone in B. We not that a pair $(u(t), v(t))$ is a solution of BVPs (2) if, and only if

$$u(t) = \lambda \int_0^1 G(t,s)a(s)f\left(\lambda \int_0^1 G(s,r)b(r)g(u(r))dr\right)ds, \quad t \in [0,1], \tag{76}$$

and

$$v(t) = \lambda \int_0^1 G(t,s)b(s)g(u(s))ds, \quad t \in [0,1]. \tag{77}$$

Now, we define an integral operator $T : K \to B$ by

$$(Tu)(t) = \lambda \int_0^1 G(t,s)a(s)f\left(\lambda \int_0^1 G(s,r)b(r)g(u(r))dr\right)ds, \quad u \in K. \tag{78}$$

We adopt the following assumptions:
$(H_1)\, a, b \in C((0,1), [0,\infty))$ and each does not vanish identically on any subinterval.
$(H_2)\, f, g \in C([0,\infty), [0,\infty))$ and each to be singular at $t = 0\, or\, t = 1$.
(H_3) All of $f_0 = \lim_{x\to 0^+} \frac{f(x)}{x}$, $g_0 = \lim_{x\to 0^+} \frac{g(x)}{x}$, $f_\infty = \lim_{x\to\infty} \frac{f(x)}{x}$, *and* $g_\infty = \lim_{x\to\infty} \frac{g(x)}{x}$ exist as real numbers.
$(H_4)\, g(0) = 0$ and f is increasing function.
Lemma 3.4 *Let λ be positive number and K be the cone defined above.*

i. If $u \in B^+$ and $v : [0,1] \to [0,\infty)$ is defined by (77), then $v \in K$.

ii. If T is the integral operator defined by (78), then $T(K) \subset K$.

iii. Assume that (H_1), (H_2) hold. Then $T : K \to B$ is completely continuous.

Proof. Let $u \in B^+$ and v be defined by (77).

i. By the nonnegativity of G, b and g it follows that $v(t) \geq 0, t \in [0,1]$. In view of (H_1), (H_2), we have

$$\int_0^1 G(t,s)b(s)g(u(s))ds \geq \int_0^1 \min_{t\in[\theta,1]} G(t,s)b(s)g(u(s))ds, \tag{79}$$

from which, we take

$$\min_{t\in[\theta,1]} \int_0^1 G(t,s)b(s)g(u(s))ds \geq \int_0^1 \min_{t\in[\theta,1]} G(t,s)b(s)g(u(s))ds. \tag{80}$$

Consequently, employing (68) and for $\lambda > 0$, we have

$$\begin{aligned} \lambda \int_0^1 G(t,s)b(s)g(u(s))ds &\geq \lambda \int_0^1 \min_{t\in[\theta,1]} G(t,s)b(s)g(u(s))ds \\ &\geq \frac{2\theta^3}{3} \lambda \int_0^1 G(t_0,s)b(s)g(u(s))ds \\ &\geq \frac{2\theta^3}{3} v(t_0), t_0 \in (0,1] \\ &\geq \frac{2\theta^3}{3} \|v\|. \end{aligned} \tag{81}$$

Therefore

$$\min_{t\in[\theta,1]} v(t) \geq \frac{2\theta^3}{3}\|v\|. \tag{82}$$

Which give that $v\in K$.

ii. Obviously, for $v\in K,\ \ T(u)\in C^+[0,1]$. For $t\in[0,1]$, we have

$$\begin{aligned}\|Tu(t)\| &= \max_{0\leq t\leq 1}\lambda\int_0^1 G(t,s)a(s)f(v(s))ds\\ &\leq \lambda\int_0^1 G(1,s)a(s)f(v(s))ds,\end{aligned} \tag{83}$$

and

$$\begin{aligned}Tu(t) &= \lambda\int_0^1 G(t,s)a(s)f(v(s))ds\\ &= \lambda\int_0^1 \frac{G(t,s)}{G(1,s)}G(1,s)a(s)f(v(s))ds\\ &\geq \frac{2\theta^3}{3}\lambda\int_0^1 G(1,s)a(s)f(v(s))ds\\ &\geq \frac{2}{3}\theta^3\|Tu(t)\|.\end{aligned} \tag{84}$$

Which give that $Tu\in K$. Therefore $T: K\to K$.

iii. By using standard arguments it is not difficult to show that the operator T : $K\to B$ is completely continuous.

The key tool in our approach is the following Krasnoselskii's fixed point theorem of cone expansion-compression type.

Theorem 1.3 (See [47]) Let B be a Banach space and $K\subset B$ be a cone in B. Assume Ω_1 and Ω_2 are open subset of B with $0\in\Omega_1$ and $\overline{\Omega_1}\subset\Omega_2$,

$T: K\cap(\overline{\Omega}_2\backslash\Omega_1)\to K$ be a completely continuous operator such that.

i. $\|Tu\|\leq\|u\|,\ \forall u\in K\cap\partial\Omega_1$ and $\|Tu\|\geq\|u\|,\ \forall u\in K\cap\partial\Omega_2$; or.

ii. $\|Tu\|\geq\|u\|,\ \forall u\in K\cap\partial\Omega_1$ and $\|Tu\|\leq\|u\|,\ \forall u\in K\cap\partial\Omega_2$.

Then, T has a fixed point in $K\cap(\overline{\Omega}_2\backslash\Omega_1)$. Throughout this section, we shall use the following notations:

$$L_1 = \max\left\{\left[\left(\frac{2\theta^3}{3}\right)^2 \int_\theta^1 \psi(r)a(r)f_\infty dr\right]^{-1}, \left[\left(\frac{2\theta^3}{3}\right)^2 \int_\theta^1 \psi(r)a(r)g_\infty dr\right]^{-1}\right\} \quad (85)$$

and

$$L_2 = \min\left\{\left[\int_0^1 \psi(r)a(r)f_0 dr\right]^{-1}, \left[\int_0^1 \psi(r)b(r)g_0 dr\right]^{-1}\right\}. \quad (86)$$

$$L_3 = \max\left\{\left[\left(\frac{2\theta^3}{3}\right)^2 \int_\theta^1 \psi(r)a(r)f_0 dr\right]^{-1}, \left[\left(\frac{2\theta^3}{3}\right)^2 \int_\theta^1 \psi(r)a(r)g_0 dr\right]^{-1}\right\} \quad (87)$$

and

$$L_4 = \min\left\{\left[\int_\theta^1 \psi(r)a(r)f_\infty dr\right]^{-1}, \left[\int_\theta^1 \psi(r)b(r)g_\infty dr\right]^{-1}\right\}. \quad (88)$$

4. Existence results

In this section, we discuss the existence of at least one positive solution for BVPs (2). We obtain the following existence results, by applying the positivity of Green's function $G(t,s)$ and the fixed-point of cone expansion-compression type.

Theorem 1.4 Assume conditions (H_1), (H_2) and (H_3) are satisfied. Then, for each λ satisfying $L_1 < \lambda < L_2$ there exists a pair (u,v) satisfying BVPs (2) such that $u(t) > 0$ and $v(t) > 0$ on $(0,1)$.

Proof. Let $L_1 < \lambda < L_2$. And let $\varepsilon > 0$ be chosen such that

$$\max\left\{\left[\left(\frac{2\theta^3}{3}\right)^2 \int_\theta^1 \psi(r)a(r)(f_\infty - \varepsilon)dr\right]^{-1}, \left[\left(\frac{2\theta^3}{3}\right)^2 \int_\theta^1 \psi(r)a(r)(g_\infty - \varepsilon)dr\right]^{-1}\right\} \le \lambda, \quad (89)$$

and

$$\lambda \le \min\left\{\left[\int_\theta^1 \psi(r)a(r)(f_0 + \varepsilon)dr\right]^{-1}, \left[\int_\theta^1 \psi(r)b(r)(g_0 + \varepsilon)dr\right]^{-1}\right\}. \quad (90)$$

From the definitions of f_0 and g_0 there exists an $R_1 > 0$ such that

$$f(u) \le (f_0 + \varepsilon)u, \quad 0 < u \le R_1, \quad (91)$$

and

$$g(u) \le (g_0 + \varepsilon)u, \quad 0 < u \le R_1, \quad (92)$$

Let $u \in K$ with $\|u\| = R_1$. From (65) and choice of ε, we have

$$\begin{aligned}\lambda\int_0^1 G(t,s)b(r)g(u(r)) &\leq \lambda\int_0^1 \psi(r)b(r)g(u(r))dr \\ &\leq \lambda\int_0^1 \psi(r)b(r)(g_0+\varepsilon)u(r)dr \\ &\leq \|u\|\lambda\int_0^1 \psi(r)b(r)dr(g_0+\varepsilon)dr \\ &\leq R_1 = \|u\|.\end{aligned} \tag{93}$$

Consequently, from (65) and choice of ε, we have

$$\begin{aligned}Tu(t) &= \lambda\int_0^1 G(t,s)a(s)f\left(\lambda\int_0^1 G(s,r)b(r)g(u(r))dr\right)ds \\ &\leq \lambda\int_0^1 \psi(s)a(s)f\left(\lambda\int_0^1 G(s,r)b(r)g(u(r))dr\right)ds \\ &\leq \lambda\int_0^1 \psi(s)a(s)(f_0+\varepsilon)\left[\lambda\int_0^1 G(s,r)b(r)g(u(r))dr\right]ds \\ &\leq \lambda\int_0^1 \psi(s)a(s)(f_0+\varepsilon)R_1 ds \\ &\leq R_1 = \|u\|.\end{aligned} \tag{94}$$

So, $\|Tu\| \leq \|u\|$. If we set $\Omega_1 = \{u \in B: \ \|u\| < R_1\}$, then

$$\|Tu\| \leq \|u\|, \ for \ u \in K \cap \partial\Omega_1 \tag{95}$$

Considering the definitions of f_∞ and g_∞ there exists an $\overline{R}_2 > 0$ such that

$$f(u) \geq (f_\infty - \varepsilon)u, \quad 0 < u \leq \overline{R}_2, \tag{96}$$

and

$$g(u) \geq (g_\infty - \varepsilon)u, 0 < u \leq \overline{R}_2. \tag{97}$$

Let $u \in K$ and $R_2 = \max\left\{2R_1, \frac{3\overline{R}_2}{2\theta^3}\right\}$ with $\|u\| = R_2$, then

$$\min_{s\in[\theta,1]} u(s) \geq \frac{2}{3}\theta^3\|u\| \geq \overline{R}_2 \tag{98}$$

Thus, from (68) and choice of ε, we have

$$\begin{aligned} \lambda\int_0^1 G(t,s)b(r)g(u(r)) &\geq \frac{2\theta^3}{3}\lambda\int_0^1 G(1,r)b(r)g(u(r))dr \\ &\geq \frac{2\theta^3}{3}\lambda\int_\theta^1 \psi(r)b(r)\left(g_\infty-\varepsilon\right)u(r)dr \\ &\geq \|u\|\left(\frac{2\theta^3}{3}\right)^2\lambda\int_\theta^1 \psi(r)b(r)dr\left(g_\infty-\varepsilon\right)dr \\ &\geq R_2=\|u\|. \end{aligned} \tag{99}$$

Consequently, from (77) and choice of ε, we have

$$\begin{aligned} Tu(t) &\geq \frac{2\theta}{3}\lambda\int_\theta^1 \psi(s)a(s)f\left(\lambda\int_\theta^1 G(s,r)b(r)g(u(r))dr\right)ds \\ &\geq \frac{2\theta^3}{3}\lambda\int_\theta^1 \psi(s)a(s)\left(f_\infty-\varepsilon\right)\left[\lambda\int_\theta^1 G(s,r)b(r)g(u(r))dr\right]ds \\ &\geq \frac{2\theta^3}{3}\lambda\int_\theta^1 \psi(s)a(s)\left(f_\infty-\varepsilon\right)H_2 ds \\ &\geq \left(\frac{2\theta^3}{3}\right)^2\lambda\int_\theta^1 \psi(s)a(s)\left(f_\infty-\varepsilon\right)H_2 ds \\ &\geq R_2=\|u\|. \end{aligned} \tag{100}$$

So, $\|Tu\|\geq\|u\|$. If we set $\Omega_2=\{u\in B:\ \|u\|<R_2\}$, then

$$\|Tu\|\geq\|u\|,\ for\ u\in K\cap\partial\Omega_2. \tag{101}$$

Applying (i) of Theorem 3.1 to (95) and (101), yields that T has a fixed point $u^*\in K\cap\left(\overline{\Omega}_2/\Omega_1\right)$. As such and with v defined by

$$v(t)=\lambda\int_0^1 G(t,s)b(s)g(u(s))ds, \tag{102}$$

the pair (u,v) is a desired solution of BVPs (2) for the given λ. The proof is complete.

Theorem 1.5 Assume conditions $(H_1),(H_2),(H_3)$ and (H_4) are satisfied. Then, for each λ satisfying $L_3<\lambda<L_4$ there exists a pair (u,v) satisfying BVPs (2) such that $u(t)>0$ and $v(t)>0$ on $(0,1)$.

Proof. Let $L_3<\lambda<L_4$ and $\varepsilon>0$ be chosen such that

$$max\left\{\left[\left(\frac{2\theta^3}{3}\right)^2\int_\theta^1 \psi(r)a(r)\left(f_0-\varepsilon\right)dr\right]^{-1},\left[\left(\frac{2\theta^3}{3}\right)^2\int_\theta^1 \psi(r)a(r)\left(g_0-\varepsilon\right)dr\right]^{-1}\right\}\leq\lambda, \tag{103}$$

and

$$\lambda \le min\left\{\left[\int_0^1 \psi(r)a(r)(f_\infty + \varepsilon)dr\right]^{-1}, \left[\int_0^1 \psi(r)b(r)(g_\infty + \varepsilon)dr\right]^{-1}\right\}. \tag{104}$$

From the definitions of f_0 and g_0 there exists an $R_1 > 0$ such that

$$f(u) \ge (f_0 - \varepsilon)u, \quad 0 < u \le R_1, \tag{105}$$

and

$$g(u) \ge (g_0 - \varepsilon)u, \quad 0 < u \le R_1, \tag{106}$$

Now $g(0) = 0$ and so there exists $0 < R_2 \le R_1$ such that

$$\lambda g(u) \le \frac{R_1}{\int_0^1 \psi(r)b(r)dr}, \quad 0 \le u \le R_2. \tag{107}$$

Let $u \in K$ with $\|u\| = R_2$. Then

$$\begin{aligned}\lambda\int_0^1 G(t,s)b(r)g(u(r)) &\le \lambda\int_0^1 \psi(r)b(r)g(u(r))dr \\ &\le \frac{\int_0^1 \psi(r)b(r)R_1 dr}{\int_0^1 \psi(s)b(s)ds} \\ &\le R_1 = \|u\|.\end{aligned} \tag{108}$$

Therefore, by (68), we have

$$\begin{aligned}Tu(t) &= \lambda\int_0^1 G(t,s)a(s)f\left(\lambda\int_0^1 G(s,r)b(r)g(u(r))dr\right)ds \\ &\ge \frac{2\theta^3}{3}\lambda\int_\theta^1 \psi(s)a(s)f\left(\frac{2\theta^3}{3}\lambda\int_\theta^1 \psi(r)b(r)g(u(r))dr\right)ds \\ &\ge \frac{2\theta^3}{3}\lambda\int_\theta^1 \psi(s)a(s)(f_0 - \varepsilon)\left[\left(\frac{2\theta^3}{3}\right)^2\lambda\int_\theta^1 \psi(r)b(r)(g_0 - \varepsilon)\|u\|dr\right]ds \\ &\ge \frac{2\theta^3}{3}\lambda\int_\theta^1 \psi(r)a(r)(f_0 - \varepsilon)\|u\| \\ &\ge \left(\frac{2\theta^3}{3}\right)^2\lambda\int_\theta^1 \psi(r)a(r)(f_0 - \varepsilon)\|u\| \\ &\ge \|u\|.\end{aligned} \tag{109}$$

So, $\|Tu\| \geq \|u\|$. If we set $\Omega_1 = \{u \in B: \ \|u\| < R_2\}$, then

$$\|Tu\| \geq \|u\|, \quad u \in K \cap (\overline{\Omega}_2 \backslash \Omega_1) \tag{110}$$

Considering the definitions of f_∞ and g_∞ there exists $\overline{R}_1 > 0$ such that

$$f(u) \leq (f_\infty + \varepsilon)u, \quad u \geq \overline{R}_1, \tag{111}$$

and

$$g(u) \leq (g_\infty + \varepsilon)u, \quad u \geq \overline{R}_1. \tag{112}$$

We consider two cases: g is bounded or g is unbounded.

Case (i). Suppose g is bounded, say $g(u) \leq N, \quad N > 0$ for all $0 < u < \infty$. Then, for $u \in K$

$$\begin{aligned} &\lambda \int_0^1 G(t,s)b(r)g(u(r)) \leq \lambda \int_0^1 \psi(r)b(r)g(u(r))dr \\ &M = \max\left\{f(u): \ 0 \leq u \leq N\lambda \int_0^1 \psi(r)b(r)dr\right\} \end{aligned} \tag{113}$$

and let

$$R_3 > \max\left\{2R_2, \ M\lambda \int_0^1 \psi(s)a(s)ds\right\}. \tag{114}$$

Then, for $u \in K$ with $\|u\| = R_3$, we have

$$\begin{aligned} Tu(t) &\leq \lambda \int_0^1 \psi(s)a(s)Mds \\ &\leq R_3 = \|u\|. \end{aligned} \tag{115}$$

So that $\|Tu\| \leq \|u\|$. If we set $\Omega_2 = \{u \in B: \ \|u\| \leq R_3\}$, then, for $u \in K \cap \partial\Omega_2$:

$$\|Tu\| \leq \|u\|, \quad u \in K \cap \partial\Omega_2 \tag{116}$$

Case(ii). g is unbounded, there exists $R_3 > max\ \{2R_2, \overline{R}_1\}$ such that $g(u) \leq g(R_3)$, for $0 < u \leq R_3$.

Similarly, there exists $R_4 > max\left\{2R_3, \ M\lambda\int_0^1 \psi(r)b(r)g(R_3)ds\right\}$ such that $f(u) \leq f(R_4)$, for $0 < u \leq R_4$.

Let $u \in K$ with $\|u\| = R_4$, from (H_4), we have

$$\begin{aligned} Tu(t) &\leq \lambda \int_0^1 \psi(s)a(s)f\left(\lambda \int_0^1 \psi(r)b(r)g(R_3)dr\right)ds \\ &\leq \lambda \int_0^1 \psi(r)a(r)f(R_4)ds \\ &\leq \lambda \int_0^1 \psi(r)a(r)(f_\infty + \varepsilon)R_4 ds \\ &\leq R_4 = \|u\|. \end{aligned} \tag{117}$$

So, $\|Tu\| \leq \|u\|$. If we set $\Omega_2 = \{x \in C[0,1] | \|x\| \leq R_4\}$, then

$$\|Tu\| \leq \|u\|, \quad for \quad u \in K \cap \partial\Omega_2. \tag{118}$$

In either of cases, application of part (ii) of Theorem 3.1 yields a fixed point u^* of T belonging to $K \cap \left(\overline{\Omega}_2/\Omega_1\right)$, which in turn yields a pair (u, v) satisfying BVPs (2) for the chosen value of λ. The proof is complete.

We construct an example to illustrate the applicability of the results presented.

Example 4.1. *Consider the two-point boundary value problem*

$$\begin{cases} u^{(4)}(t) = \lambda t v(t)\left(v(t)e^{-v(t)} + \dfrac{v(t)+K}{1+\eta v(t)}\right), & 0 < t < 1, \\ v^{(4)}(t) = \lambda t u(t)\left(u(t)e^{-u(t)} + \dfrac{u(t)+K}{1+\eta u(t)}\right), & 0 < t < 1, \end{cases} \tag{119}$$

and satisfying two-point boundary conditions

$$\begin{cases} u(0) = 0, u'(0) = 0, u''(1) = 0, u'''(1) = 0, \\ v(0) = 0, v'(0) = 0, v''(1) = 0, v'''(1) = 0, \end{cases} \tag{120}$$

where $a(t) = b(t) = t,\ f(v) = v\left(ve^{-v} + \frac{v+K}{1+\eta v}\right),\ g(u) = u\left(1 + \frac{u+K}{1+\eta u}\right)$.

By simple calculations, we find $g(0) = 0, f_\infty = g_\infty = \frac{1}{\eta}, f_0 = g_0 = K$.

Choosing $\theta = \frac{1}{3}, \eta = 100,$ *and* $K = 10^4$, *we obtain* $L_3 \cong 1,1817237, L_4 \cong 9,1666667$.

By Theorem 4, *it follows that for every* λ *such that* $1,1817237 < \lambda < 9,1666667$, *there exists a pair* (u, v) *satisfying* BVPs (25–2526).

5. Conclusions

This chapter concerns the boundary value problem of a class of fractional differential equations involving the Caputo fractional derivative with nonlocal boundary conditions. By using the Leray-Shauder nonlinear alternative and Banach contraction principle, one shows that the problem has at least one positive solutions and has unique solution. Secondly, we derive explicit eigenvalue intervals of λ for the existence of at least one positive solution for the second problem by using Krasnosel'skii fixed point theorem. The results of the present chapter are significantly contribute to the existing literature on the topic.

Acknowledgements

The authors want to thank the anonymous referee for the thorough reading of the manuscript and several suggestions that help us improve the presentation of the chapter.

Conflict of interest

The authors declare no conflict of interest.

Author details

Noureddine Bouteraa[1,2*†] and Habib Djourdem[1,3*†]

1 Laboratory of Fundamental and Applied Mathematics of Oran (LMFAO), University of Oran1, Ahmed Benbella, Algeria

2 Oran Graduate School of Economics, Oran, Algeria

3 University of Ahmed Zabbana, Relizane, Algeria

*Address all correspondence to: bouteraa-27@hotmail.fr and djourdem.habib7@gmail.com

† These authors contributed equally.

References

[1] Agarwal RP, Baleanu D, Hedayati V, Rezapour SH. Two fractional derivative inclusion problems via integral boundary condition. Applied Mathematics and Computation. 2015;**257**:205-212

[2] Kac V, Cheung P. Quantum Calculus. New-York: Springer; 2002

[3] Lakshmikantham V, Vatsala AS. General uniqueness and monotone iterative technique for fractional differential equations. Applied Mathematics Letters. 2008;**21**(8): 828-834

[4] Miller S, Ross B. An Introduction to the Fractional Calculus and Fractional Differential Equations. New-York: John Wiley and Sons, Inc.; 1993

[5] Rudin W. Functional analysis. In: International Series in Pure and Applied Mathematics. 2nd ed. New York: Mc Graw-Hill; 1991

[6] Samko SG, Kilbas AA, Marichev OI. Fractional Integrals and Derivatives: Theory and Applications. Yverdon: Gordon & Breach; 1993

[7] Jarad F, Abdeljaw T, Baleanu D. On the generalized fractional derivatiives and their Caputo modification. Journal of Nonlinear Sciences and Applications. 2017;**10**(5):2607-2619

[8] Tian Y. Positive solutions to m-point boundary value problem of fractional differential equation. Acta Mathematica Sinica, English Series. 2013;**29**:661-672

[9] EL-Sayed AMA, NESREEN FM, EL-Haddade. Existence of integrable solutions for a functional integral inclusion. Differential Equations & Control Processes. 2017;**3**:15-25

[10] Bouteraa N, Benaicha S. Existence of solutions for nonlocal boundary value problem for Caputo nonlinear fractional differential inclusion. Journal of Mathematical Sciences and Modelling. 2018;**1**(1):45-55

[11] Bouteraa N, Benaicha S. Existence results for fractional differential inclusion with nonlocal boundary conditions. Ri. Mat. Parma. 2020;**11**:181-206

[12] Cernia A. Existence of solutions for a certain boundary value problem associated to a fourth order differential inclusion. International Journal of Analysis and Applications. 2017;**14**:27-33

[13] Ntouyas SK, Etemad S, Tariboon J, Sutsutad W. Boundary value problems for Riemann-Liouville nonlinear fractional diffrential inclusions with nonlocal Hadamard fractional integral conditions. Mediterranean Journal of Mathematics. 2015;**2015**:16

[14] Bouteraa N, Benaicha S. Triple positive solutions of higher-order nonlinear boundary value problems. Journal of Computer Science and Computational Mathematics. 2017;**7**(2):25-31

[15] Bouteraa N, Benaicha S. Existence of solutions for three-point boundary value problem for nonlinear fractional equations, Analele Universitatii Oradia Fasc. Mathematica. 2017;**Tom XXIV**(2): 109-119

[16] Benaicha S, Bouteraa N. Existence of solutions for three-point boundary value problem for nonlinear fractional differential equations. Bulltin of the Transilvania University of Brasov, Serie III: Mathematics, Informtics, Physics. 2017;**10**(59):2

[17] Bouteraa N, Benaicha S. Existence of solutions for third-order three-point

boundary value problem. Mathematica. 2018;**60**(83):21-31

[18] Bouteraa N, Benaicha S. The uniqueness of positive solution for higher-order nonlinear fractional differential equation with nonlocal boundary conditions. Advances in the Theory of Nonlinear and it Application. 2018;**2**(2):74-84

[19] Bouteraa N, Benaicha S, Djourdem H. Positive solutions for nonlinear fractional differential equation with nonlocal boundary conditions. Universal Journal of Mathematics and Applications. 2018;**1**(1):39-45

[20] Bouteraa N, Benaicha S. The uniqueness of positive solution for nonlinear fractional differential equation with nonlocal boundary conditions. Analele universitatii Oradia Fasc. Matematica. 2018;**Tom XXV**(2):53-65

[21] Bouteraa N, Benaicha S, Djourdem H, Benatia N. Positive solutions of nonlinear fourth-order two-point boundary value problem with a parameter. Romanian Journal of Mathematics and Computer Science. 2018;**8**(1):17-30

[22] Bouteraa N, Benaicha S. Positive periodic solutions for a class of fourth-order nonlinear differential equations. Siberian Journal of Numerical Mathematics. 2019;**22**(1):1-14

[23] Bouteraa N. Existence of Solutions for Some Nonlinear Boundary Value Problems [thesis]. Ahmed Benbella, Algeria: University of Oran 1; 2018

[24] Bouteraa N, Benaicha S, Djourdem H. ON the existence and multiplicity of positive radial solutions for nonlinear elliptic equation on bounded annular domains via fixed point index. Maltepe Journal of Mathematics. 2019;**I**(1):30-47

[25] Bouteraa N, Benaicha S, Djourdem H. Positive solutions for systems of fourth-order two-point boundary value problems with parameter. Journal of Mathematical Sciences and Modeling. 2019;**2**(1):30-38

[26] Bouteraa N, Benaicha S. Existence and multiplicity of positive radial solutions to the Dirichlet problem for the nonlinear elliptic equations on annular domains. Universitatis Babeş-Bolyai. 2020;**65**(1):109-125

[27] Benaicha S, Bouteraa N, Djourdem H. Triple positive solutions for a class of boundary value problems with integral boundary conditions. Bulletin of Transilvania University of Brasov, Series III: Mathematics, Informatics, Physics. 2020;**13**(62):51-68

[28] Bouteraa N, Djourdem H, Benaicha S. Existence of solution for a system of coupled fractional boundary value problem. Proceedings of International Mathematical Sciences. 2020;**II**(1):48-59

[29] Bouteraa N, Benaicha S. Existence results for second-order nonlinear differential inclusion with nonlocal boundary conditions. Numerical Analysis and Applications. 2021;**14**(1): 30-39

[30] Bouteraa N, In M, Akinlar MA, Almohsen B. Mild solutions of fractional PDE with noise. Mathematical Methods in the Applied Sciences. 2021:1-15. DOI: 10.1002/mma.7139

[31] Bouteraa N, Benaicha S. A study of existence and multiplicity of positive solutions for nonlinear fractional differential equations with nonlocal boundary conditions. Universitatis Babeş-Bolyai. 2021;**66**(2):361-380

[32] Djourdem H, Benaicha S. Existence of positive solutions for a nonlinear three-point boundary value problem with integral boundary conditions. Acta Mathematica Universitatis Comenianae. 2018;**87**(2):167-177

[33] Djourdem H, Benaicha S, Bouteraa N. Existence and iteration of monotone positive solution for a fourth-order nonlinear boundary value problem. Fundamental Journal of Mathematics and Applications. 2018; **1**(2):205-211

[34] Djourdem H, Benaicha S, Bouteraa N. Two positive solutions for a fourth-order three-point BVP with sign-changing Green's function. Communications in Advanced Mathematical Sciences. 2019;**II**(1):60-68

[35] Djourdem H, Bouteraa N. Mild solution for a stochastic partial differential equation with noise. WSEAS Transactions on Systems. 2020;**19**:246-256

[36] Ghorbanian R, Hedayati V, Postolache M, Rezapour SH. On a fractional differential inclusion via a new integral boundary condition. Journal of Inequalities and Applications. 2014; **2014**:319. 20 pages

[37] Inc M, Bouteraa N, Akinlar MA, Chu YM, Weber GW, Almohsen B. New positive solutions of nonlinear elliptic PDEs. Applied Sciences. 2020;**10**:4863. DOI: 10.3390/app10144863. 17 pages

[38] Bouteraa N, Benaicha S. Nonlinear boundary value problems for higher-order ordinary differential equation at resonance. Romanian Journal of Mathematic and Computer Science. 2018;**8**(2):83-91

[39] Bouteraa N, Benaicha S. A class of third-order boundary value problem with integral condition at resonance. Maltepe Journal of Mathematics. 2020; **II**(2):43-54

[40] Zhang X, Zhong Q. Multiple positive solutions for nonlocal boundary value problems of singular fractional differential equations. Boundary Value Problems. 2016;**2016**:65

[41] Lin X, Zhao Z, Guan Y. Iterative Technology in a Singular Fractional Boundary Value Problem With q-Difference. Applications of Mathematics. 2016;7:91-97

[42] Alsulami HH, Ntouyas SK, Ahmad B, Alsaedi A. A study of third-order single-valued and multi-valued problems with integral boundary conditions. Boundary Value Problems. 2015:20

[43] Rezapour SH, Hedayati V. On a Caputo fractional differential inclusion with integral boundary condition for convex-compact and nonconvex-compact valued multifunctions. Kragujevac Journal of Mathematics. 2017;**41**:143-158

[44] Kilbas AA, Srivastava HM, Trijull JJ. Theory and Applications of Fractional Differential Equations. Amsterdam: Elsevier Science B. V; 2006

[45] Annaby MH, Mansour ZS. q-Fractional calculus and equations. In: Lecture Notes in Mathematics. Vol. 2056. Berlin: Springer-Verlag; 2012

[46] Agrawal O. Some generalized fractional calculus operators and their applications in integral equations. Fractional Calculus and Applied Analysis. 2012;**15**:700-711

[47] Krasnosel'skii MA. Positive Solutions of Operator Equations. Groningen: Noordhoff; 1964

Chapter 6

A System of Singularly Perturbed Parabolic Equations with a Power Boundary Layer

Asan Omuraliev and Peil Esengul Kyzy

Abstract

The work is devoted to the construction of the asymptotics of the solution of a singularly perturbed system of equations of parabolic type, in the case when the limit equation has a regular singularity as the small parameter tends to zero. The developed algorithm allows construction of the asymptotics of solutions containing power, parabolic, and angular boundary layer functions.

Keywords: singularly perturbed problems, power boundary layer, parabolic boundary layer, angular boundary-layer functions, regularized asymptotics

1. Introduction

1.1 A singularly perturbed system of parabolic equations

This article studies the problem

$$L_\varepsilon u(x,t,\varepsilon) \equiv (\varepsilon + t)\partial_t u - \varepsilon^2 a(x)\partial_x^2 u - B(t)u = f(x,t), \ (x,t) \in \Omega,$$
$$u|_{t=0} = u|_{x=0} = u|_{x=1} = 0, \tag{1}$$

where $\varepsilon > 0$ is a small parameter, $\Omega = \{(x,t) : x \in (0,1), t \in (0,T]\}$, $u = (u_1, u_2, \ldots, u_n)$. Suppose the following conditions:

$$0 < a(x) \in C^\infty[0,1], B(t) \in C^\infty\left([0, \ T], C^{n^2}\right), f(x,t) \in C^\infty\left(\overline{\Omega}, C^n\right);$$

The eigenvalues $\lambda_i(t)$ of the matrix $B(t)$ satisfy the conditions: $B(t)\psi_i(t) = \lambda_i(t)\psi_i(t), i = 1,2,\ldots,n, \Re\lambda_i(t) < 0, \lambda_j(t) \neq \lambda_i(t), \forall i \neq j, t \in [0, \ T]$ ($\Re$ is the real part of the complex number).

Singularly perturbed parabolic problems in various statements are devoted to Ref. [1–23]. In Ref. [6, 9–17], by the method of boundary layer functions, various boundary value problems for parabolic equations with a small parameter are studied. The regularization asymptotic for solving parabolic problems [7, 8, 18–23] with different occurrences of a small parameter in the equations and limit operators has different structures using the regularization method for singularly perturbed problems.

The papers [16, 23] are devoted to systems of singularly perturbed parabolic equations. In [16], the method of boundary functions, and in Ref. [23], the method of regularization. For singularly perturbed problems, equations are studied when the spatial derivative is preceded by a scalar function, and [24] is a matrix. In the latter case, the structure of the solution asymptotic is greatly multiplied. The order of the equation does not go down.

Problems with power boundary layers, that is, problems where, when a small parameter tends to zero, it acquires a regular feature were studied in Ref. [9, 10, 25, 26]. Ordinary differential equations with a power boundary layer are studied in Ref. [8–10, 26]. For equations of parabolic type, when a small parameter does not enter the factors of the spatial derivative, the asymptotic of the power boundary layer is constructed. In contrast to Ref. [9], in our equation, there is a small parameter in front of the spatial derivative and we will improve the algorithm for constructing the asymptotic.

1.2 Regularization of problems

We introduce regularizing variables

$$\xi_l = \frac{\varphi_l(x)}{\sqrt{\varepsilon^3}},\quad \varphi_l(x) = (-1)^{l-1}\int_{l-1}^{x}\frac{ds}{\sqrt{a(s)}},\quad \eta_i = \lambda_i(0)\ln(1+t/\varepsilon),\quad \tau = \varepsilon^{-1}\ln(1+t/\varepsilon) \tag{2}$$

along with the independent variables (x, t) of the function $\tilde{u}\,(M, \varepsilon)$, $M = (x, t, \xi, \tau, \eta)$, $\xi = (\xi_1, \xi_2)$, $\eta = (\eta_1, \eta_2, \ldots, \eta_n)$ such that

$$\tilde{u}(M,\ \varepsilon)|_{\zeta=\gamma(x,\ t,\ \varepsilon)} \equiv u(x, t, \varepsilon),$$

$$\begin{aligned}\zeta &= (\xi, \tau, \eta), \gamma(x, t, \varepsilon)\\ &= \left(\frac{\varphi_1(x)}{\sqrt{\varepsilon^3}}, \frac{\varphi_2(x)}{\sqrt{\varepsilon^3}}, \lambda_1(0)\ln(1+t/\varepsilon), \ldots, \lambda_n(0)\ln(1+t/\varepsilon), \varepsilon^{-1}\ln(1+t/\varepsilon)\right).\end{aligned}$$

Hence, on the basis of (2), we find

$$\partial_t u \equiv \left(\partial_t\tilde{u} + \sum_{i=1}^{n}\frac{\lambda_i(0)}{t+\varepsilon}\partial_{\eta_i}\tilde{u} + \frac{1}{\varepsilon t+\varepsilon}\partial_\tau\tilde{u}\right)_{\zeta=\gamma(x,\ t,\ \varepsilon)}, \partial_x u \equiv \left(\partial_x\tilde{u} + \sum_{l=1}^{2}\frac{\varphi_l'(x)}{\sqrt{\varepsilon^3}}\partial_{\xi_l}\tilde{u}\right)_{\zeta=\gamma(x,\ t,\ \varepsilon)},$$

$$\partial_x^2 u \equiv \left(\partial_x^2\tilde{u} + \sum_{l=1}^{2}\left[\left(\frac{\varphi_l'(x)}{\sqrt{\varepsilon^3}}\right)^2\partial_{\xi_l}^2\tilde{u} + \frac{1}{\sqrt{\varepsilon^3}}D_{\xi,l}\tilde{u}\right]\right)_{\zeta=\gamma(x,\ t,\ \varepsilon)},\quad D_{\xi,l} \equiv 2\varphi_l'(x)\partial_{\xi_l x}^2 + \varphi_l''(x)\partial_{\xi_l},$$

then, according to (1), we set the extended problem

$$\begin{aligned}&L_\xi \equiv \varepsilon^{-1}T_0\tilde{u} + T_1\tilde{u} - \sqrt{\varepsilon}L_\xi\tilde{u} + \varepsilon\partial_t\tilde{u} - \varepsilon^2 L_x\tilde{u} = f(x, t),\ M \in Q,\\ &\tilde{u}|_{t=\tau=\eta_i=0} = \tilde{u}|_{x=0,\xi_1=0} = \tilde{u}|_{x=1,\xi_2=0} = 0,\ Q = \Omega\times(0, \infty)^3.\end{aligned} \tag{3}$$

The notation is entered here

$$T_0 \equiv \partial_\tau - \Delta_\xi,\ T_1 \equiv \sum_{i=1}^{n} \lambda_i(0)\partial_{\eta_i} + t\partial_t - B(t),\ \ \Delta_\xi \equiv \sum_{l=1}^{2} \partial^2_{\xi_l}, L_\xi \equiv \sum_{l=1}^{2} a(x) D_{\xi,l},\ L_x \equiv a(x)\partial_x^2.$$

Note that the identity holds

$$\left(\tilde{L}_\varepsilon \tilde{u}\right)_{\zeta=\gamma(x,\ t,\ \varepsilon)} \equiv L_\varepsilon u, \tag{4}$$

solutions of problem (3) will be defined as

$$\tilde{u}(M,\ \varepsilon) = \sum_{k=0}^{\infty} \varepsilon^{\frac{k}{2}} u_k(M), \tag{5}$$

then for the coefficients of this series we obtain the following iterative problems:

$$T_0 u_\nu = 0,\ \nu = 0{,}1,\ T_0 u_2 = f(x,t) - T_1 u_0,\ T_0 u_k = -T_1 u_{k-2} + L_\xi u_{k-3} - \partial_t u_{k-4} + L_x u_{k-6},$$
$$u_k\big|_{t=\tau=\eta_i=0} = 0,\ u_k\big|_{x=0,\xi_1=0} = u_k\big|_{x=1,\xi_2=0} = 0,\ k \geq 3. \tag{6}$$

1.3 Solvability of iterative problems

We introduce the space of functions in which the iterative problems will be solved:

$$U_1 = \{u_1(N_1) : u_1(N_1) = \ < v(x,t) + [c(x,\ t) + \Lambda(P(x))]e^\eta, \psi(t) >,$$
$$v(x,t) \in C^\infty\left(\overline{\Omega},\ C^n\right),\ P(x) \in C^\infty([0,\ 1],\ C^n),\ c(x,t) \in C^\infty\left(\overline{\Omega},\ C^{n^2}\right),\ \psi(t) \in C^\infty([0,\ T],\ C^n)\},$$

$$U_2 = \left\{ u_2\left(N^l\right) : u_2\left(N^l\right) = \sum_{l=1}^{2} < Y\left(N^l\right) + z\left(N^l\right)e^\eta, \psi(t) >, \left\| Y\left(N^l\right)\right\| < ce^{-\frac{\xi_l^2}{8\tau}}, \right.$$
$$\left. \left\| z\left(N^l\right)\right\| < ce^{-\frac{\xi_l^2}{8\tau}} \right\},$$

$$Y\left(N^l\right) = col\left(Y_1\left(N^l\right),\ Y_2\left(N^l\right),\ \dots,\ Y_n\left(N^l\right)\right),\ z\left(N^l\right) = \left(z_{ij}\left(N^l\right)\right),\ i,j = \overline{1,n},$$

$$N^l = \left(x, t, \xi^l, \tau\right),\ \ \eta = col(\eta_1, \eta_2,\ \dots, \eta_n),\ \Lambda(P(x)) = diag(P_1(x), P_2(x),\ \dots, P_n(x)),$$

$$< [c(x,\ t) + \Lambda(P(x))]e^\eta, \psi(t) > \ = \sum_{i,j=1}^{n} c_{ij}(x,t)e^{\eta_j}\psi_i(t) + \sum_{i=1}^{n} P_i(x)e^{\eta_i}\psi_i(t).$$

$$< Y\left(N^l\right) + z\left(N^l\right)e^\eta, \psi(t) > \ = \sum_{i=1}^{n}\left[Y_i\left(N^l\right) + \sum_{j=1}^{n} z_{ij}\left(N^l\right)e^{\eta_j}\right]\psi_i(t).$$

Here $col(\eta_1, \eta_2,\ \dots, \eta_n)$ is the vector, $diag(P_1(x), P_2(x),\ \dots, P_n(x))$ is a diagonal matrix. Element $u_k(M) \in U = U_1 \oplus U_2$ has an idea

$$\begin{aligned} u_k(M) &= <v_k(x,t) + \left[c^k(x,\ t) + \Lambda\left(P^k(x)\right)\right]e^\eta, \psi(t)> \\ &+ \sum_{l=1}^{2} <Y^k\left(N^l\right) + z^k\left(N^l\right)e^\eta, \psi(t)>. \end{aligned} \tag{7}$$

We calculate the action of the operators included in the extended equation on the function $u_k(M) \in M$, for which we first decompose $\psi_{i'}(t) = \sum_{j=1}^{n} \alpha_{ij}(t)\psi_j(t)$, or we write by entering the notation

$$T_0 u_k(M) = \sum_{l=1}^{2} <\partial_\tau Y^k\left(N^l\right) - \partial_{\xi_l}^2 Y^k\left(N^l\right) + \left[\partial_\tau z^k\left(N^l\right) - \partial_{\xi_l}^2 z^k\left(N^l\right)\right]e^\eta, \psi(t)>,$$

$$\begin{aligned} T_1 u_k(M) &= <t\partial_t v_k(x,t) + tA^T(t)v_k(x,t) - B(t)v_k(x,t) + t\partial_t c^k(x,t) \\ &+ tA^T(t)c^k(x,t) - B(t)c^k(x,t) + c^k(x,t)\Lambda(0) - B(t)\Lambda\left(P^k(x)\right) + tA^T(t)\Lambda\left(P^k(x)\right) \\ &+ \Lambda\left(P^k(x)\right)\Lambda(0), \psi(t)> + \sum_{l=1}^{2} <\left[t\partial_t Y^k\left(N^l\right) + tA^T(t)Y^k\left(N^l\right) - B(t)Y^k\left(N^l\right)\right. \\ &\left. + t\partial_t z^k\left(N^l\right) + tA^T(t)z^k\left(N^l\right) - B(t)z^k\left(N^l\right) + z^k\left(N^l\right)\Lambda(0)\right]e^\eta, \psi(t)> \\ &= <D_1 v_k(x,t) + D_2\left[c^k(x,\ t) + \Lambda\left(P^k(x)\right)\right]e^\eta + \sum_{l=1}^{2}\left[<D_1 Y^k\left(N^l\right) + D_2 z^k\left(N^l\right)e^\eta\right], \psi(t)>, \end{aligned} \tag{8}$$

$$D_1 \equiv t\partial_t + tA^T(t) - B(t),\ D_2 \equiv D_1 + D_\lambda,\ D_\lambda = \sum_{i=1}^{n} \lambda_i(0)\partial_{\eta_i},$$

$$\Lambda(0) = diag(\lambda_1(0),\ \lambda_2(0),\ \dots,\ \lambda_n(0)),$$

$$A(t) = \left(\alpha_{ij}\right),\quad \alpha_{ij} = \left(\psi_i'(t),\ \psi_j(t)\right);$$

$$\begin{aligned} L_\xi u_k &= <a(x)\sum_{l=1}^{2}\left\{\left[2\varphi_l'(x)\partial_{x\xi_l}^2 + \varphi_l''(x)\partial_{\xi_l}\right]Y^k\left(N^l\right)\right. \\ &\left. + \left[2\varphi_l'(x)\partial_{x\xi_l}^2 + \varphi_l''(x)\partial_{\xi_l}\right]z^k\left(N^l\right)e^\eta, \psi(t)\right\}> \end{aligned}$$

$$\begin{aligned} L_x u_k(M) &= <L_x v_k(x,t) + \sum_{l=1}^{2} L_x Y^k\left(N^l\right) \\ &+ \left[L_x c^k(x,\ t) + L_x\Lambda\left(P^k(x)\right) + \sum_{l=1}^{2} L_x z^k\left(N^l\right)\right]e^\eta, \psi(t)>. \end{aligned}$$

Iterative equations are written in the form.

$$T_0 u_k(M) = h^k(M). \tag{9}$$

Theorem 1: Let conditions (1), (2), and $h^k(M) \in U_2$ hold. Then Eq. (9) is solvable in U.

Proof: Let $h^k(M) \in U_2$:

$$h^k(M) = \sum_{l=1}^{2} <\left[h^{k,1}\left(N^l\right) + h^{k,2}\left(N^l\right)e^\eta\right], \psi(t)>. \tag{10}$$

Function (7) will be the solution of Eq. (9) in U. If the functions $Y^k(N^l), z^k(N^l)$ are the solution of the equations

$$\partial_\tau Y^k(N^l) - \partial^2_{\xi_l} Y^k(N^l) = h^{k,1}(N^l), \quad \partial_\tau z^k(N^l) - \partial^2_{\xi_l} z^k(N^l) = h^{k,2}(N^l). \tag{11}$$

These equations are obtained by substituting (7) into Eq. (9), taking into account calculations (8) and representation (10). Eqs. (11), with appropriate boundary conditions, have solutions that satisfy the estimates [25]:

$$\left\|Y^k(N^l)\right\| < ce^{-\frac{\xi_l^2}{8\tau}}, \quad \left\|z^k(N^l)\right\| < ce^{-\frac{\xi_l^2}{8\tau}}.$$

Theorem 2: Let conditions (1), (2), and $\overline{h^{k,3}(x,\ 0)} = 0$ hold. Then problem

$$< D_2\left[c^k(x,\ t) + \Lambda(P^k(x))\right], \psi(t) > \ = \ < h^{k,3}(x,t), \psi(t) > \tag{12}$$

$$\overline{c^k(x,\ 0)} = -\Lambda(v_k(x,\ 0)) - \Lambda(P^k(x)) - \Lambda\left(\overline{\overline{c^k(x,\ 0)}}\cdot \mathbf{1}\right), \ \mathbf{1} = col(1,\ 1,\ \dots,\ 1), \tag{13}$$

$$\left(c^k_{ii}(x,\ t)\big|_{t=0} = -v_{k,i}(x,\ 0) - P^k_i(x) - \sum_{j=1(i\neq j)}^{n} c^k_{ij}(x,\ 0) \right),$$

has a unique solution. Hereinafter, $\overline{c}$ denotes a matrix with nonzero diagonal, and $\overline{\overline{c}}$ with zero diagonal elements, that is, $c = \overline{c} + \overline{\overline{c}}$.

Proof: We write the relation (12) in the coordinate form

$$\sum_{i,j=1}^{n}\left\{ t\left[\partial_t c^k_{ij}(x,\ t) + \sum_{\mu=1}^{n}\alpha_{\mu,i}(t)\left(c^k_{\mu,i}(x,\ t) + P^k_\mu(x)\right)\right] + (\lambda_j(0) - \lambda_i(t))c^k_{ij}(x,t)\right\}\psi_i(t)e^{\eta_j}$$
$$+\sum_{i=1}^{n}(\lambda_i(0) - \lambda_i(t))P^k_i(x)e^{\eta_i}\psi_i(t) = \sum_{i,j=1}^{n} h^{k,3}_{ij}(x,t)e^{\eta_j}\psi_i(t).$$

Removing the degeneracy of this equation, assuming that

$$\left.(\lambda_j(0) - \lambda_i(t))c^k_{ij}(x,t)\right|_{t=0} = h^{k,3}_{ij}(x,\ 0), \ \forall i\neq j. \tag{14}$$

Equating the coefficients ψ, we get

$$t\left[\partial_t c^k_{ii}(x,\ t) + \alpha_{ii}(t)c^k_{ii}(x,t)\right] = (\lambda_i(t) - \lambda_i(0))P^k_i(x) + h^{k,3}_{ii}(x,t) - t\sum_{\mu=1}^{n}\alpha_{\mu,i}(t)P^k_\mu(x), \ i = \overline{1,n}, \tag{15}$$

$$t\left[\partial_t c^k_{ij}(x,\ t) + \sum_{\mu=1(i\neq\mu)}^{n}\alpha_{\mu,i}(t)c^k_{\mu,j}(x,t)\right] + (\lambda_j(0) - \lambda_i(t))c^k_{ij}(x,t) = h^k_{ij}(x,t), \ i\neq j, \ i,j = \overline{1,n}. \tag{16}$$

Eq. (15), by virtue of condition $h^k_{ii}(x,\ 0) = 0$, under the initial condition (13), and Eq. (16) with condition (14) have one-to-one solutions.

Remarks. *In iterative problems, the condition* $h^k_{ii}(x,0) = 0$ *will be provided by the choice of the function* $P^k_i(x)$.

Theorem 3: Let conditions (1) and (2) be fulfilled. Then Eq. (9) has a unique solution satisfying the conditions: (a) $u_k(M)|_{t=\tau=\eta=0} = 0$, $u_k(M)|_{x=l-1,\xi_l=0} = 0$; (b) $T_1u_k(M) + h^k(M) \in U_2$; (c) $L_\xi u_k(M) = 0$.

Proof: Let satisfy the function (7) to the boundary conditions (a):

$$< v_k(x,0) + \left[c^k(x,\ 0) + \Lambda(P^k(x))\right]1 + \sum_{l=1}^{2}\left[Y^k(N^l) + z^k(N^l)1\right]_{t=\tau=0}, \psi(0) > = 0,$$

$$< v_k(l-1,\ t) + \left[c^k(l-1,t) + \Lambda(P^k(x))\right]e^\eta + \sum_{l=1}^{2}\left[Y^k(N^l)\right|_{x=l-1,\xi_l=0}$$

$$+ z^k(N^l)|_{x=l-1,\xi_l=0}e^\eta\right], \psi(t) > = 0,$$

Hence, we define

$$\overline{c^k(x,0)} = -\Lambda(v_k(x,\ 0)) - \Lambda(P^k(x)) - \Lambda\left(\overline{\overline{c^k(x,\ 0)}}1\right),$$

$$Y^k(N^l)|_{t=\tau=0} = 0,\ z^k(N^l)|_{t=\tau=0} = 0, \tag{17}$$

$$z^k(N^l)|_{\xi_l=0} = W^{k,l}(x,t), W^{k,l}(x,t)|_{x=l-1} = c^k(l-1,t) - \Lambda(P^k(x)),$$

$$Y^k(N^l)|_{\xi_l=0} = d^{k,l}(x,t),$$

$$d^{k,l}(x,\ t)\Big|_{x=l-1} = -v_k(l-1,t).$$

Ensuring the solvability of Eq. (9) with the right side of

$$F_k(M) = T_1u_k(M) + h^k(M) \in U_2,$$

based on the calculations (8) and the representation of

$$h^k(M) = < h^{k,1}(x,t) + h^{k,2}(x,t)e^\eta + \sum_{l=1}^{2}\left[h^{k,3}(N^l) + h^{k,4}(N^l)e^\eta\right], \psi(t) >$$

assuming that

$$< D_2\left[c^k(x,\ t) + \Lambda(P^k(x))\right]e^\eta, \psi(t) > = - < h^{k,2}(x,t)e^\eta, \psi(t) >, \tag{18}$$

$$< D_1v_k(x,t) - h^{k,1}(x,t), \psi(t) > = 0. \tag{19}$$

Eq. (18) under the initial condition (17), on the basis of Theorem 2, are uniquely solvable. Eqs. (19) are solved without an initial condition and have a bounded solution [27].

Eqs. (11) with boundary conditions (17) have solutions that can be represented as

$$Y^k_i(N^l) = d^{k,l}_i(x,t)erfc\left[\xi_l/\left(2\tau^{1/2}\right)\right] + h^{k,3}_i(x,t)I_1(\xi_l,\tau),$$

$$z_{ij}^{k}\left(N^{l}\right) = W_{ij}^{k,l}(x,t)erfc\left[\xi_{l}/\left(2\tau^{1/2}\right)\right] + h_{ij}^{k,4}(x,t)I_{2}(\xi_{l},\tau),$$

where $erfc(x) = 2\pi^{-1/2}\int_{0}^{x}e^{-t^{2}}dt$ is integral of the additional function and describes a parabolic boundary layer, $d^{k,l}(x,t)$,$W^{k,l}(x,t)$ are arbitrary functions that are chosen, like solving the equations

$$D_{x}d_{i}^{k,l}(x,t) = -D_{x}h_{i}^{k,3}(x,t),\ D_{x}W_{ij}^{k,3}(x,t) = -D_{x}h_{ij}^{k,4}(x,t),\quad D_{x} \equiv 2\varphi_{l'}(x)\partial_{x} + \varphi_{l''}(x), \tag{20}$$

with boundary conditions from (17). Eqs. (20) are obtained by satisfying condition c) and taking into account that the functions $erfc\left[\xi_{l}/\left(2\tau^{1/2}\right)\right]$ and $I_{l}(\xi_{l},\tau)$ have single estimates.

Thus, a unique solution to Eq. (9) is obtained that satisfies conditions (a)–(c).

1.4 Construction of solutions of iterative equations

For $\nu = 0{,}1$ the equations for $u_{\nu}(M)$ are homogeneous; therefore, the condition of Theorem 1 holds; therefore, the solution of these equations exists and can be represented in the form (7).

The following iterative equation, on the grounds (8), has a free term

$$F_{2}(M) = f(x,t) - T_{1}u_{0} = f(x,t) - \langle D_{1}v_{0}(x,t) + D_{2}\left[c^{0}(x,\ t) + \Lambda\left(P^{0}(x)\right)\right]e^{\eta}$$
$$+ \sum_{l=1}^{2}\left[D_{1}Y^{0}\left(N^{l}\right) + D_{2}z^{0}\left(N^{l}\right)e^{\eta}\right],\psi(t)\rangle.$$

We decompose $f(x,t)$ by the system $\psi_{i}(t)$ and substitute it with the previous relation. Further, providing the conditions of Theorem 1, we set

$$D_{1}v_{0}(x,t) = f(x,t),\ t\left[\partial_{t}v_{0,i}(x,\ t) + \sum_{\mu=1}^{n}\alpha_{i\mu}(t)v_{0,\mu}(x,t)\right] - \lambda_{i}(t)v_{0i}(x,t) = (f(x,\ t),\psi_{i}),$$
$$\langle D_{2}\left[c^{0}(x,\ t) + \Lambda\left(P^{0}(x)\right)\right],\psi(t)\rangle = 0.$$

The first system has a smooth solution, and the second system by Theorem 2 is solvable if

$$t\left[\partial_{t}c_{ii}^{0} + \alpha_{ii}(t)c_{ii}^{0}(x,t)\right] = (\lambda_{i}(t) - \lambda_{i}(0))P_{i}^{0}(x) - t\sum_{\mu=1}^{n}\alpha_{\mu,i}(t)P_{\mu}^{0}(x),$$
$$c_{ii}^{0}(x,\ t)\big|_{t=0} = -v_{i0}(x,0) - P_{i}^{0}(x), \tag{21}$$

$$t\left[\partial_{t}c_{ij}^{0} + \sum_{\mu=1(\mu\neq i)}^{n}\alpha_{\mu,i}(t)c_{\mu i}^{0}(x,t)\right] + \left(\lambda_{j}(0) - \lambda_{i}(t)\right)c_{ij}^{0}(x,t) = 0,$$
$$\left(\lambda_{j}(0) - \lambda_{i}(t)\right)c_{ij}^{0}(x,0) = 0,\ i \neq j,\ i,j = \overline{1,n}. \tag{22}$$

Problem (21) is uniquely solvable, and problem (22) has a trivial solution.

For $k = 3$, by Theorem 3, ensuring condition (c) for $d^{0,l}(x, t)$ and $W^{0,l}(x, t)$, we obtain problem

$$D_x d_i^{0,l}(x, t) = 0, \quad d_i^{0,l}(x,\ t)\Big|_{x=l-1} = -v_{0,i}(l-1, t),$$
$$D_x W_{ij}^{0,l}(x, t) = 0, \quad W_{ij}^{0,l}(x,\ t)\Big|_{x=l-1} = -c_{ij}^{0}(l-1, t).$$

By this, we have determined the main term of the asymptotic. In addition, conditions (b) of Theorem 3 gives

$$D_1 v_{1,i} = 0, \quad t\left[\partial_t c_{ii}^1 + \alpha_{ii}(t)c_{ii}^1\right] = (\lambda_i(t) - \lambda_i(0))P_i^1(x) - t\sum_{\mu=1}^{n}\alpha_{\mu,i}(t)P_\mu^1(x),$$
$$c_{ii}^1(x,\ t)\big|_{t=0} = -v_{1i}(x, 0) - P_i^1(x),$$
$$t\left[\partial_t c_{ij}^1 + \sum_{\mu=1(\mu\neq i)}^{n}\alpha_{\mu,i}(t)c_{\mu j}^1(x, t)\right] + \left(\lambda_j(0) - \lambda_i(t)c_{ij}^1(x, t) = 0, \quad c_{ij}^1(x,\ t)\Big|_{t=0}.\right.$$

From here we define $v_{1i}(x, t) = 0$, below it will be shown that $P_i^1(x) = 0$, therefore $c_{ii}^1(x, t) = 0$, and from the last equation we find $c_{ij}^1(x, t) = 0, \ i \neq j$.

In the next $k = 4$ step free member

$$\partial_t,$$
$$F_4(M) = -T_1 u_2 + L_\xi u_1 - \partial_t u_0.$$

Satisfying condition (c) of Theorem 3, we obtain the problems

$$D_x d_i^{1,l}(x, t) = 0, \ d_i^{1,l}(x,\ t)\Big|_{x=l-1} = -v_{1,i}(l-1, t) = 0,$$
$$D_x W_{ij}^{1,l}(x, t) = 0, \ W_{ij}^{1,l}(x,\ t)\Big|_{x=l-1} = -c_{ij}^1(l-1, t) = 0.$$

That is, $d_i^{1,l}(x, t) = 0$, $W_{ij}^1, l(x, t) = 0$, and, therefore, $u_1(M) = 0$. Regarding $Y_i^3(N^l)$,$W_{ij}^3(N^l)$ we obtain homogeneous equations, therefore

$$Y_i^3(N^l) = d_i^{3l}(x, t)erfc\left[\xi_l/\left(2\tau^{1/2}\right)\right], \ W_{ij}^3(N^l) = W_{ij}^{3,l}(x, t)erfc\left[\xi_l/\left(2\tau^{1/2}\right)\right].$$

We calculate

$$F_4(M) = -<D_1 v_2(x, t) + D_2\left[c^2(x,\ t) + \Lambda(P^2(x))\right]e^\eta$$
$$+\sum_{l=1}^{2}\left[Y^2(N^l) + z^2(N^l)e^\eta\right],\psi(t)> \ - \ <\partial_t v_0(x, t) + A^T(t)v_0(x, t)$$
$$+\left[\partial_t c^0(x,\ t) + A^T(t)c^0(x, t) + A(t)\Lambda(P^0(x))e^\eta\right.$$
$$+\sum_{l=1}^{2}\left[\partial_t Y^0(N^l) + A^T(t)Y^0(N^l) + (\partial_t z^0(N^l) + A^T(t)z^0(N^l))e^\eta\right],\psi(t)>$$

and ensuring the solvability in U of the iteration equation for $k = 4$, we assume

$$< D_t v_2(x, t), \psi(t) > \; = - < \partial_t v_0(x, t) + A(t) v_0(x, t), \psi(t) >, \tag{23}$$
$$D_2\left[c^2(x, t) + \Lambda\left(P^2(x)\right)\right], \psi(t) > \; = - < \partial_t c^0(x, t) + A(t) c^0(x, t) + A(t)\Lambda\left(P^0(x)\right), \psi(t) >.$$

The solvability of the second system is ensured by the choice of $P^0(x)$:

$$\Lambda\left(P^0(x)\right) = -\overline{A^{-1}(t)\left[\partial_t c^0(x, t) + A(t) c^0(x, t)\right]},$$

$$P_i^0(x) = \alpha_{ii}^{-1}\left[\partial_t c_{ii}^0(x, t) + \sum_{k=1}^{n} \alpha_{ki}(t) c_{ki}^0(x, t)\right]_{t=0},$$

as well as

$$< \overline{c^2(x, t)\Lambda(0) - \Lambda(t) c^2(x, t), \psi(t)} > |_{t=0}$$
$$= - < \overline{\overline{\partial_t c^0(x, t) + A(t) c^0(x, t) + A(t)\Lambda\left(P^0(x)\right), \psi(t)}} >$$

$$c_{ij}^2(x, t)\Big|_{t=0} = \frac{1}{\lambda_j(0) - \lambda_i(0)}\left[\partial_t c_{ij}^0(x, t) + \sum_{k=1}^{n} \alpha_{ki}(t) c_{kj}^0(x, t) + \alpha_{ji}(t) P_j^0(x)\right]_{t=0}, \quad i \neq j. \tag{24}$$

Under such assumptions, the second system of (23) is solvable. It is solved under the initial conditions (24) and the initial condition $c_{ii}^2(x, t)\big|_{t=0} = -v_{2i}(x, 0) - P_i^2(x)$, which is obtained from (17). This completely defines the main term of the asymptotic.

Further, repeating the process described above, we find all the coefficients of the partial sum (5), and the coefficients with odd indices are zero.

1.5 Estimates of the remainder term

Denote by $R_{\varepsilon n}(M) = \tilde{u}(M, \varepsilon) - u_{\varepsilon n}(M)$, where $u_{\varepsilon n}(M) = \sum_{k=0}^{n} \varepsilon^k u_{2k}(M)$. Substituting $\tilde{u}(M, \varepsilon) = u_{\varepsilon n}(M) + R_{\varepsilon n}(M)$ into the extended problem (3), then, taking into account (6), with respect to $R_{\varepsilon n}(M)$, we obtain

$$\tilde{L}_\varepsilon R_{\varepsilon n(M)} = \varepsilon^{n+1} g_{\varepsilon n}(M), \; R_{\varepsilon n}(M)|_{t=\tau=0} = R_{\varepsilon n}(M)|_{x=l-1, \xi_l=0} = 0, \; l = 1,2,$$

where the function $g_{\varepsilon n}(M)$ is expressed through $u_{\varepsilon,k}(M)$. Producing constriction this problem, taking into account the identity (4), with respect to $R_{\varepsilon,n}(x, t, \varepsilon) \equiv (R_{\varepsilon n}(M))|_{\zeta=\gamma(x, t, \varepsilon)}$ we get the problem

$$L_\varepsilon R_{\varepsilon,n}(x, t, \varepsilon) = \varepsilon^{n+1} g_{\varepsilon n}(x, t, \varepsilon), \; R_{\varepsilon n}|_{t=0} = R_{\varepsilon n}|_{x=l-1} = 0, \; l = 1,2.$$

From the construction of solutions to iterative problems, it can be seen that the function $g_{\varepsilon n}(x, t, \varepsilon)$ is uniformly bounded in $\overline{\Omega}$. Applying the maximum principle [28], we can establish a uniform estimate in $\overline{\Omega}$

$$\|R_{\varepsilon n}(x, t, \varepsilon)\| < c\varepsilon^{n+1}. \tag{25}$$

Theorem 4: Let conditions (1) and (2) be fulfilled. Then the function $u_{\varepsilon n}(x, t, \varepsilon)$ is a uniform asymptotic solution of problem (1), that is, the estimate (25) holds.

2. A singularly perturbed parabolic equation system

We consider the first boundary-value problem for a system of singularly perturbed parabolic equations

$$L_\varepsilon u \equiv (\varepsilon + t)\partial_t u - \varepsilon^2 A(x)\partial_x^2 u - D(t)u = f(x, t), \ (x, t) \in \Omega, \tag{26}$$

$$u(x, 0, \varepsilon) = h(x), \ u(0, t, \varepsilon) = u(1, t, \varepsilon) = 0,$$

where $\Omega = (0, 1) \times (0, T], \ \varepsilon > 0$ is a small parameter,

$$u = u(x, t, \varepsilon) = col(u_1(x, t, \varepsilon), u_2(x, t, \varepsilon), \ u_n(x, t, \varepsilon)), A(x) \in C^\infty\left([0, 1], \mathbb{C}^{n^2}\right),$$

$$D(t) \in C^\infty\left([0, T], \mathbb{C}^{n^2}\right), f(x, t) \in C^\infty\left(\overline{\Omega}, \mathbb{C}^n\right).$$

The work is a continuation of [29], where instead of the matrix-function $A(x)$, there was a scalar function and an asymptotic of the solution was constructed containing two functions describing the boundary layers along $x = 0$ and $x = 1$. Lomov was the first to introduce the concept of a power-law boundary layer based on the study of the Lighthill equation and he based his method on it [29]. In this case, the asymptotics contains $2m$ parabolic boundary layer functions describing the boundary layers along $x = 0$ and $x = 1$.

The asymptotics of the solution to this problem, along with the parabolic boundary layer function (the parabolic boundary layer is described by the function), $erfc\left(\frac{\varphi(x)}{2\varepsilon\sqrt{t}}\right)$, also contain the power boundary layer function

$$\Pi_\varepsilon(t) = \left(\frac{\varepsilon}{\varepsilon + t}\right)^\lambda, \ \lambda > 0$$

as well as their products, which describe the corner boundary layer in the vicinity of $(0, 0)$.

Construction of the asymptotic solution of a singularly perturbed system of parabolic equations is devoted to works [8–10] and [30]. In Ref. [8], a regularized asymptotic is constructed in the case when the matrix of coefficients for the desired function has zero multiple eigenvalues. A similar problem was studied in [9] and an asymptotic of the boundary layer type was constructed. The method of boundary functions in [10] studied the bisingular problem for systems of parabolic equations, which is characterized by the presence of non-smoothness of the asymptotic terms and a singular dependence on a small parameter. In Ref. [30] and [26], various problems for split systems of two equations of parabolic type were studied, and asymptotics of the boundary layer type were constructed. The problems of differential equations of parabolic type with a small parameter were studied in Ref. [24, 31, 32].

2.1 Statement of the problem

We consider the first boundary-value problem (26). The problem is solved under the following assumptions:

1. For n-dimensional vector functions $f(x, t)$ and $h(x)$, the inclusions

$$f(x, t) \in C^{\infty}\left(\overline{\Omega}, \mathbb{C}^{n}\right), \quad h(x) \in C^{\infty}([0,\ 1], \mathbb{C}^{n}),$$

are fulfilled for $n \times n$-matrix-valued functions $D(t)$ and $A(x)$-inclusions

$$D(t) \in C^{\infty}([0,\ T], \mathbb{C}^{n\times n}), \quad A(x) \in C^{\infty}([0,\ 1], \mathbb{C}^{n\times n});$$

2. The real parts of all roots $\lambda_i(x),\ i = \overline{1,n}$, of the equation $det(A(x) - \lambda E) = 0$ are positive and $\lambda_i(x) \neq \lambda_j(x)$ for all $x \in [0, 1], i \neq j,\ \ i,j = \overline{1,n}$;

3. The real parts of the eigenvalues $\beta_j(t),\ j = \overline{1,n}$ of the matrix $D(t)$ are negative, that is, $Re\,\beta_j(0) < 0$ and $\beta_i(0) \neq \beta_j(t)\ \ \forall t \in [0, T],\ i \neq j,\ \ i,j = \overline{1,n}$;

4. Completed the conditions of approval of the initial and boundary conditions $h(0) = h(1) = 0$.

2.2 Regularization of the problem

Following [29], p. 316; 30, P. 18], we introduce regularizing variables

$$\xi_{i,l} = \frac{\varphi_{i,l}(x)}{\sqrt{\varepsilon^3}};\varphi_{i,l}(x) = (-1)^{l-1}\int_{l-1}^{x} \frac{ds}{\sqrt{\lambda_i(s)}},l = 1,2,i = \overline{1,n}, \tag{27}$$

$$\tau = \frac{1}{\varepsilon}\ln\left(\frac{t+\varepsilon}{\varepsilon}\right),\ \mu_j = \beta_j(0)\ln\left(\frac{t+\varepsilon}{\varepsilon}\right) \equiv K_j(t, \varepsilon),j = \overline{1,n},$$

and an extended function such that

$$\begin{gathered}\tilde{u}(M,\ \varepsilon)|_{\xi=\varphi(x)/\varepsilon} \equiv u(x, t, \varepsilon),M = (x, t, \xi, \tau, \mu),\ \ \xi = (\xi_1, \xi_2),\\ \xi_l = \left(\xi_{1,l}, \xi_{2,l}, \ \dots, \xi_{n,l}\right),\varphi(x) = (\varphi_1(x), \varphi_2(x)),\\ \varphi_l(x) = \left(\varphi_{1,l}, \varphi_{2,l}(x),\ \dots, \varphi_{n,l}(x)\right),\ l = 1,2,\ \ \mu = (\mu_1, \mu_2,\ \dots, \mu_n).\end{gathered} \tag{28}$$

Based on (27), we find the derivatives from (28):

$$\partial_t u \equiv \left.\left(\partial_t \tilde{u} + \frac{1}{\varepsilon(t+\varepsilon)}\partial_\tau \tilde{u} + \sum_{j=1}^{n}\frac{\beta_j(0)}{t+\varepsilon}\partial_{\mu_j}\tilde{u}\right)\right|_{\theta=\chi(x,\ t,\ \xi,\ \tau,\ \mu)},$$

$$\partial_x u \equiv \left.\left(\partial_x \tilde{u} + \sum_{l=1}^{2}\sum_{i=1}^{n}\left[\frac{1}{\sqrt{\varepsilon^3}}\varphi'_{i,l}(x)\partial_{\xi_{i,l}}\tilde{u}\right]\right)\right|_{\theta=\chi(x,\ t,\ \xi,\ \tau,\ \mu)},$$

$$\partial_x^2 u \equiv \left.\left(\partial_x^2 \tilde{u} + \sum_{l=1}^{2}\sum_{i=1}^{n}\left[\frac{1}{\varepsilon^3}\left(\varphi'_{i,l}(x)\right)^2\partial^2_{\xi_{i,l}}\tilde{u} + \frac{1}{\sqrt{\varepsilon^3}}L^{\xi}_{i,l}\tilde{u}\right]\right)\right|_{\theta=\chi(x,\ t,\ \xi,\ \tau,\ \mu)},$$

$$L_{i,l}^{\xi}\tilde{u} = 2\varphi'_{i,l}(x)\partial^2_{x\xi_{i,l}}\tilde{u} + \varphi''_{i,l}(x)\partial_{\xi_{i,l}}\tilde{u}, \ \chi(x, t, \varepsilon) = \left(\frac{\varphi(x)}{\sqrt{\varepsilon^3}}, \frac{1}{\varepsilon} ln\left(\frac{t+\varepsilon}{\varepsilon}\right), K(t, \varepsilon)\right),$$

$$K(t, \varepsilon) = (K_1(t, \ \varepsilon), K_2(t, \varepsilon), \ \dots, K_n(t, \varepsilon)), \ \ \theta = (\tau, \mu, \xi).$$

According to these calculations, as well as (26) and (28), we pose the following extended problem

$$\tilde{L}_\varepsilon\tilde{u} \equiv \varepsilon\partial_t\tilde{u} + \frac{1}{\varepsilon}T_0\tilde{u} + T_1\tilde{u} - \sqrt{\varepsilon}L_\xi\tilde{u} - t\partial_t\tilde{u} - \varepsilon^2 L_x\tilde{u} = f(x, t), \tag{29}$$

$$\tilde{u}|_{t=0} = 0, \ \tilde{u}|_{x=0,\xi_{i,1}=0} = \tilde{u}|_{x=1,\xi_{i,2}=0} = 0, \ i = \overline{1,n},$$

$$T_0\tilde{u} = \partial_\tau\tilde{u} - \sum_{l=1}^{2}\sum_{i=1}^{n}\left(\varphi'_{i,l}(x)\right)^2 A(x)\partial^2_{\xi_{i,l}}\tilde{u}, \ T_1\tilde{u} = t\partial_t\tilde{u} + \sum_{j=1}^{\infty}\beta_j(0)\partial_{\mu_j}\tilde{u} - D(t)\tilde{u},$$

$$L_\xi\tilde{u} = \sum_{l=1}^{2}\sum_{i=1}^{n}A(x)L_{i,l}^{\xi}\tilde{u}, \ L_x\tilde{u} = A(x)\partial_x^2\tilde{u}.$$

In this case, the identity holds

$$\left(\tilde{L}_\varepsilon\tilde{u}(M, \ \varepsilon)\right)_{\theta=\chi(x, \ t, \ \xi, \ \tau, \ \mu)} \equiv L_\varepsilon u(x, t, \varepsilon). \tag{30}$$

The solution to the extended problem (29) will be defined as a series

$$\tilde{u}(M, \varepsilon) = \sum_{k=0}^{\infty}\varepsilon^{k/2}u_k(M). \tag{31}$$

Substituting (31) into problem (29) and equating the coefficients for the same powers of P, we obtain the following equations:

$$T_0u_0 = 0, \ T_0u_1 = 0, \ T_0u_2 = f(x, t) - T_1u_0, \ T_0u_3 = L_\xi u_0 - T_1u_1, \tag{32}$$

$$T_0u_k = L_\xi u_{k-3} + L_x u_{k-6} - \partial_t u_{k-4} - T_1 u_{k-2}.$$

The initial and boundary conditions for them are set in the form

$$u_k(M)|_{t=\mu=\tau=0} = 0,$$

$$u_k(M)|_{x=l-1,\xi_{i,l}=0} = 0, \ k \geq 0, \ i = \overline{1,n}, \ l = 1,2.$$

2.3 Solvability of iterative problems

Each of the problems (32) has innumerable solutions, therefore, we single out a class of functions in which these problems were uniquely solvable. We introduce the following function classes:

$$U_1 = \left\{V(x, \ t) : V(x, t) = \sum_{i=1}^{n} v_i(x, t)\psi_i(t), \ \ v_i(x, t) \in C^\infty\left(\overline{\Omega}\right)\right\},$$

$$U_2 = \left\{ Y(N) : Y(N) = \sum_{l=1}^{2} \sum_{i=1}^{n} y_i^l(N_i^l) b_i(x), \ \left| y_i^l(N_i^l) \right| < c \exp\left(-\frac{\xi_{i,l}}{8\tau} \right) \right\},$$

$$U_3 = \left\{ C(x,\ t) : C(x,t) = \sum_{i=1}^{n} \left[\sum_{j=1}^{n} c_{ij}(x,\ \ t) \exp\left(\mu_j\right) + p_i(x) \right] \psi_i(t), \ \ c_{ij}(x,t) \in C^\infty\left(\overline{\Omega}\right) \right\},$$

$$U_4 = \left\{ Z(N) : Z(N) = \sum_{l=1}^{2} \sum_{i,\,j=1}^{n} z_{i,j}^l(N_i^l) b_i(x) \exp\left(\mu_j\right), \ \left| z_{i,j}^l(N_i^l) \right| < c \exp\left(-\frac{\xi_{i,l}^2}{8\tau} \right) \right\},$$

where $N_i^l = \left(x,\ t,\ \xi_{i,l},\ \tau,\ \mu_i\right),\ i = 1,2, \dots, n,\ l = 1,2$. From these classes of functions we construct a new class as a direct sum: $U = U_1 \oplus U_2 \oplus U_3 \oplus U_4$. The function $u_k(M) \in U$ is representable in vector form

$$u_k(M) = \Psi(t)\left[V_k(x,\ t) + C^k(x,t)\exp(\mu)\right]$$

$$+ \sum_{l=1}^{2} B(x)\left[Y^{k,l}(N^l) + Z^{k,l}(N^l)\exp(\mu)\right],\ C^k(x,t) = C_1^k(x,t) + \Lambda(P(x)),$$

$$C_1^k(x,t) = \left(c_{ij}(x,\ t)\right),\ V_k(x,t) = col(v_{k1}, v_{k2}, \ \dots, v_{kn}),$$

$$Y^{k,l}(N^l) = col\left(y_1^{k,l}(N_1^l), y_2^{k,l}(N_2^l),\ \dots, y_n^{k,l}(N_n^l)\right),\ Z^{k,l}(N^l) = \left(z_{ij}^{k,l}(N_i^l)\right),$$

$$\Psi(t) = (\psi_1(t), \psi_2(t),\ \dots, \psi_n(t)),\ B(x) = (b_1(x), b_2(x),\ \dots, b_n(x)),$$

$$\exp(\mu) = col(\exp(\mu_1),\ \exp(\mu_2),\ \dots,\ \exp(\mu_n))$$

or in coordinate form

$$u_k(M) = \sum_{i=1}^{n} v_{k,i}(x,t)\psi_i(t) + \sum_{l=1}^{2}\sum_{i=1}^{n} y_i^{k,l}(N_i^l) b_i(x) \tag{33}$$

$$+ \sum_{i,j=1}^{n} \left\{ c_{i,j}^k(x,\ t)\psi_i(t) + \sum_{l=1}^{2} z_{ij}^{k,l}(N_i^l) b_i(x) \right\} \exp\left(\mu_j\right) + \sum_{i=1}^{n} p_i^k(x)\psi_i(t)\exp(\mu_i).$$

The vector-functions $b_i(x)$,$\psi_i(t)$ included in these classes are eigenfunctions of the matrices $A(x)$ and $D(t)$, respectively

$$A(x)b_i(x) = \lambda_i(x) b_i(x),\ D(t)\psi_i(t) = \beta_i(t)\psi_i(t),\ i = \overline{1,n}. \tag{34}$$

Moreover, according to condition (1), they are smooth in their arguments.

Along with the eigenvectors $b_i(x)$ and $\psi_i(t)$, the eigenvectors $b_i^*(x)$,$\psi_i^*(t),\ i = \overline{1,n}$ of the conjugate matrices $A^*(x),\ D^*(t)$ will be used

$$A^*(x) b_i^*(x) = \overline{\lambda}_i(x) b_i^*(x),\ \ D^*(t)\psi_i^*(t) = \overline{\beta}_i(t)\psi_i^*(t)$$

and they are selected biorthogonal

$$\left(b_i(x), b_j^*(x)\right) = \delta_{i,j},\ \ \left(\psi_i(t), \psi_j^*(t)\right) = \delta_{i,j},\ i,j = \overline{1,n},$$

where $\delta_{i,j}$ is the Kronecker symbol.

We calculate the action of the operators T_0, T_1, L_ξ, L_x on the function $u_k(M, \varepsilon)$ from (34), taking into account relations (35) and

$$\varphi'^2_{i,l}(x) = \frac{1}{\lambda_i(x)}, i = \overline{1,n}$$

we have

$$T_0 u_k(M) \equiv \sum_{i=1}^{n} \sum_{l=1}^{2} \left\{ \partial_\tau y_i^{k,l}(N_i^l) - \partial^2_{\xi_{i,l}} y_i^{k,l}(N_i^l) \right. \quad (35)$$

$$\left. + \sum_{j=1}^{n} \left[\partial_\tau z_{i,j}^{k,l}(N_i^l) - \partial^2_{\xi_{i,l}} z_{i,j}^{k,l}(N_i^l) \right] \exp\left(\mu_j\right) \right\} b_i(x);$$

or vector form

$$T_0 u_k(M) \equiv \sum_{l=1}^{2} B(x) \left\{ \partial_\tau Y^{k,l}(N^l) - \partial^2_{\xi_l} Y^{k,l}(N^l) + \left[\partial_\tau Z^{k,l}(N^l) - \partial^2_{\xi_l} Z^{k,l}(N^l) \right] exp\,(\mu) \right\},$$

$Y^{k,l}(N^l)$ is n-vector, $Z^{k,l}(N^l)$ is $n \times n$-matrix. Here $B(x)$ is a matrix function $(n \times n)$ whose columns are the eigenvectors $b_i(x)$ of the matrix $A(x)$. We calculate

$$T_1 u_k = t\partial_t u_k + \sum_{j=1}^{n} \beta_j(0) \partial_{\mu_j} u_k - D(t) u_k$$

$$= t \sum_{i=1}^{n} \left\{ \left[\partial_t v_{k,i} + \sum_{r=1}^{n} \alpha_{r,i}(t) v_{k,r}(x, t) \right] \psi_i(t) + \sum_{l=1}^{2} \partial_t y_i^{k,l}(N_i^l) b_i(x) \right.$$

$$+ \sum_{j=1}^{n} \left[\left(\partial_t c_{ij}^k(x, t) + \sum_{r=1}^{n} \alpha_{ri}(t) c_{r,j}^k(x, t) + \alpha_{ji}(t) p_j^k(x) \right) \psi_i(t) \right. \quad (36)$$

$$\left. \left. + \sum_{l=1}^{2} z_{ij}^{k,l}(N_i^l) b_i(x) \right] \exp\left(\mu_j\right) \right\}$$

$$+ \sum_{i,j=1}^{n} \beta_j(0) c_{ij}^k(x, t) \psi_i(t) \exp\left(\mu_j\right) + \sum_{i=1}^{n} \beta_i(0) p_i^k(x) \psi_i(t) \exp\left(\mu_i\right)$$

$$+ \sum_{l=1}^{2} \sum_{i,j=1}^{n} \beta_j(0) z_{i,j}^{k,l}(x, t) b_i(x) \exp\left(\mu_j\right)$$

$$- \sum_{i=1}^{n} \left\{ \beta_i(t) v_{ki}(x, t) \psi_i(t) + \sum_{l=1}^{2} \sum_{r=1}^{n} \gamma_{r,i}(x, t) y_i^{k,l}(N_i^l) b_i(x) \right\}$$

$$- \sum_{i,j=1}^{n} \beta_j(t) c_{i,j}^k(x, t) \psi_i(t) \exp\left(\mu_j\right) - \sum_{i=1}^{n} \beta_i(t) p_i^k(x) \psi_i(t) \exp\left(\mu_i\right)$$

$$- \sum_{l=1}^{2} \sum_{i,j=1}^{n} \sum_{r=1}^{n} \gamma_{r,i}(x, t) z_{r,j}^{k,l}(N_i^l) b_i(x) \exp\left(\mu_j\right).$$

Here $\alpha_{i,r} = \left(\psi'_i(t), \psi^*_r(t)\right)$, $\gamma_{i,r}(x, t) = \left(D(t) b_i(x), b^*_r(x)\right)$.

It will be shown below that the scalar functions $y_i^{k,l}(N^l)$ and $z_{i,j}^{k,l}(N^l)$ are representable in the form

$$y_i^{k,l}(N_i^l) = d_i^{k,l}(x,t)y_i^{k,l}(\xi_l,\tau),\ z_{i,j}^{k,l}(N_i^l) = \omega_{i,j}^{k,l}(x,t)z_{i,j}^{k,l}(\xi_l,\tau).$$

Given these representations, we calculate

$$L_\xi u_k(M) = \sum_{l=1}^{2} A(x)\sum_{i=1}^{n}\Big\{\Big[2\varphi_i'(x)\Big(b_i(x)d_i^{k,l}(x,\ t)\Big)_x \tag{37}$$

$$+\varphi''_i(x)\Big(b_i(x)d_i^{k,l}(x,\ t)\Big)\Big]\partial_{\xi_{i,l}}y_i^{k,l}(\xi_{i,l},\tau)$$

$$+\sum_{j=1}^{n}\Big[2\varphi_i'(x)\Big(b_i(x)\omega_{i,j}^{k,l}(x,\ t)\Big)_x+\varphi''_i(x)\Big(b_i(x)\omega_{i,j}^{k,l}(x,\ t)\Big)\Big]\partial_{\xi_{i,l}}z_{i,j}^{k,l}(\xi_{i,l},\ \tau)\exp\left(\mu_j\right)\Big\},$$

$$L_x u_k(M) = A(x)\Bigg[\partial_x^2(V_k(x,\ t)) + \sum_{l=1}^{2}\partial_x^2(B(x)Y^{l,k}(N^l))$$

$$+\partial_x^2(\Psi(t)C^{k,0}(x,\ t))\exp(\mu) + \sum_{l=1}^{2}\partial_x^2(B(x)Z^{l,k}(N^l))\exp(\mu)\Bigg].$$

Satisfy the function (34) of the boundary conditions from (29)

$$y_i^{k,l}(N_i^l)\Big|_{t=\tau=\mu=0} = 0,\ \ z_{i,j}^{k,l}(N_i^l)\Big|_{t=\tau=\mu=0} = 0, \tag{38}$$

$$c_{ii}^k(x,0) = -v_{k,i}(x,0) - p_i^k(x) - \sum_{i\neq j}c_{ij}^k(x,0),$$

$$y_i^{k,l}(N_i^l)\Big|_{\xi_{i,l}=0} = d_i^{k,l}(x,t),\ \ d_i^{k,l}(x,t)b_i(x)\Big|_{x=l-1} = -v_i(l-1,t)\psi_i(t),$$

$$z_{i,j}^{k,l}(N_i^l)\Big|_{\xi_{i,l}=0} = \omega_{i,j}^{k,l}(x,t),\quad \omega_{i,j}^{k,l}(x,t)b_i(x)\Big|_{x=l-1} = -\left[c_{i,j}(l-1,\ t)+p_i(l-1)\right]\psi_i(t).$$

In general form, the iterative Eqs. (32) are written as

$$T_0u_k(M) = h_k(M). \tag{39}$$

Theorem 1: Let $h_k(M)\in U$ and conditions (2) and (3) on hold. Then Eq. (40) has a solution $u_k(M)\in U$, if the equations are solvable

$$\partial_\tau y_i^{k,l}(N_i^l) - \partial_{\xi_{i,l}}^2 y_i^{k,l}(N_i^l) = h_i^{k,1}(N_i^l) \equiv \overline{h}_i^{k,1}(x,t)\overline{\overline{h}}_i^{k,1}(\xi_{i,1},\tau), \tag{40}$$

$$\partial_\tau z_{i,j}^{k,l}(N_i^l) - \partial_{\xi_{i,l}}^2 z_{i,j}^{k,l}(N_i^l) = h_{ij}^{k,2}(N_i^l) \equiv \overline{h}_{ij}^{k,2}(x,t)\overline{\overline{h}}_i^{k,2}(\xi_{i,2},\tau).$$

Proof: Let $h_k(M) = \sum_{l=1}^2\sum_{i=1}^n\left[h_i^{k,1}(N_i^l) + \sum_{j=1}^n h_{ij}^{k,2}(N_i^l)\exp\left(\mu_j\right)\right]b_i(x)\in U$. Pose (34) into Eq. (40), then, on the basis of calculations (38), with respect $y_i^{k,l}(N_i^l)$, $z_{ij}^{k,l}(N_i^l)$, we obtain Eqs. (41). These equations, under appropriate boundary value conditions

$$y_i^{k,l}(N_i^l)\Big|_{t=\tau=\mu=0}=0,\;\; y_i^{k,l}(N_i^l)\Big|_{\xi_{i,l}=0}=d_i^{k,l}(x,t),$$

$$z_{i,j}^{k,l}(N_i^l)\Big|_{t=\tau=\mu=0}=0,\;\; z_{i,j}^{k,l}(N_i^l)\Big|_{\xi_{i,l}=0}=\omega_{i,j}^{k,l}(x,t).$$

Solutions are represented in the form

$$y_i^{k,l}(N_i^l)=d_i^{k,l}(x,t)erfc\left(\frac{\xi_{i,l}}{2\sqrt{\tau}}\right)+\overline{h}_i^{k,1}(x,t)I_1(\xi_{i,l},\tau),$$

$$z_{i,j}^{k,l}(N_i^l)=\omega_{i,j}^{k,l}(x,t)erfc\left(\frac{\xi_{i,l}}{2\sqrt{\tau}}\right)+\overline{h}_{ij}^{k,2}(x,t)I_2(\xi_{i,l},\tau),$$

$$I_r(\xi_{i,l},\tau)=\frac{1}{2\sqrt{\pi}}\int_0^\tau\int_0^\infty\frac{\overline{\overline{h}}_i^{k,r}(\eta,s)}{\sqrt{\tau-s}}\left[\exp\left(-\frac{(\xi_{i,l}-\eta)^2}{4(\tau-s)}\right)\right.$$

$$\left.-\exp\left(-\frac{(\xi_{i,l}+\eta)^2}{4(\tau-s)}\right)\right]d\eta ds,\;\; r=1,2,$$

where $\overline{h}_i^{k,r}(x,t),\overline{\overline{h}}_i^{k,r}(\eta,s)$ are known functions. Evaluation of the integral

$$|I_r(\xi_{i,l},\tau)|\le c\exp\left(-\frac{\xi_{i,l}^2}{8\tau}\right).$$

Theorem 2: Let conditions (1)–(4) be satisfied, then Eq. (32) under additional conditions

1. $u_k\big|_{t=\mu=\tau=0}=0,\;\; u_k\big|_{x=l-1,\xi_{i,l}=0}=0,\;\; l=1,2;$

2. $-T_1u_k-\partial_tu_{k-2}+L_xu_{k-4}\in U_2\oplus U_4;$

3. $L_\xi u_k=0,$

has a unique solutionin U.

Proof: Satisfying the function $u_k(M)\in U$ with the boundary conditions from (26) we obtain (39). Based on calculations (36–38) we have

$$-T_1u_k-\partial_tu_{k-2}+L_xu_{k-4}=-t\sum_{i=1}^n\left\{\partial_tv_{k,i}(x,t)\psi_i(t)\right.$$

$$+\sum_{r=1}^n\alpha_{ri}(t)v_{k,r}(x,t)\psi_i(t)+\sum_{r=1}^n\alpha_{ri}(t)p_r^k(x)\psi_i(t)\exp(\mu_r)$$

$$+\sum_{j=1}^n\left[\partial_tc_{ij}^k+\sum_{r=1}^n\alpha_{r,i}(t)c_{r,j}^k(x,t)\right]\psi_i(t)\exp(\mu_j)$$

$$\left.+\sum_{l=1}^2\left[\partial_ty_i^{k,l}(N^l)+\sum_{j=1}^n\partial_tz_{ij}^k(N^l)\exp(\mu_j)\right]b_i(x)\right\}$$

$$-\sum_{i,j=1}^{n}\psi_i(t)c_{i,j}^k(x,t)\left(\beta_j(0)-\beta_j(t)\right)\exp\left(\mu_j\right)$$

$$-\sum_{i=1}^{n}\psi_i(t)p_i^k(x)(\beta_i(0)-\beta_i(t))\exp\left(\mu_i\right)$$

$$+\sum_{l=1}^{2}\sum_{i=1}^{n}\left\{\sum_{r=1}^{n}\gamma_{ir}(x,\ t)b_r(x)y_i^{k,l}\left(N^l\right)+\sum_{j=1}^{n}\sum_{r=1}^{n}\gamma_{i,r}(x,t)b_r(x)z_{i,j}^{k,l}\left(N^l\right)\exp\left(\mu_j\right)\right\}$$

$$-\sum_{i=1}^{n}\sum_{r=1}^{n}\alpha_{r,i}(t)p_r^{k-2}(x)\exp\left(\mu_r\right)\psi_i(t)$$

$$-\sum_{i=1}^{n}\left\{\partial_t v_{k-2,i}+\sum_{r=1}^{n}\alpha_{r,i}(t)v_{k-2,r}+\sum_{j=1}^{n}\left[\partial_t c_{i,j}^{k-2}+\sum_{r=1}^{n}\alpha_{r,i}(t)c_{r,j}^{k-2}(x,\ t)\right]\exp\left(\mu_j\right)\right\}\psi_i(t)$$

$$-\sum_{l=1}^{2}\sum_{i=1}^{n}\left\{\partial_t y_i^{k,l}\left(N^l\right)+\sum_{j=1}^{n}\partial_t z_{i,j}^{k,l}\left(N^l\right)\exp\left(\mu_j\right)\right\}b_i(x)$$

$$+A(x)\sum_{i=1}^{n}\left\{\partial_x^2 v_{k-4,i}+\sum_{j=1}^{n}\partial_x^2\left[c_{i,j}^{k-4}(x,\ \ t)+p_i^{k-4}(x)\right]\exp\left(\mu_j\right)\right\}\psi_i(t)$$

$$+A(x)\sum_{l=1}^{2}\sum_{i=1}^{n}\partial_x^2\left(\left[y_i^{k,l}\left(N^l\right)+\sum_{j=1}^{n}z_{i,j}^{k,l}\left(N^l\right)\exp\left(\mu_j\right)\right]b_i(x)\right).$$

Providing the condition of Theorem 1, we set

$$t\left[\partial_t V_k+A^T(t)V_k\right]=-\partial_t V_{k-2}-L_x V_{k-4}(x,t),\tag{41}$$

$$\begin{aligned}t\left[\partial_t C^k+A^T(t)\left(C^k(x,\ \ t)+\Lambda\left(P^k(x)\right)\right)\right]+\left[C^k(x,\ \ t)+\Lambda\left(P^k(x)\right)\right]\Lambda(\beta(0))\\ -\Lambda(\beta(t))\left[C^k(x,\ \ t)+\Lambda\left(P^k(x)\right)\right]\\ =-\partial_t C^{k-2}-A^T(t)\left[C^{k-2}+\Lambda\left(P^{k-2}(x)\right)\right]+L_x\left[C^{k-4}+\Lambda\left(P^{k-4}(x)\right)\right],\end{aligned}\tag{42}$$

Eq. (43) is solved without an initial condition and uniquely determines the function $V_k(x,t)$ [27].

Providing the solvability of Eq. (44) we set

$$\overline{\overline{C^k(x,\ \ t)\Lambda(\beta(0))-\Lambda(\beta(t))C^k(x,\ \ t)}}\Big|_{t=0}$$

$$=-\overline{\overline{\partial_t C^{k-2}-A^T(t)C^{k-2}(x,\ \ t)-A^T(t)\Lambda\left(P^{k-2}\right)+L_x\left(C^{k-4}\right)}}\Big|_{t=0}$$

$$\overline{A^T(t)\Lambda\left(P^{k-2}(x)\right)}=\overline{\left[-\partial_t C^{k-2}-A^T(t)C^{k-2}-L_x\left(C^{k-4}(x,\ \ t)+\Lambda\left(P^{k-4}(x)\right)\right)\right]}$$

or in coordinate form

$$c_{ij}^{k}(x,\ t)\Big|_{t=0} = -\frac{1}{\beta_j(0)-\beta_i(t)}\left[\partial_t c_{ij}^{k-2} + \sum_{r\neq j}\alpha_{ri}(t)c_{rj}^{k-2}(x,t) - q_{ij}(x,t)\right]_{t=0},$$

$$p_i^{k-2}(x) = -\frac{1}{\alpha_{ii}(t)}\left[\partial_t c_{ii}^{k-2} + \alpha_{ii}(t)c_{ii}^{k-2} - q_{ii}(x,t)\right]_{t=0},$$

where $q_{ij}(x,t)$ is known function included in $A(x)\partial_x^2\left[C^{k-4}(x,\ t) + \Lambda\left(P^{k-4}(x)\right)\right]$.

On the basis of (38), condition (3), Theorem 2 is ensured if arbitrary functions $d_i^{l,k}(x,t)b_i(x)$,.

$\omega_{ij}^{k,l}(x,t)b_i(x)$ are solutions to the problems

$$2\varphi_{i,l}'(x)\left(d_i^{l,k}(x,\ t)b_i(x)\right)_{x'} + \varphi_{i,l}''(x)\left(d_i^{l,k}(x,\ t)b_i(x)\right) = 0,$$

$$d_i^{l,k}(x,\ t)b_i(x)\Big|_{x=l-1} = -v_k(l-1,t)\psi_i(t),\ i=\overline{1,n},$$

$$2\varphi_{i,l}'(x)\left(\omega_{i,j}^{l,k}(x,\ t)b_i(x)\right)_{x'} + \varphi_{i,l}''(x)\left(\omega_{i,j}^{l,k}(x,\ t)b_i(x)\right) = 0,$$

$$\omega_{i,j}^{l,k}(x,\ t)b_i(x)\Big|_{x=l-1} = -\left[c_{ij}^{k}(l-1,\ t) + p_i(l-1)\right]\psi_j(t).$$

Thus, arbitrary functions $d_i^{l,k}(x,t),\ \omega_{ij}^{k,l}(x,t),\ v_{ki}(x,t),\ c_{ij}^{k,l}(x,t)$ included in (34) are uniquely determined.

Iterative Eq. (32) for $k = 0{,}1$ is homogeneous; therefore, by Theorem 1, it has a solution $u_k(M)\in U$ if the functions $y_i^{l,k}(N_i^l), z_{i,j}^{l,k}(N_i^l)$ are solutions of the equations

$$\partial_\tau y_i^{k,l}(N_i^l) = \partial_{\xi_{i,l}^2} y_i^{k,l}(N_i^l),\quad \partial_\tau z_{i,j}^{k,l}(N_i^l) = \partial_{\xi_{i,l}^2} z_{i,j}^{k,l}(N_i^l)$$

for boundary value conditions

$$y_i^{k,l}(N_i^l)\Big|_{t=\tau=0} = 0,\ \ y_i^{k,l}(N_i^l)\Big|_{\xi_{i,l}=0} = d_i^{k,l}(x,t),$$

$$z_{i,j}^{k,l}(N_i^l)\Big|_{t=\tau=0} = 0,\ \ z_{i,j}^{k,l}(N_i^l)\Big|_{\xi_{i,l}=0} = \omega_{i,j}^{k,l}(x,t).$$

From this problem, we find

$$y_i^{0,l}(N_i^l) = d_i^{0,l}(x,t)erfc\left(\frac{\xi_{i,l}}{2\sqrt{\tau}}\right),\quad z_{i,j}^{0,l}(N_i^l) = \omega_{i,j}^{0,l}(x,t)erfc\left(\frac{\xi_{i,l}}{2\sqrt{\tau}}\right).$$

The functions $d_i^{0,l}(x,t), \omega_{i,j}^{0,l}(x,t)$ are determined from problems (44) which ensure that condition $L^{\xi}u_0 = 0$ is satisfied. Using calculations (37), the free term of iterative Eq. (32) at $k = 2$ is written as $F_2(M) = -T_1u_0(M) + f(x,t)$ by Theorem 1, an equation with such a free term is solvable in U, if

$$t\sum_{i=1}^{n}\left\{\left[\partial_t v_{0,i}(x,\ t) + \sum_{r=1}^{n}\alpha_{ri}(x)v_{0,r}(x,\ t)\right] - \beta_i(t)v_{0,i}(x,t)\right\} = f(x,t)$$

$$t\sum_{i,j=1}^{n}\left\{\left[\partial_t c_{ij}^0(x,\ t)+\sum_{r=1}^{n}\alpha_{ri}(t)c_{rj}^0(x,\ t)\right]+\alpha_{ji}(t)p_j^0(x)\right\}$$

$$+\sum_{i,j=1}^{n}\left[\beta_j(0)-\beta_i(t)\right]c_{ij}^0(x,t)+\sum_{i=1}^{n}[\beta_i(0)-\beta_i(t)]p_i^0(x)=0. \qquad (43)$$

From (47) we uniquely determine $c_{i,j}^0(x,t)=0,\forall i\neq j$ and the function $c_{i,i}^0(x,t)$ is determined from the equation

$$t\left[\partial_t c_{ii}^0(x,\ t)+\alpha_{ii}(t)c_{ii}^0(x,t)\right]+(\beta_i(0)-\beta_i(t))c_{ii}^0(x,t)+[\beta_i(0)-\beta_i(t)]p_i^0(x)=0$$

under the initial condition

$$c_{ii}^0(x,0)=-v_{0,i}(x,0)-p_i^0(x).$$

The first equation from (47), by virtue of condition (2), has a solution satisfying the condition [27] $\|v^0(x,\ 0)\|<\infty$. We calculate the free term of Eq. (32) at $k=3$:

$$F_3(M)=-T_1u_1,$$

which has the same view as T_1u_0. Providing the solvability of equation $T_0u_3=-T_1u_1$ in U, with respect to $c_{ij}^1(x,t)$,$v_{1i}(x,t)$ we obtain Eqs. (47).

In the next step, $(k=4)$ the free term has the view

$$F_4(M)=-T_1u_2-\partial_t u_0+L_\xi u_1.$$

The functions $d_i^{1,l}(x,t)$,$\omega_{i,j}^{1,l}(x,t)$ entering the $u_1(M)$ provide the condition $L_\xi u_1=0$. Providing the solvability of the iterative equation at $k=4$, we set

$$t\left[\partial_t v_{2i}(x,\ t)+\sum_{k=1}^{n}\alpha_{ki}(x)v_{2,k}(x,t)\right]-\beta_i(t)v_{2,i}(x,t)=-\partial_t v_{0,i}(x,t).$$

For $c_{ij}^2(x,t)$ we obtain the same equation of the form (47), but with the right-hand side $\partial_t c_{i,j}^0(x,t)+\sum_{k=1}^{n}\alpha_{k,i}(x)\left(c_{k,j}^0(x,\ t)+\alpha_{j,i}(x)p_j^0(x)\right)$. Taking off the degeneracy of this equation as $t=0$, we set $p_i^0(x)=-\frac{1}{\alpha_{ii}(t)}\left[\partial_t c_{ii}^0+\alpha_{ii}(x)c_{ii}^0\right]_{t=0}$. Further repeating the described process, using Theorems 1 and 2, sequentially determining $u_k(M),\ k=0,1,\dots,n$, we construct a partial sum

$$u_{\varepsilon n}(M)=\sum_{k=0}^{n}\varepsilon^{k/2}u_k(M). \qquad (44)$$

2.4 Estimate for the remainder

For the remainder term

$$R_{\varepsilon n}(M)=u(M,\varepsilon)-u_{\varepsilon n}(M)=u(M,\varepsilon)-\sum_{k=0}^{n+2}\varepsilon^{k/2}u_k(M)+\sum_{l=1}^{2}\varepsilon^{\frac{n+l}{2}}u_{n+l}$$

we get the problem

$$\tilde{L}_\varepsilon R_{\varepsilon n}(M) = \varepsilon^{\frac{n+1}{2}} g_{\varepsilon n}(M),$$

$$R_{\varepsilon n}(M)\big|_{t=\tau=\mu=0} = R_{\varepsilon n}\big|_{x=l-1,\xi=0} = 0,\ l = 1{,}2,$$

where

$$g_{n\varepsilon}(M) = -\sum_{l=0}^{1}(T_1 u_{n+1+l})\varepsilon^{l/2} - \sum_{l=0}^{3}\varepsilon^{l/2}\partial_t u_{n-1+l} + \sum_{l=0}^{5}\varepsilon^{l/2}L_x u_{n-3+l} - \sum_{l=0}^{1}\varepsilon^{l/2}\tilde{L}_\varepsilon u_{n+1+l}.$$

From the form (34) of the function U, based on conditions (1)–(3) and the form of regularizing variables from (27), we conclude that

$$\|g_{n\varepsilon}(M)\| < C.$$

In the equation for R, we make the restriction by means of regularizing functions, then, based on (30), we obtain the problem

$$L_\varepsilon R_{\varepsilon n}(x, t, \varepsilon) = \varepsilon^{(n+1)/2} g_{n\varepsilon}(x, t, \varepsilon),$$

$$R_{\varepsilon n}|_{t=0} = R_{\varepsilon n}|_{x=0} = R_{\varepsilon n}|_{x=1} = 0.$$

We divide both sides of this equation by $(t+\varepsilon)$, while the properties of the matrices A are preserved. Therefore, using Theorem 5.5 of [33], we obtain the estimate

$$\|R_{\varepsilon n}(x,\ t,\ \varepsilon)\| < c\varepsilon^{(n+1)/2}.$$

Passing to Euclidean norms, like [28], or the same estimate can be obtained using the maximum principle [33], p. 23.

By narrowing this problem by means of regularizing functions. Following [33] and [28], passing to Euclidean norms, we obtain a problem that is limited by the maximum principle

$$\|R_{\varepsilon n}(x,\ t,\ \varepsilon)\| < c\varepsilon^{\frac{n+1}{2}}. \tag{45}$$

Theorem 3: Let conditions (1)–(4) be satisfied. Then the restriction of the constructed solution (44) is an asymptotic solution to the problem (26), that is, $\forall n = 0{,}1, \ldots$ the estimate (45) holds at $\varepsilon \to 0$.

Proof: Let us rewrite the problem (31)

$$\partial_t u - \varepsilon^2 \frac{1}{t+\varepsilon} A(x)\partial_x^2 u - \frac{1}{t+\varepsilon} D(t)u = \frac{1}{t+\varepsilon} f(x, t)$$

here, the expression $(t+\varepsilon)$ does not affect sufficiently small ε the properties of the matrices $A(x)$,$D(t)$ for which the conditions of the maximum principle theorem are valid [28], p. 20]. Therefore, on the basis of this theorem, it is easy to establish an estimate (45).

Author details

Asan Omuraliev and Peil Esengul Kyzy*
Kyrgyz-Turkish Manas University, Bishkek, Kyrgyzstan

*Address all correspondence to: peyil.esengul@manas.edu.kg

References

[1] Aronson DG. Linear parabolic differential equations containing a small parameter. Journal of Rational Mechanical Analysis. 1956;**5**(6): 1003-1014

[2] Vishik MI, Lusternik LA. Regular degeneration and boundary layer for linear differential equations with a small parameter. Russian Mathematical Surveys. 1957;**12**(5):3-122

[3] Vishik MI, Lusternik LA. On the asymptotics of the solution of boundary value problems for quasilinear differential equations. Doklady Mathematics. 1958;**121**(5):778-782

[4] Gevrey D. Sur les equations aux derivees partielles du type parabolique. Journal of Mathematics. 19313;**6**(9):305

[5] Isakova EK. Asymptotics of the solution of a second-order partial differential equation of parabolic type with a small parameter with the highest derivative. Doklady Mathematics. 1957; **117**(6):935-938

[6] Trenogin VA. Asymptotic behavior of solutions of almost linear parabolic equations with parabolic boundary layers. Uspekhi Mat. Nauk. 1961;**16**(1): 164-169

[7] Lomov SA. Vvedenie v obshchuyu teoriyu singulyarnykh vozmushchenii [Introduction to the General Singular Perturbation Theory]. Moscow: Nauka; 1981

[8] Lomov SA, Lomov IS. Osnovy matematicheskoi teorii pogranichnogo sloya [Fundamentals of the Mathematical Theory of the Boundary Layer]. Moscow: Moscow State Univ; 2011

[9] Lomov SA. Power boundary layer in problems with singular perturbation. Izvestiya. 1966;**30**(3):525-572

[10] Koniev UA. Construction of the exact solution of some singularly perturbed problems for linear ordinary differential equations with a power boundary layer. Mathematical Notes. 2006;**79**(6):950-954

[11] Vasil'eva AB, Butuzov VF. Asimptoticheskie metody v teorii singulyarnykh vozmushchenii [Asymptotic Methods in Singular Perturbation Theory]. Moscow: Vysshaya Shkola; 1990

[12] Butuzov VF. Asymptotic properties of the solution of a finite-difference equation with small steps in a rectangular region. Journal of Computational Mathematics and Mathematical Physics. 1973;**12**(3):14-34

[13] Butuzov VF, Buchnev VY. Asymptotics of a solution of a singularly perturbed parabolic problem in the two-dimensional case. Differential Equations. 1989;**25**(3):315-322

[14] Butuzov VF, Buchnev VY. Asymptotic approximation of the solution of a singularly perturbed parabolic two-dimensional problem with a nonsmooth limit solution. Differential Equations. 1989;**25**(12): 1514-1518

[15] Butuzov VF, Buchnev VY. The asymptotic behavior of the solution of a singularly perturbed two-dimensional problem of the reaction-diffusion-transport type. Journal of Computational Mathematics and Mathematical Physics. 1991;**31**(4):37-44

[16] Butuzov VF, Kalachev LV. Asymptotic approximation of solution of boundary-value problem for a singularly perturbed parabolic equation in the critical case. Mathematical Notes. 1986; **39**(6):445-451

[17] Butuzov VF, Levashova NT. On a singularly perturbed reaction-diffusion-transfer system in the case of slow diffusion and fast reactions. Fundam. Prikl. Mat. 1995;**1**(4):907-922

[18] Butuzov VF, Urazgil'dina TA. Asymptotics of solutions to a boundary value problem for the heat equation with a powerful nonlinear source in a thin rod. Differential Equations. 1995;**31**(3): 435-445

[19] Nesterov AV. The asymptotic behavior of the solution to a parabolic equation with singularly perturbed boundary conditions. Journal of Computational Mathematics and Mathematical Physics. 1997;**37**(9): 1051-1057

[20] Omuraliev AS. Regularization of a two-dimensional singularly perturbed parabolic problem. Journal of Computational Mathematics and Mathematical Physics. 2006;**46**(8)

[21] Omuraliev AS, Kulmanbetova S. Asymptotics of the solution of a parabolic problem in the absence of a limit operator. Journal of Computational Mathematics and Mathematical Physics. 2012;**52**(8)

[22] Omuraliev AS, Kyzy MI. Singularly perturbed parabolic problems with multidimensional boundary layers. Differential Equations. 2017;**53**(12):1-5

[23] Omuraliev AS, Kulmanbetova S. Singularly perturbed system of parabolic equations in the critical case. Journal of Mathematical Sciences. 2018;**230**(5): 728-731

[24] Omuraliev AS. Asymptotics of the solution of a parabolic linear system with a small parameter. Differential Equations. 2019;**55**(6):1-5

[25] Omuraliev AS. Asymptotics of the Solution of Singularly Perturbed Parabolic Problems. LAMBERT Academic Publishing; 2017

[26] Omuraliev A, Abylaeva E. Ordinary differential equations with power boundary layers. Ukrainian Mathematical Bulletin. 2018;**15**(4): 536-542

[27] Vazov V. Asymptotic Expansions of Solutions of Ordinary Differential Equations. Moscow: Nauka; 1968

[28] Ladyjenskaya OA, Solonnikov VA, Ural'seva NN. Linear and Quasilinear Parabolic Equations. Moscow: Nauka; 1967

[29] Lomov SA. About the Lithill model equation. MO SSSR. 1964;**54**:74-83

[30] Koniev UA. Singularly perturbed problems with a double singularity math. Zametki. 1997;**62**(4):494-501

[31] Omuraliev AS, Abylaeva E, Kyzy PE. A system of singularly perturbed parabolic equations with a power boundary layer Lobachevskii. Journal of Mathematics. 2020;**41**(1):71-79

[32] Omuraliev AS, Abylaeva E, Kyzy PE. Parabolic problem with a power boundary layer Diff. Equation. 2021; **57**(1):67-77

[33] Hartman F. Ordinary Differential Equations. 1970

Section 3

Boundary Layer Theory in Aerodynamic Flows

Chapter 7

Wind Tunnel Experiments on Turbulent Boundary Layer Flows

Adrián R. Wittwer, Acir M. Loredo-Souza, Jorge O. Marighetti and Mario E. De Bortoli

Abstract

The knowledge and experimental development of boundary layer turbulent flows is extremely important in applications related to the building aerodynamics, wind comfort, atmospheric dispersion, and even aeronautics. The Aerodynamic Laboratories of the UFRGS and UNNE have been making joint activities related to wind engineering such as those mentioned earlier for more than 25 years. In this work, a compilation of different experiments on turbulent boundary layer flows realized in these both laboratories is carried out. The characteristics of flows that develop on a smooth surface of the wind tunnel are experimentally evaluated. Then, reduced-scale models of atmospheric boundary layer flows are analyzed including the effects of turbulence generators and surface roughness. Special attention on the behavior of the turbulent parameters in the case of experimental studies using low mean velocity is paid. Finally, some comments referring to recent studies on thermal effects in turbulent boundary layer flows and the development of reduced-scale models of convective flows are included.

Keywords: wind engineering, aerodynamics, turbulence, scale models, smooth and roughness surface effects

1. Introduction

The wind tunnel is an experimental device for aerodynamic studies. Turbulent boundary layer flows are modeled within the test section of a wind tunnel in order to apply them to complex experimental studies. The physical simulation of atmospheric boundary layer (ABL) flows is one of the bases of wind engineering. That type of experimental procedure requires a careful investigation of the characteristic parameters of mean flow and turbulence. Different measurements must be carried out with specific equipment to be able to evaluate the velocity fluctuations of the turbulent flow.

Boundary layer flows are commonly characterized by their mean velocity and turbulent intensity profiles. The Prandtl logarithmic law (Eq. (1)), proposed from similarity theories, can be used can be used to compare experimental results. In this mathematical expression, U is the mean velocity, u^* is the friction velocity, k is 0,4 (Kármán constant), and z_0 is known as the roughness height:

$$U(z)/u^* = (1/k)\ln(z/z_0) \tag{1}$$

IntechOpen

Effects of the surface roughness, temperature, and others properties are transmitted by turbulent flows in case of the atmospheric boundary layer (ABL). Turbulent exchanges are very weak when there are very stable thermal stratification wind conditions [1]. The atmospheric boundary layer over nonhomogeneous terrain is not well defined because topographical features generate highly complex flows. The depth of the atmospheric boundary layer is only 100 m during stable nighttime conditions and could reach 1 km during unstable daytime conditions [2]. The aforementioned expression of Prandtl's law is modified when applied near the surface in the case of a neutral boundary layer (Eq. (2)). The zero-plane displacement z_d is then included for very rough surface:

$$U(z)/u^{*} = (1/k)\ln\left[(z - z_d)/z_0\right] \tag{2}$$

Another widely used expression used to characterize the vertical profile is the power law (Eq. (2)). The exponent α varies between 0.10 and 0.43, and the thickness of the boundary layer z_g ranges from 250 to 500 m, depending on the type of terrain [2]. This type of vertical velocity distribution is verifiable under conditions of neutral stability, which are those to be considered for the analysis of wind loads:

$$U(z)/U\left(h_g\right) = \left(z/h_g\right)^{\alpha} \tag{3}$$

The turbulence spectrum is the analytical tool used to investigate the velocity fluctuation characteristics of laboratory turbulent flows. Experimental spectra are fitting to some functional form to define dimensionless turbulence spectra in accordance with a similarity theory [3, 4]. Using similarity theory will allow the dimensionless spectra of atmospheric and laboratory flows to collapse, if the dimensionless spectra were formulated with the appropriate parameters [5].

2. Boundary layer wind tunnels

The most efficient tool to experimental study of aerodynamic phenomena is the wind tunnel. Basically, there are three types of wind tunnels. Aerodynamic tunnels, called the "first generation" of wind tunnels, are mainly used in aeronautics and vehicular applications. Boundary layer tunnels, called "second generation," are used for studies involving atmospheric boundary layer (ABL) flows. Finally, the "third generation" of wind tunnels is beginning, the three-dimensional tunnels, in which different types of flow can be simulated. An example of the "third generation" of wind tunnels is the WindEEE Dome [6, 7], a hexagonal wind tunnel with 25 m in diameter and 40 m in diameter for the external return built in recent years in Canada.

In this work, measurements of flow velocity realized in two different boundary layer wind tunnels will be used for the experimental analysis. The closed-return wind tunnel "Prof. Joaquim Blessmann" of the Laboratório de Aerodinâmica das Construçoes, Universidade Federal de Rio Grande do Sul, UFRGS, Brazil [8] and the "Jacek Gorecki" open-circuit wind tunnel of the Laboratorio de Aerodinámica, Universidad Nacional del Nordeste, UNNE [9] are the boundary layer tunnels employed to develop the experiments. Both wind tunnels can be considered as low-speed tunnels.

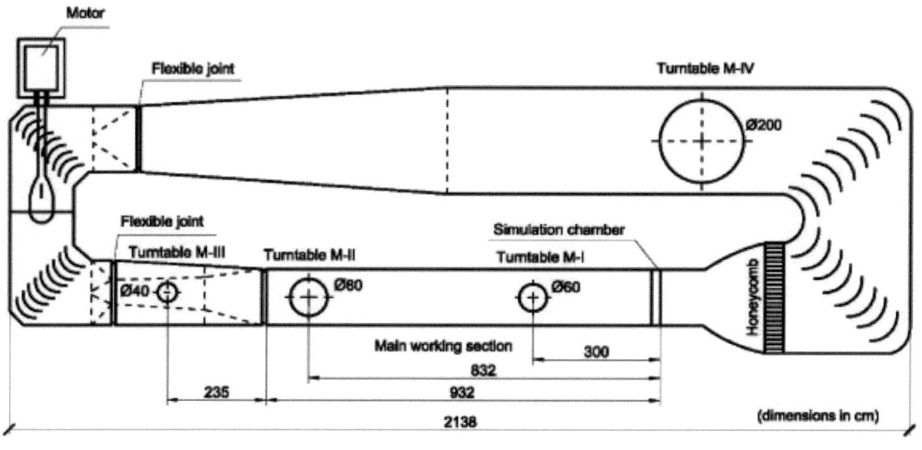

Figure 1.
The "Prof. Joaquim Blessmann" boundary-layer wind tunnel of the UFRGS.

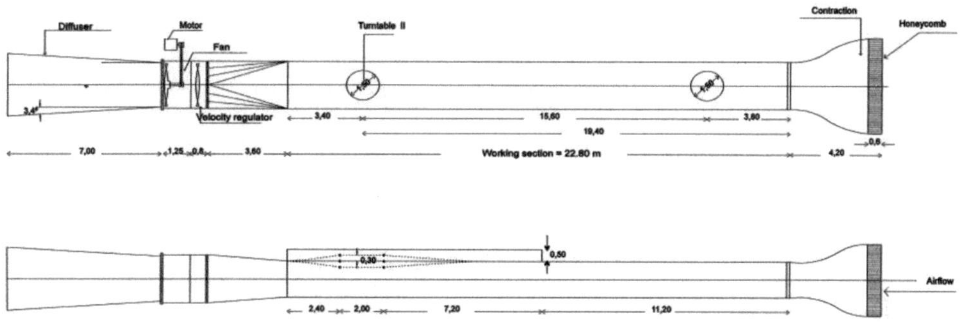

Figure 2.
The "Jacek Gorecki" boundary-layer wind tunnel of the UNNE.

The "Prof. Joaquim Blessmann" wind tunnel of UFRGS has a cross section of 1.30 m 0.90 m at downstream end of the main working section that has 9.32 m long (**Figure 1**). Characteristics and performance of this wind tunnel are described by Blessmann [9]. The "Jacek Gorecki" UNNE wind tunnel is a 39.56 m-long channel, and the test section is a 22.8-m-long rectangular chamber (2.40 m width × 1.80 m height) with two rotating tables to place test models (**Figure 2**).

3. Boundary layer flows on smooth surface

In this first part, we show the behavior of the boundary layer that develops on the smooth surface (floor) of a wind tunnel. Different measurements were made in both wind tunnels, the J. Blessmann tunnel at UFRGS and the J. Gorecki tunnel at UNNE, where different wind speeds were used. **Figure 3** shows the general characteristics of the interior of both tunnels, with the dimensions of the test chamber being 1.30 m width × 0.90 m high × 9.32 m long for the UFRGS tunnel and 2.40 m width × 1.80 m high × 22.8 m long for the UNNE tunnel, respectively.

Figure 3.
Test sections for smooth surface experiments at the UFRGS wind tunnel (left) and at the UNNE wind tunnel (right).

Experiments.	Position	U_{ref} [m/s]	H_{BL} [m]	Re_H
UFRGS -TI	Center line	38.9	0.075	1.95×10^5
UNNE –TI	Center line	22.7	0.130	1.97×10^5
UFRGS -TII	Center line	39.5	0.125	3.29×10^5
UNNE –TII – 00	Center line	24.5	0.300	4.90×10^5
UNNE –TII – 06(L)	Left side	24.5	0.300	4.90×10^5
UNNE –TII – 06(R)	Right side	24.5	0.300	4.90×10^5
UFRGS-LV1	Center line	0.9	0.175	1.05×10^4
UFRGS-LV2	Center line	2.1	0.150	2.10×10^4
UFRGS-LV5	Center line	4.7	0.150	4.70×10^4

Table 1.
Characteristic parameters of the smooth surface experiments.

The different experiments used to characterize the smooth surface boundary layer flow are described in **Table 1**. TI and TII refer to rotating Tables I and II of each tunnel (positions of rotating tables are indicated in **Figures 1** and **2**). Most of the measurements were made in the central position of the test chamber, but for the UNNE tunnel, two measurements are indicated at 0.6 m to the left (L) and right (R) of the central position in order to illustrate the lateral homogeneity. Each experiment is considered stationary, and it is characterized by the reference velocity U_{ref} measured at the top of the boundary layer, the boundary layer height H_{BL}, and the Reynolds number Re_H.

All measurements were made with a constant temperature hot-wire anemometer. Acquisition time and sampling frequency of the velocity records were modified in each case according to the type of boundary layer flow and the reference speed of the experiment. These values are indicated later in the text where the description and analysis of the different experiments are made. In the case of the experiments carried out at UFRGS, the CTA System Data Acquisition System—DANTEC-Stream-Line Pro [10] was used. On the other hand, the measurements at UNNE were made by a Dantec 56C constant temperature hot-wire anemometer connected to a Stanford amplifier with low- and high-pass analogic filters. Hot-wire signals were digitalized by a DAS-1600 A/D converter board controlled by a computer which was also used for the analysis of the results. Voltage output from hot wires was converted in mean velocity and velocity fluctuations [11, 12] by the probe calibration curves previously determined.

3.1 High-velocity measurements

Firstly, the mean speed and turbulence intensity profiles obtained with relatively high speeds, between 20 and 30 m/s, shown in **Figure 4** are analyzed. Velocity profiles are dimensionless with the reference velocity measured "outside" the boundary layer, where the mean velocity remains approximately constant. The vertical coordinate z is not dimensionless in order to show how the height of the boundary layer varies in each case. In this way, we can observe that the UNNE tunnel, where the cross section of the test section has larger dimensions, the thicknesses of the boundary layer generated are relatively greater. Likewise, it is observed that in the rotating Table I, located at the beginning of the test section, the boundary layer thicknesses developed in both tunnels are less than the thicknesses obtained in the rotating Table II.

The fluctuating speed records measured with a hot-wire anemometer have a duration of 60 seconds in the tests carried out in the UFRGS tunnel and 90 seconds in the tests carried out in the UNNE tunnel. The acquisition frequency of the velocity records is 2048 Hz for the UFRGS measurements and 3000 Hz for the UNNE measurements. Finally, low-pass filtering was implemented for the analog signals measured at UNNE, while low-pass filtering was not used for the UFRGS measurements.

Velocity profiles indicate a fairly harmonic velocity distribution and even good lateral homogeneity, with respect to flow quality. The maximum turbulence intensities

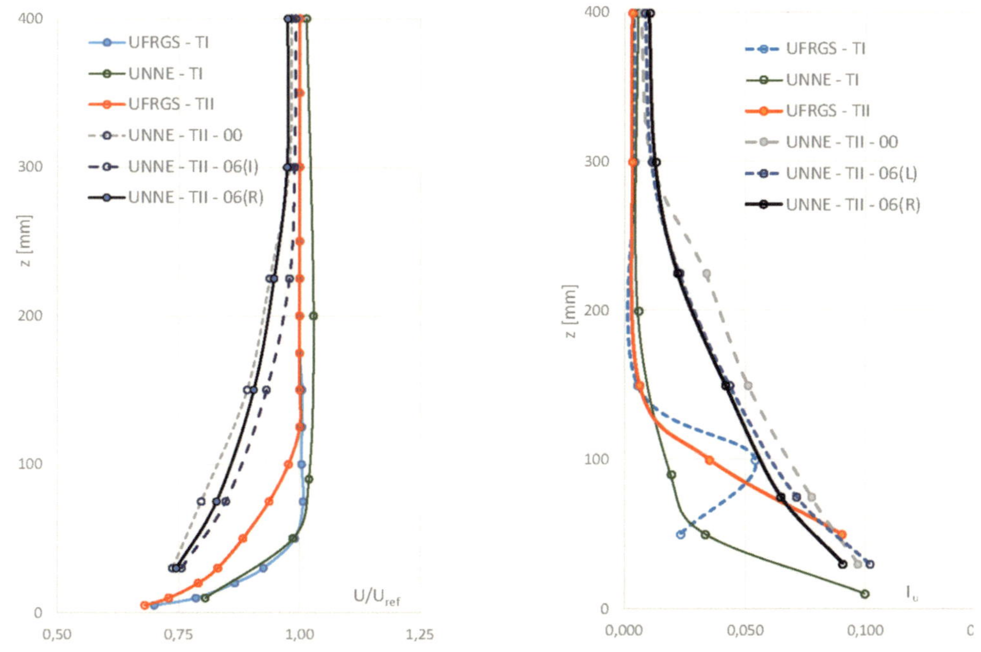

Figure 4.
Mean velocity and turbulence intensity profiles for smooth surface experiments (high velocity).

measured at distances of 15 to 30 mm from the floor do not exceed values of 10% and "outside" the boundary layer are less than 1%. In general, the vertical distribution shows a fairly harmonic behavior, except in the measurements corresponding to rotating Table I—UFRGS wind tunnel, where a localized increase is observed for height z = 100 mm, possibly due to a vortex shedding phenomena located at the convergent.

3.2 Low-velocity measurements

The mean velocity and turbulence intensity profiles for relatively lower mean velocities are now analyzed. All fluctuating speed records used for the low-speed analysis have a duration of 180 seconds, and the acquisition frequency for the speed records is 2048 Hz. These records come from the tests carried out on the rotating Table II of the UFRGS tunnel, and the corresponding analog signals do not include low-pass filtering.

The analysis of the vertical distribution of the mean velocity (**Figure 5**) shows an increase in the thickness of the boundary layer when the velocity decreases from 150 mm (Uref =4.7 m/s - LV5) to approximately 200 mm (Uref =0.9 m/s - LV1). The profile obtained with the lowest speed (LV1) presents a more irregular behavior than the others, probably because the effects of viscosity are more noticeable. **Figure 5** also includes the profile obtained on Table II of the UFRGS tunnel at high speed (Uref =39.5 m/s – TII - HV) to compare and verify a more harmonic behavior when the mean flow speed is higher.

The turbulence intensity profiles accompany the behavior shown by the average speeds. The highest values of I_u obtained up to 200 mm height correspond to the measurements of lower speed (LV1). However, the measurement point closest to the tunnel floor shows a slight decrease in I_u with respect to the one immediately above it (from 0.13 to 0.10) as a result of viscous effects.

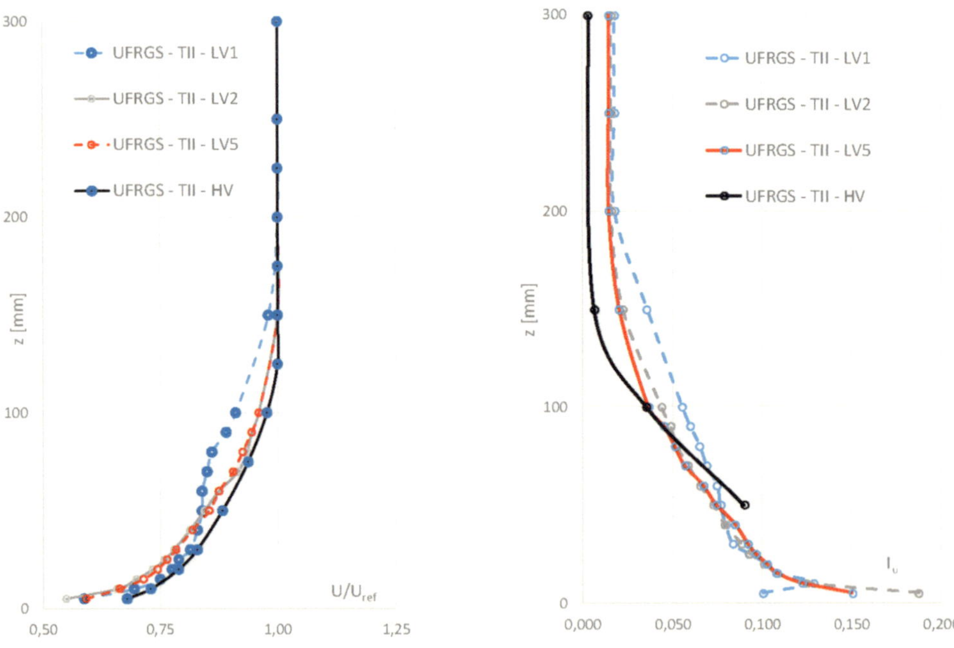

Figure 5.
Mean velocity and turbulence intensity profiles for smooth surface experiments (low velocity).

3.3 Spectral analysis

Spectral results from longitudinal velocity fluctuations were obtained by Fourier analysis applied to different sampling series, obtained in the UNNE wind tunnel (high velocity) and UFRGS wind tunnel (low velocity). Time acquisition and sampling frequency are given in **Table 2**. The obtained series were divided in blocks to which a fast Fourier transform (FFT) algorithm was applied [13]. Three spectra obtained at the UFRGS wind tunnel and a spectrum measured at the UNNE wind tunnel are shown in **Figure 6**. Values of the spectral function decrease as the mean velocity also decreases. The spectrum obtained at the measurement point closest to the tunnel floor with low speed allows verification of the behavior observed in the turbulence intensity profile (it shows a slight decrease in I_u at this point with respect to the immediately higher one). Viscous attenuation effects start around 10 Hz at this measurement point. An important characteristic of the spectra measured with high velocity is the presence of a clear region with a $-5/3$ slope, characterizing Kolmogorov's inertial subrange.

4. Physical models of the atmospheric boundary layer

Jensen [14] established in 1958 that: "The correct model test with phenomena in the wind must be carried out in a turbulent boundary layer, and the model-law requires that this boundary layer be to scale as regards the velocity profile." Later studies on physical reduced-scale models of atmospheric boundary layer attempted to reproduce as closely as possible the mean flow and turbulence structure of the atmospheric flow, in particular, those properties related to neutral atmospheric boundary layer and wind loading effects.

In the preceding section, it was verified that the boundary layer thickness achieved with a smooth surface is relatively small. Structural model tests require higher boundary layer thicknesses and turbulence structures representative of atmospheric winds. This problem gave rise to different methods of physical simulation of atmospheric flows in a wind tunnel.

The general similarity requirements for reproducing ABL flows in wind tunnels are geometric, kinematic, dynamic, thermal, and boundary conditions similarity. In terms of dimensionless parameters, it is necessary to maintain the equality of Reynolds, Richardson, Rossby, and Prandtl numbers, in addition to geometric similarity and boundary conditions [15].

The fluid used in wind tunnels for ABL simulation is air, so the Prandtl number is automatically the same in the model and prototype. On the other hand, Rossby number equality cannot be considered in conventional tunnels, so applications are restricted to flow conditions where the effects of terrestrial rotation are negligible.

Experiments	Time [s]	Sampling frequency [Hz]	Low-pass filter [Hz]
UFRGS -TII – LV1	180	2048	—
UFRGS -TII – LV5	180	2048	—
UNNE – TII – HV	60	3000	1000

Table 2.
Characteristic parameters of time series for spectral analysis.

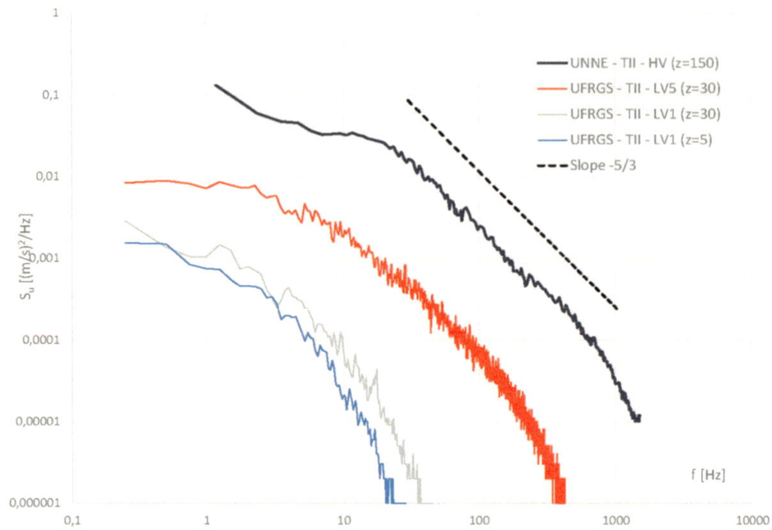

Figure 6.
Spectra for smooth surface experiments (low and high velocity).

The Reynolds number is usually much smaller in the wind tunnel than the real scale, so the equality condition is relaxed. The aim is simply to reach values high enough to obtain the aerodynamic similarity of the phenomenon. Finally, the Richardson number describes the relationship between thermal and mechanical turbulence [16], establishing the difference between thermally stratified and neutral boundary layers.

4.1 Wind tunnel simulation of atmospheric neutral boundary layer

Counihan [17] and Standen [18] have developed simulation techniques particularly suitable for reproducing ABL's in neutral stability employing the roughness, barrier, and blending methods. These techniques make it possible to obtain representations of the boundary layer flows that are generated over rural and urban terrains. Particularly, Standen technique allows the full-depth and part-depth simulations of ABL. Thus, it is possible to change the model scales for the tests.

These physical models of ABL under neutral stability conditions behave well at the relatively high speeds used in wind tunnels for aerodynamic loading studies. Atmospheric dispersion tests must be carried out at lower speeds, and it is normal for exaggerated speed fluctuations to occur in the low-frequency region of the turbulence spectrum. These fluctuations do not reproduce the characteristics of atmospheric wind [19]. For this reason, it is necessary to make a more thorough evaluation of the wind tunnel performance before carrying out diffusion studies [20]. Most studies in wind tunnels are carried out using simulations of neutral ABL; however, atmospheric thermal stratification must be considered in some cases. The drawback of modeling thermal stratification is the requirement of a large investment in equipment for its development. A relatively simple way of considering some non-neutral flow characteristics was proposed by Janssen [21] by modifying the layout and roughness elements of a simulation of neutral characteristics.

Next, some experimental results obtained in ABL simulations carried out in the UFRGS and UNNE wind tunnels are shown and analyzed.

4.2 Experimental analysis on atmospheric boundary layer models

A first ABL simulation method (UFRGS – S1) was applied at the UFRGS wind tunnel. Four perforated spires, a barrier and surface roughness elements were used to simulate a full-depth boundary layer. The arrangement of the simulation hardware is shown in **Figure 7** (left).

Then, an ABL full-depth simulation (UNNE – S2) was implemented in the UNNE tunnel based on the Counihan method. Four high elliptic vortex generators and a barrier were used, together with prismatic roughness elements placed on the test section floor along 17 m (see **Figure 7**—center). Finally, two Irwin-type generators separated 1.5 m were used to simulate the part-depth boundary layer by means of the Standen method (UNNE – S3). The windward plate of the simulator has a trapezoidal shape. The roughness elements distributed on the test section floor are the same that were used for the Counihan method (**Figure 7**—right).

Five experiments were analyzed to characterize the modeled boundary layer flow in each simulation. Different average speeds were evaluated only for simulation UFRGS – S1. Characteristic parameters of the different experiments on ABL simulations are indicated in **Table 3**.

Figure 8 shows the mean speed and turbulence intensity profiles for the five cases analyzed. The vertical z positions are dimensionless (the reference height H_{ref} is the boundary layer thickness). The vertical non-dimensionalization allows a better comparison of the profiles obtained in both wind tunnels and also of the different simulation methods used. The velocity profiles show a fairly harmonic behavior and the values remain consistent with those obtained by the power law using exponents 0.23 and 0.25. The I_u profiles indicate a distortion of the values in the two low-speed cases,

Figure 7.
Simulation hardware: Perforated spires, barrier, and roughness (UFRGS – S1); Counihan method (UNNE – S2); Standen method (UNNE – S3).

Experiment	Position	U_{ref} [m/s]	H_{ref} [m]	Re_H
UFRGS –S1 - HV	Center line	35	0.60	1.40×10^6
UFRGS –S1 – LV1	Center line	1	0.60	4.00×10^4
UFRGS –S1 – LV2	Center line	3.5	0.60	1.40×10^5
UNNE – S2 – HV	Center line	27.5	1.17	2.15×10^6
UNNE – S3 – HV	Center line	25.3	1.21	2.04×10^6

Table 3.
Characteristic parameters of the experiments on ABL models.

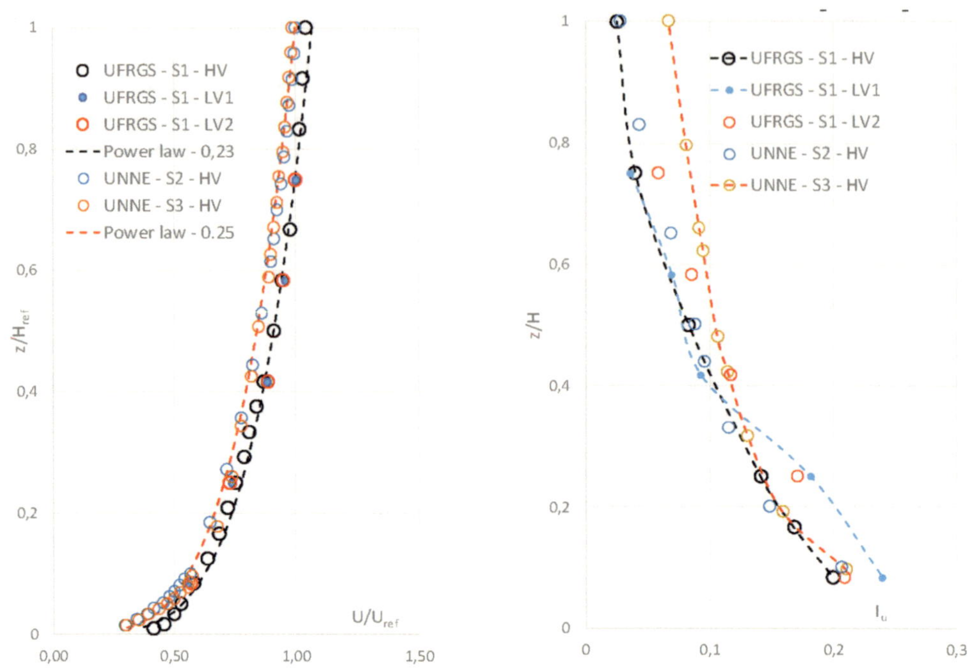

Figure 8.
Mean velocity and turbulence intensity profiles for atmospheric boundary layer models.

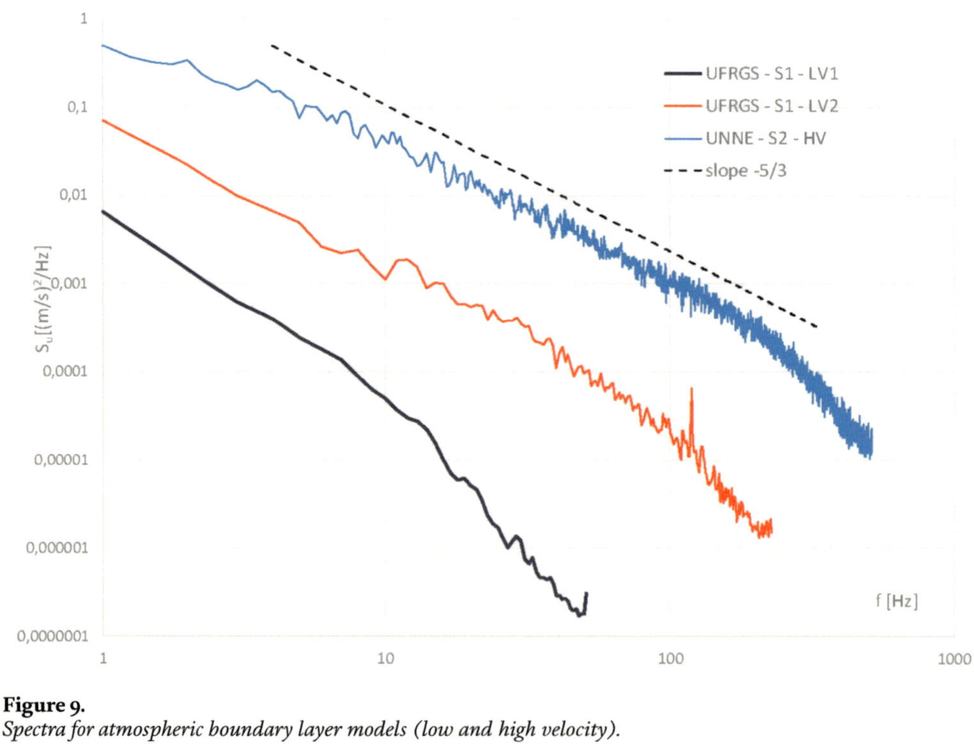

Figure 9.
Spectra for atmospheric boundary layer models (low and high velocity).

mainly in the lowest speed one (UFRGS – S1 – LV1). On the other hand, an increase in turbulence levels is observed from z/H ≈ 0.3 for the profile corresponding to part-depth simulation. This is consistent with the proposed model (part-depth) that aims to simulate the lower part of the ABL where the turbulence intensities are higher.

Two spectra obtained at low speed at the position z/H = 0.25 are shown in **Figure 9** to verify the energy distribution when the mean velocity decreases. Fluctuating velocity records used for the spectral analysis were acquired with a sampling frequency of 1024 Hz. These two spectra are compared with the spectrum obtained at high speed at the position z/H = 0.15. A poor definition of the Kolmogorov's inertial subrange is observed for the spectra measured at velocity U_{ref} = 1 m/s.

The objective of the evaluation of the mean flow, the turbulent parameters, and the spectral analysis is the general comparison between simulations carried out in two different tunnels, implemented with different methods and analyzed for different mean speeds. A comparison with data measured in the atmosphere is not intended. This type of analysis requires the non-dimensionalization of the spectra for their comparison with atmospheric spectra (see [22, 23]).

4.3 Some comments about stratified boundary layer

Most of the wind tunnel simulations do not include some effects measured in field experiments, such as vertical thermal flows due to the presence of solar radiation and other thermal sources [24–26]. The physically relevant quantities in turbulent winds are horizontal and vertical wind speeds, vertical temperature, pressure profile, and humidity distribution. These quantities are subject to fluctuations due to the flow dynamics and due to the complex conditions of the boundary layer [27]. The experiments in the wind tunnel use mechanical and thermal forces to simulate realistic ABL as observed in the literature [28, 29].

A recent study carried out in the UFRGS wind tunnel includes thermal effects in experiments that simulate a boundary layer. To carry out this work, a sector of the wind tunnel test section was equipped with a metal sheet and Peltier elements attached to it. Peltier elements heat or cool the floor of the wind tunnel depending on the polarity. Thus, heating the floor simulates the effects of solar radiation. Thermal effects generating new flow patterns become feasible to realize comparisons of wind tunnel simulations to stratified boundary layer conditions [30].

This work realized in the UFRGS Aerodynamics Laboratory presents and discusses the turbulent energy spectrum of the velocity fluctuations obtained with different mean incident wind velocities [31]. Smooth wind tunnel floor measurements were repeated under the same conditions with floor roughness. The time series of the longitudinal velocity component is determined with hot-wire anemometry and analyzed using the Hilbert-Huang transform method. Both neutral and convective wind tunnel boundary layers are considered in this study.

5. Final considerations

This work presents a compilation of different experiments on turbulent boundary layer flows realized in the UFRGS and UNNE wind tunnels. First, boundary layer flows developed on a smooth surface of the wind tunnel test section are experimentally evaluated; next, some scale models of neutral ABL developed by different methods are analyzed. General characteristics and turbulence spectra are determined considering the effects of the average velocity variation and the influence of the dimensions of the test section. Finally, some studies on thermal effects in turbulent boundary layer flows developed in wind tunnels are analyzed. Specifically, a work recently carried out in the UFRGS tunnel to evaluate the thermal and roughness effects is analyzed and commented.

Acknowledgements

The authors acknowledge the technical personnel of the Laboratório de Aerodinâmica das Construções, Universidade Federal de Rio Grande do Sul (UFRGS, Brazil) and of the Laboratorio de Aerodinámica, Universidad Nacional del Nordeste (UNNE, Argentina).

This research was partially funded by Conselho Nacional de Desenvolvimento Científico e Tecnológico (CNPq, Brazil), Secretaría General de Ciencia y Técnica of the Universidad Nacional del Nordeste (SGCYT-UNNE, Argentina) and Facultad de Ingeniería of the Universidad Nacional del Nordeste (FI-UNNE, Argentina).

Author details

Adrián R. Wittwer[1]*, Acir M. Loredo-Souza[2], Jorge O. Marighetti[1] and Mario E. De Bortoli[1]

1 Facultad de Ingeniería, Universidad Nacional del Nordeste (UNNE), Resistencia, Argentina

2 Universidade Federal de Rio Grande do Sul (UFRGS), Laboratório de Aerodinâmica das Construções, Porto Alegre, Brazil

*Address all correspondence to: a_wittwer@yahoo.es

References

[1] Arya S. In: Plate EJ, editor. Atmospheric Boundary Layers over Homogeneous Terrain. Engineering Meteorology. Amsterdam: Elsevier Scientific Publishing Company; 1982. pp. 233-266

[2] Vento BJO. na Engenharia Estrutural. Editora da Universidade. Porto Alegre. Brazil: UFGRS; 1995

[3] Kaimal JC, Wyngaard JC, Izumi Y, Cote OR. Spectral characteristics of surface-layer turbulence. Quarterly Journal of the Royal Meteorological Society. 1972;**98**:563-589

[4] Kaimal JC. Atmospheric Boundary Layer Flows: Their Structure and Measurement. New York: Oxford University Press, Inc.; 1994

[5] Cook NJ. Determination of the model scale factor in wind-tunnel simulations of the adiabatic atmospheric boundary layer. Journal of Industrial Aerodynamics. 1978;**2**:311-321

[6] Hangan H. The WindEEE research institute and the WindEEE dome. In: Wind EEE Scientific Symposium. London, Canadá: Western University; 2013, 2013

[7] Loredo-Souza AM, Rocha MM, Wittwer AR, Oliveira MK. Modelagem de edifícios altos em túnel de vento. Concreto & Construções. Revista do IBRACON - Instituto Brasileiro do Concreto. 2020;**99**:48-63

[8] Blessmann J. The boundary layer TV-2 wind tunnel of the UFGRS. Journal of Wind Engineering and Industrial Aerodynamics. 1982;**10**:231-248

[9] Wittwer AR, Möller SV. Characteristics of the low speed wind tunnel of the UNNE. Journal of Wind Engineering & Industrial Aerodynamics. 2000;**84**:307-320

[10] Single Sensor Miniature Wire Probes. Dantec Dynamics. 2019. Available online: https://www.dantecdynamics.com/productsand-services/single-sensor-miniature-wire-probes. [Accessed: September 6, 2019]

[11] Vosáhlo L. Computer Programs for Evaluation of Turbulence Characteristics from Hot-Wire Measurements. Karlsruhe: KfK 3743, Kernforschungszentrum Karlsruhe; 1984

[12] Möller S. Experimentelle Untersuchung der Vorgänge in engen Spalten zwischen den Unterkanälen von Stabbündeln bei turbulenter Strömung. Dissertation. Karlsruhe. RFA. 1988; KfK 4501: Universität Karlsruhe (TH); 1989

[13] Press WH, Flannery BP, Teukolsky SA, Vetterling WT. Numerical Recipes: The Art of Scientific Computing. New York: Cambridge University Press; 1990

[14] Jensen M. The model law phenomena in natural wind. Ingeniøren. 1958;**2**: 121-128

[15] Cermak JE. Laboratory simulation of the atmospheric boundary layer. AIAA Journal. 1971;**9**(9):1746-1754

[16] Ruscheweyh H, Fisher K. Aerodinamic effects of large natural-draught cooling towers on the atmospheric dispersion from a stack. Journal of Industrial Aerodynamics. 1979;**4**:399-413

[17] Counihan J. An improved method of simulating an atmospheric boundary layer in a wind tunnel. Atmospheric Environment. 1969;**3**:197-214

[18] Standen NM. A Spire Array for Generating Thick Turbulent Shear Layers for Natural Wind Simulation in Wind Tunnels. National Research Council of Canada, National Aeronautical Establishment (NAE), Laboratory Technical Report LTR-LA-94. 1972

[19] Isymov N, Tanaka H. Wind tunnel modelling of stack gas dispersion – Difficulties and aproximations. Wind engineering. In: Cermak JE, editor. Proceedings of the Fifth International Conference. Fort Collins, Colorado, USA: Pergamon Press Ltd; 1979

[20] Meroney R, Neff D. Laboratory simulation of liquid natural gas vapour dispersion over land or water. Wind engineering. In: Cermak J, editor. Proceedings of the Fifth International Conference. Vol. 2. Fort Collins, Colorado, USA: Pergamon Press Ltd; 1980. pp. 1139-1149

[21] Janssen LAM. Wind tunnel modelling of dispersion of Odours in the neighborhood of pig houses. Journal of Industrial Aerodynamics. 1979;**4**:391-398

[22] Wittwer A, Welter G, Loredo-Souza A. Statistical analysis of wind tunnel and atmospheric boundary layer turbulent flows. In: Wind Tunnel Designs and their Diverse Engineering Applications, book edited by N. A. Ahmed, ISBN 978-953-51-1047-7. London, UK: INTECH; 2013

[23] Wittwer A, Loredo-Souza A, De Bortoli M, Marighetti J. Physical models of atmospheric boundary layer flows: Some developments and recent applications. Book chapter in: Boundary Layer Flows - Theory, Applications and Numerical Methods (ISBN 978-1-83968-186-8). Book edited by: Vallamapati Ramachandra Prasad. London, UK: INTECH OPEN; 2019

[24] Chamorro LP, Porté-Agel F. Effects of thermal stability and incoming boundary-layer flow characteristics on wind-turbine wakes: A wind-tunnel study. Boundary-Layer Meteorology. 2010;**136**:515-533

[25] Demarco G, Puhales F, Acevedo OC, Costa FD, Avelar AC, Fisch G. Dependence of turbulence-related quantities on the mechanical forcing for wind tunnel stratified flow. American Journal of Environment Enginerring. 2015;**5**:15-26

[26] Puhales FS, Demarco G, Martins LGN, Acevedo OC, Degrazia GA, Welter GS, et al. Estimates of turbulent kinetic energy dissipation rate for a stratified flow in a wind tunnel. Physica A: Statistical Mechanics and its Applications. 2015;**431**:175-187

[27] Arya SPS. Buoyancy effects in a horizontal flat-plate boundary layer. Journal of Fluid Mechanics. 1975;**68**:321-343

[28] Williams O, Hohman T, Van Buren T, Bou-Zeid E, Smits AJ. The effect of stable thermal stratification on turbulent boundary layer statistics. Journal of Fluid Mechanics. 2017;**812**:1039-1075

[29] Doosttalab A, Araya G, Newman J, Adrian RJ, Jansen K, Castillo L. Effect of small roughness elements on thermal statistics of a turbulent boundary layer at moderate Reynolds number. Journal of Fluid Mechanics. 2016;**787**:84-115

[30] Kabir IFSA, Ng E. Effect of different atmospheric boundary layers on the wake characteristics of NREL phase VI wind turbine. Renewable Energy. 2019;**130**:1185-1197

[31] Demarco G, Martins LGN, Bodmann BEJ, Puhales FS, Acevedo OC, Wittwer AR, et al. Analysis of thermal and roughness effects on the turbulent characteristics of experimentally simulated boundary layers in a wind tunnel. International Journal of Environmental Research and Public Health. 2022;**19**:5134

Chapter 8

Hypersonic Flow over Closed and Open Nose Missile Bodies: Raw and SVD-Enhanced Schlieren Imaging, Numerical Modeling, and Physical Analysis

Tyler Watkins, Jesse Redford, Franklin Green, Jerry Dahlberg, Peter Tkacik and Russell Keanini

Abstract

Understanding and mitigating against high heat loads at leading and blunt aerodynamic surfaces during hypersonic flight represents an ongoing technological challenge. Recent work has shown that the commercial software package, STAR CCM+, can provide reliable predictions of hypersonic aerothermodynamic flow and heating, under a wide range of complex, but common conditions. This chapter presents a preliminary experimental and numerical investigation of hypersonic flow over closed- and open-nose missile bodies, where the latter have been proposed as a means of reducing leading edge heat transfer. Four contributions are presented. First, a novel singular value decomposition (SVD)-based image processing technique is introduced, which significantly enhances the quality of raw schlieren images obtained in high-speed compressible flows. Second, numerically predicted hypersonic flow about a scale-model missile body, obtained using STAR-CCM+, is validated against experimental schlieren image data, an empirical correlation connecting bow shock stand-off distance and shock density ratio, and estimated drag forces. Third, scaling and physical arguments are presented as a means of choosing appropriate gas equations of state and for interpreting results of numerical simulations and experiments. Last, numerical experiments show that the forward facing cavity used in our wind tunnel experiments functions as a heat sink, reducing heat fluxes on the missile body downstream of the cavity.

Keywords: hypersonics, aerothermodynamics, forward facing cavity, hypersonic heat loads

1. Introduction

Hypersonic flight is often described as flight at free stream Mach numbers exceeding five. However, from a physical standpoint, hypersonic flight refers to

Figure 1.
Hypersonic glide vehicle, leading edges experiencing high friction and heat loading [1].

conditions in near-body boundary layers where a variety of complex, Mach-number dependent processes can emerge, including non-equilibrium flow of gas constituents, ionization and recombination of gas species, emission and absorption of radiation both within the gas and at the aerodynamic surface, difficult-to-predict laminar-to-turbulent boundary layer flow transition, conversion of high gas enthalpy and kinetic energy into high intensity surface heating, thermal ablation of aerodynamic surfaces, generation of high frequency screech modes, excitation of high frequency in-solid flutter, and various shock-boundary layer interactions.

Predicting and mitigating against intense surface heating and surface stress, particularly at the leading edges of aerodynamic bodies, remains a central challenge in hypersonic aerodynamics. High heat loads and stresses can ablate the leading edge, destabilizing the high speed flow, threatening the integrity of the aerodynamic body, and placing the body well outside its design envelope.

These phenomena are visually represented in **Figure 1**. As shown, large heat loads exist along all leading edges, being highest at the nose and decaying, due to boundary layer thickening and reduction of viscous heating, with distance from the nose.

The aerothermodynamic phenomena that arise during hypersonic flight make for very difficult mechanical design. This study focuses on the problem of leading edge heating. Here, a potential approach for mitigating against leading edge heating, introduction of a forward facing cavity, placed at the nose of a missile-shaped test body, is investigated in preliminary fashion.

1.1 Chapter overview and main results

Due to the complexity of hypersonic aerodynamic flows, numerical modeling and experimental diagnostics remain active areas of research. This Chapter presents a preliminary experimental and numerical investigation of heat transfer and flow structure produced by hypersonic flow over a missile shaped body, having both a solid nose and an on-nose forward-facing cavity.

An overview of the work performed and the results obtained is as follows:

a. Hypersonic flow experiments are carried out in the Hypersonic Wind Tunnel facility recently designed and built at UNC Charlotte [2–4]. Time-dependent schlieren images are obtained for Mach 4.5 flow about a small, solid nosed, missile-shaped body, as well as for Mach 3.5 flow over the same body, modified with an on-nose forward facing cavity.

b. A numerical model simulating these experimental flows is developed using STAR CCM+, a commercial software package that has recently been validated [5] against experimental heat transfer measurements in a range of hypersonic flows, including laminar and turbulent boundary layer flows [5], flows dominated by large separation bubbles [5], and flows featuring both boundary layer-shock interactions [5] and shock-shock interactions [5]. The present study appears to be one of the first to implement the modeling best practices documented in Cross and West's wide ranging investigation [5]. Following [5], we highlight procedures for setting up high fidelity adaptive grids, setting up grids for resolving turbulent boundary layers, and for diagnosing solution convergence.

c. The Chapter introduces a new technique for significantly enhancing Schlieren images obtained in high speed flows. The technique replaces an array of raw, time-dependent, pixelated Schlieren images with a k-rank singular value image decomposition (SVD) [6], which suppresses static background and enhances time-varying flow features, such as oscillating bow shocks and body-generated Mach waves.

d. The numerical model is validated, in preliminary fashion, via three tests: (i) the computed drag force on the solid nose test article is compared against the drag estimated via Newton's drag law, (ii) computed schlieren images of the near-body density gradient field are compared against experimental schlieren images, and (iii) computed bow shock standoff distances are compared against a correlation that connects the standoff distance to the cross-shock density ratio.

e. The complexity of hypersonic aerodynamic flows suggests that order of magnitude analyses can play an important role in interpreting experimental results, and in testing the physical consistency of numerical solutions. Here, we use scaling arguments to: (i) explain Newton's drag law for hypersonic flow [7], and (ii) show that time scales for relaxation of excited vibration modes in shock layer molecular N_2 and O_2 exceed (N_2), and are of the same order of magnitude (O_2) as the shock layer flow time scale.

2. Experiment: hypersonic flow over solid nose and open nose missile shapes

Between 2019 and 2021, a blow down hypersonic wind tunnel was designed, built and tested at UNC Charlotte [2–4]; see **Figure 2**. The wind tunnel facility, pictured in **Figure 2**, consists of a modular/interchangeable bench top de Laval nozzle, a rectangular 25 in^2 cross-section test section, a diffuser, and near-exit silencer/noise baffle. The test section, 49.5 cm in length, 27.3 cm wide and 25.4 cm in height, has a removable top plate, allowing convenient placement of aerodynamic models within the test section. The test chamber is fitted with a pressure sensor, two thermocouples,

Figure 2.
UNC Charlotte hypersonic wind tunnel [2, 3, 8].

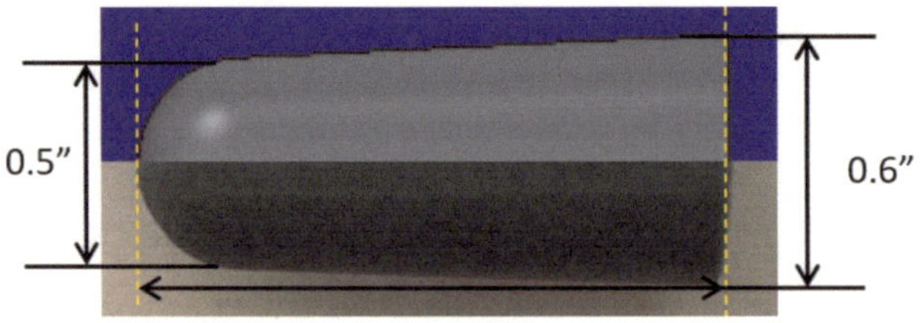

Figure 3.
Solid nose model geometry used in wind tunnel experiments and numerical models.

and a reinforced static pressure Pitot tube which can be swept across the test chamber inlet. To allow visual access, the lateral sides of the test chamber are fitted with square (15 cm × 15 cm) high strength glass windows.

Compressed room air is stored in twelve high pressure cylinders, each connected to a single high pressure manifold, located immediately upstream of the nozzle plenum. The cylinders, rated for pressures up to 10,000 psi, are charged before each run using a commercial compressor. Prior to any given experimental run, a pressure-actuated, fast response ball valve, separating the manifold and plenum, is closed and manifold pressure is set at a magnitude that produces a pre-specified run time plenum pressure. Plenum/stagnation pressure and temperature are monitored using an in-plenum pressure sensor and thermocouple.

A schlieren system is used to image density gradient fields within the test section; experimental details are available in [3, 4, 8]. For this study, density gradient fields were imaged about a missile body having a solid nose, shown in **Figure 3**, and an open nose missile body, shown in **Figure 4**.

2.1 Test article geometry

As depicted in **Figure 3**, the solid nose test piece investigated in this study is a pseudo-blunt body that mimics a missile-shaped body. The total length is 1.35 inches (3.43 cm), the forward diameter is 0.5 inches (1.26 cm), and the aft diameter is 0.6 inches (1.52 cm). The same body dimensions and shape are used in construction of all numerical models. Geometric specifications for the open-nose/forward-facing cavity body are shown in **Figure 4**. A simple cavity shape was chosen in which the depth and diameter of the circular cavity are the same, 0.2 inches (0.508 cm). Again, these dimensions are used in all numerical models of flow over the open-nose body.

Figure 4.
Open nose/forward facing cavity geometry used in experiments and numerical models.

3. Experimental results

Density gradient fields and test chamber pressure histories captured during Mach 4.5 flow over the solid nose missile body are shown in **Figure 5**. The rapid rise and fall in chamber Mach number is produced by opening and closure of the fast response (0.25 s response time) ball valve separating the high pressure manifold and nozzle plenum. The noisy variations in pre- and post-run Mach numbers reflect leakage of high pressure air from the pressurized manifold through the ball valve into the nozzle plenum.

The raw schlieren image clearly indicates the presence of a bow shock, as well as Mach waves emanating from the quasi-cylindrical portion of the missile body. The latter are induced by small grooves in the body surface, produced during machining of the shape. In addition, a (conical) expansion fan is indicated at the trailing (circular) base of the missile body. The somewhat coarse nature of the images, here and below, reflects the nature of these early experiments, designed to shake down and optimize the wind tunnel. Note, that the solid body on the aft side of the test body is a stinger to which the body is attached.

Similar results, shown in **Figure 6**, are observed for Mach 3.5 flow past the open-nose missile body. Here, the schlieren image is darker and more monochrome than in **Figure 5**, reflecting slight misalignment of the schlieren system. Due to failure of an early Pitot tube at Mach 4.5, these tests were run at Mach 3.5.

3.1 Singular value decomposition (SVD) compression for improved schlieren imaging

Singular value decomposition (SVD) has found wide application in image processing [9], where the common theme centers on obtaining a sparse

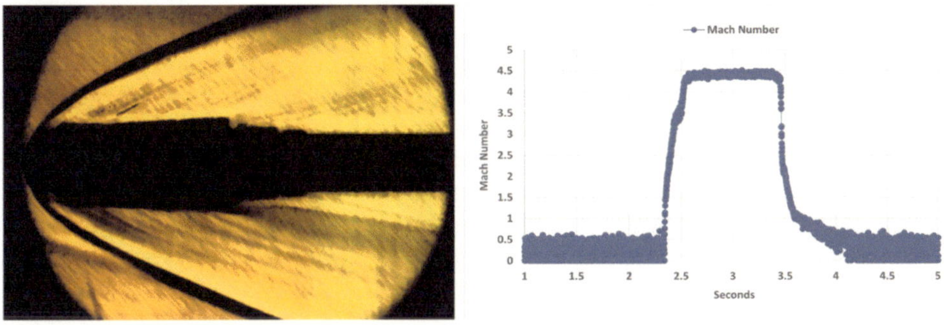

Figure 5.
Schlieren imaging and Mach number measurements in Mach 4.5 flow over solid nose missile body.

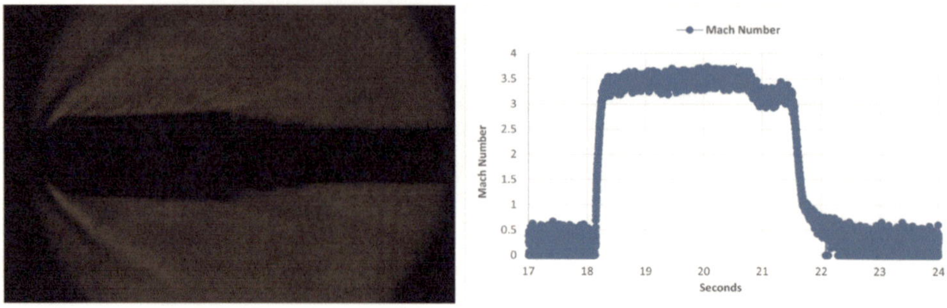

Figure 6.
Schlieren imaging and Mach number measurements in Mach 3.5 flow over open nose missile body.

representation, via SV decomposition, of a raw image or a time sequence of raw images, i.e., a video. This study introduces what we believe to be the first application of SVD to the enhancement of raw schlieren images, here obtained in our hypersonic wind tunnel experiments.

The technique, described in [6], proceeds as follows:

a. Raw, digitized schlieren video images, taken during any given experimental run, are sequentially arranged into a time series matrix, $\tilde{A}$, of dimension $M^*N \times T$, where each (nominally instantaneous) vectorized digital video frame contains $M \times N$ pixels, and where T is the total number of frames obtained during the experiment.

b. Perform SV decomposition on matrix $\tilde{A}$.

c. Remove static (time-invariant) features from $\tilde{A}$ by performing a rank-k background subtraction,

$$\hat{A} = \tilde{A} - \tilde{A}_k \tag{1}$$

where $\tilde{A}_k$ is the rank-k approximation of $\tilde{A}$, and where $k \le T \le M^* N$. For imaged processes in which a 'visually dominant' static background obscures time-varying events, this step suppresses the background, effectively enhancing obscured, time-varying features [6].

Applying this image enhancement technique to our raw schlieren video images, single frames of which are shown in **Figures 5** and **6**, we obtain the images in **Figures** 7 and 8. Roughly speaking, in the present application, choosing a larger rank k effectively removes more static background. Comparing **Figure 5** with **Figures** 7 and 8, we observe that an appropriate selection of rank k significantly improves resolution of the density gradient fields extant in our experiments. This feature is particularly apparent in **Figure 8**, where large scale unsteadiness and turbulence—apparently reflecting high-frequency fluid oscillations in and near the on-nose cavity (see below)—appears to produce an oscillating bow shock, as well as turbulent flow downstream of the shock.

Further details and results obtained by this method will be reported in a separate publication.

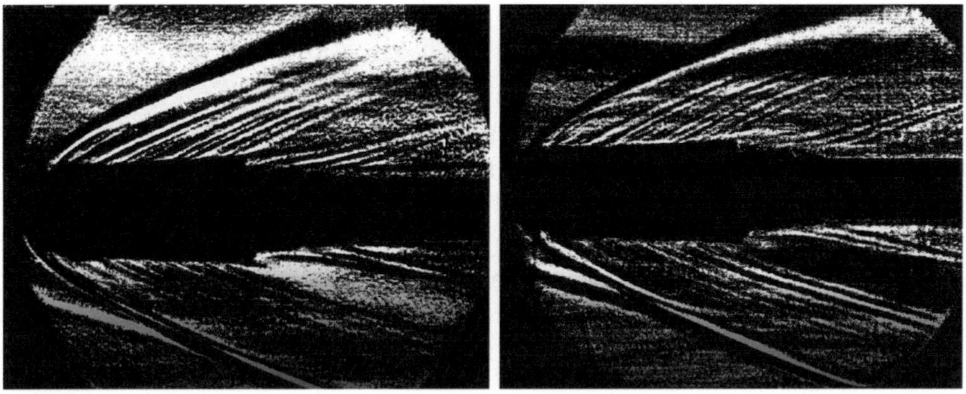

Figure 7.
SVD-enhanced imaging of Mach 4.5 flow over closed nose missile body; left image is a rank 1 representation of a raw schlieren image and the right is a rank 20 representation.

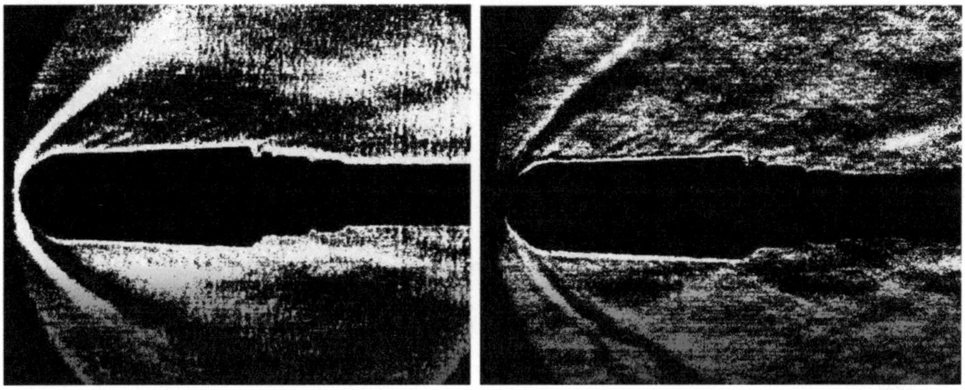

Figure 8.
SVD-enhanced imaging of Mach 3.75 flow over open nose missile body; left image is a rank 1 representation of a raw schlieren image and the right is a rank 20 representation.

4. Computational model

Computational models of unsteady, three-dimensional hypersonic flow over the closed and open nose missile bodies used in our experiments were developed. As noted, and based on the recent validation studies reported by Cross and West [5], the commercial CCM+ package was employed. The objectives of this effort were two-fold:

a. Determine appropriate turbulence models and adaptive meshing strategies that provide consistent predictions of experimentally observed density gradient fields, as revealed by schlieren imaging.

b. Use the experimentally validated computational model to investigate, in preliminary fashion, surface heat transfer to closed and open nose missile bodies.

A secondary objective centered on investigating the effects of three gas models – ideal gas, equilibrium real gas, and non-equilibrium, two-specie (nitrogen and oxygen) models – on computed flow fields and surface heat transfer.

In this section, we briefly describe the general flow model, then highlight three modeling features that are crucial to obtaining physically reasonable computed results, and finally discuss a limitation associated with both the present model and the model developed by Cross and West [5].

4.1 Model description

The model solution domain is depicted in **Figure 9**. The missile body is placed forward of center in a spherical domain. As is often the case in aerodynamic flow models, the size of the solution domain is somewhat arbitrarily chosen; the goal in this study is to make the domain large enough that free stream boundary conditions can be reasonably imposed on the far field boundary. [Model validation against experimental data, here, experimental schlieren images, implies that the chosen domain size is appropriate.]

No slip and no penetration conditions are imposed on the missile body surface. Mirror flow (and thermal transport) symmetry is assumed about any plane passing through the center of the sphere. Thus, on the (deep blue) circular symmetry boundary, derivatives of all field variables with respect to the azimuthal angle, ϕ, are zero. Based on the experimentally validated STAR-CCM+ hypersonic flow simulations in a study conducted by Cross and West [5], the full, variable property Navier-Stokes equations, including the continuity and energy equations, are solved. Technical details concerning model set-up, which followed the best practices outlined in [5], can be found in [8].

4.2 Adaptive meshing and resolution of shocks and turbulent boundary layers

STAR-CCM+ provides a number of utilities that allow high fidelity modeling of stationary hypersonic aerodynamic flows. The most important of these, adaptive meshing, iteratively refines meshes in regions where pressure gradients are high, e.g., in and near shocks. Similarly, in high velocity gradient regions, e.g., turbulent boundary layer viscous sublayers, buffer layers, and logarithmic regions, STAR CCM+ continuously monitors and alters mesh thicknesses.

STAR-CCM+ constructs stationary turbulent solutions of the Navier-Stokes equations in two steps. First, beginning from a specified initial condition and a user-

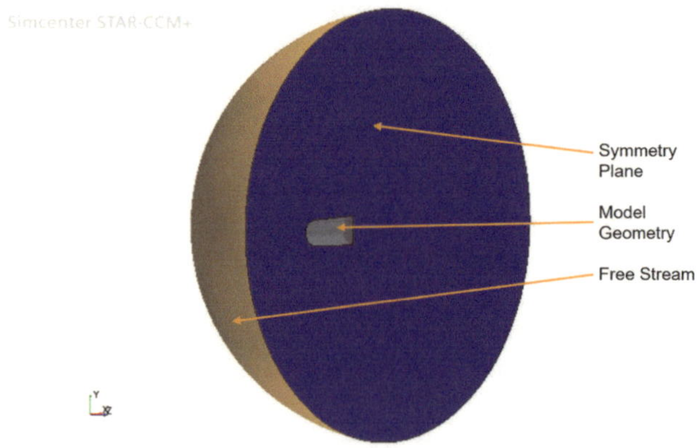

Figure 9.
Solution domain.

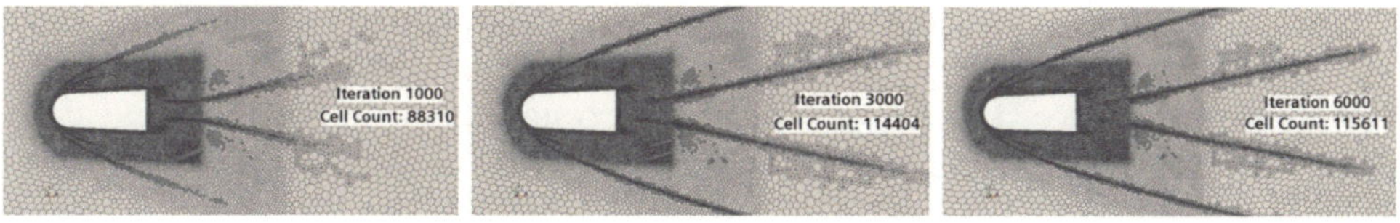

Figure 10.
Automated mesh refinement based on inviscid flow solution, obtained at 1000, 3000 and 6000 iterations, respectively; the final inviscid mesh provides the initial mesh for the complete viscous solution.

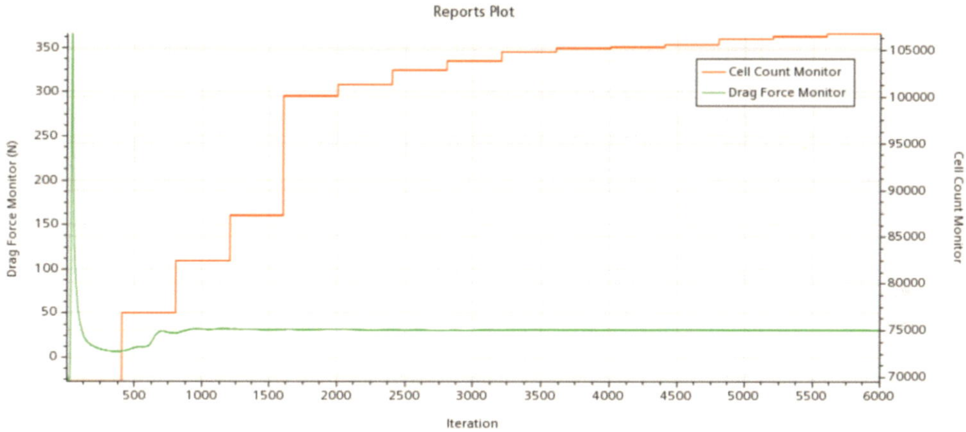

Figure 11.
Solution convergence is determined by monitoring mesh cell count and net computed drag during iterative solution of the full viscous flow problem.

specified initial coarse mesh, STAR CCM+ incrementally increases the free stream Mach number, solves the inviscid Navier-Stokes equations, and, using computed pressure gradient and velocity fields, refines the mesh in high pressure and velocity gradient regions. The initial condition and mesh can be somewhat arbitrarily specified: a global zero or low speed velocity can be imposed, for example, while the initial mesh is constructed via a few user-specified mesh parameters [5]. Examples of inviscid construction of the initial mesh are shown in **Figure 10**.

Once an inviscid solution is obtained at the desired free stream Mach number, STAR CCM+ then iteratively solves the full Navier-Stokes equations, continuing to adapt the mesh as a converged stationary solution is approached. Regarding convergence, in all numerical experiments, we monitor both computed drag force and mesh cell count, stopping a simulation when these have reached nominally steady magnitudes. **Figure 11** shows an example.

Cross and West [5] provide detailed guidance on choosing software settings designed to ensure third order solution accuracy (except within shocks, where second order accuracy is achieved), solution stability, and proper resolution of boundary layers.

4.3 Qualitative validation of computational model

Comparing model predictions against experimental data represents the gold standard in code validation. In this study, we are limited to presenting three semi-quantitative checks and one consistency check on our computational model.

First, based on the experimental stagnation temperature and pressures, T_{01} and P_{01}, and the missile body geometry used in the model, the predicted drag force on the body, shown in **Figure 11**, is $D_{model} \approx 30$ N. As a rough check on this result, we can estimate the actual drag, D_{est}, as follows. First, we note that the pressure at the base of the body is approximately equal to the free stream/test section pressure, $P_{base} \sim P_1$. This is shown by recognizing that the flow over the trailing edge of the quasi-cylindrical missile produces a thin shear layer immediately downstream of the body. A (conical) expansion fan emanating from the trailing edge bends the shear layer inward toward the body's axial centerline. Thus, define a local coordinate system within the shear layer, labeling the cross-layer coordinate as n, and layer-parallel coordinate as t, where n and t are orthogonal. Consider the time-average n-component of the momentum equation. Since the average velocity in the n-direction is negligible, and the turbulent flow is stationary, time-average n-direction inertia and time-average viscous forces are likewise negligible. Thus, the time-average pressure gradient in the n-direction—across the shear layer—is small, on the order of the n-direction gradient in (turbulent) Reynolds stresses: $P_{base} \sim P_1$.

Next, estimate the drag force on the body as

$$D_{exp} \sim \left(P_1 + \rho_1 u_1^2\right) \cdot A - P_{base} \cdot A \tag{2}$$

or

$$D_{exp} \sim P_{01} \cdot A - P_1 \cdot A \tag{3}$$

where P_1, ρ_1, and u_1 are the free stream/test section pressure, density, and velocity upstream of the bow shock, and A is the projected area of the missile body (in the direction of the free stream flow). For isentropic flow between the nozzle plenum and upstream face of the bow shock, and for the test section Mach number of 4.5, $P_{01}/P_1 \approx 289$. Thus, $D_{est} \sim P_{01} \cdot A \approx 27$ N, or,

$$D_{model} \sim D_{est} \tag{4}$$

As a second check, and as shown in **Figure 12**, we compare a computed density gradient field, as indicated by the numerical schlieren image shown, against a corresponding (raw) experimental schlieren image, both obtained for Mach 4.5 flow over the solid nose missile body. Comparing the distances from the bow shock nose to the locations on the upper and lower image boundaries where the bow shock meets the boundary, we find that

$$\frac{|L_{upper,exp} - L_{upper,model}|}{L_{upper,exp}} \approx 0.02 \tag{5}$$

and

$$\frac{|L_{lower,exp} - L_{lower,model}|}{L_{lower,exp}} \approx 0.02 \tag{6}$$

Since computed density gradient fields require accurate solutions for velocity, pressure, temperature, and density fields, this rough comparison indicates the physical fidelity of computed results. Note that the validation case implemented the ideal

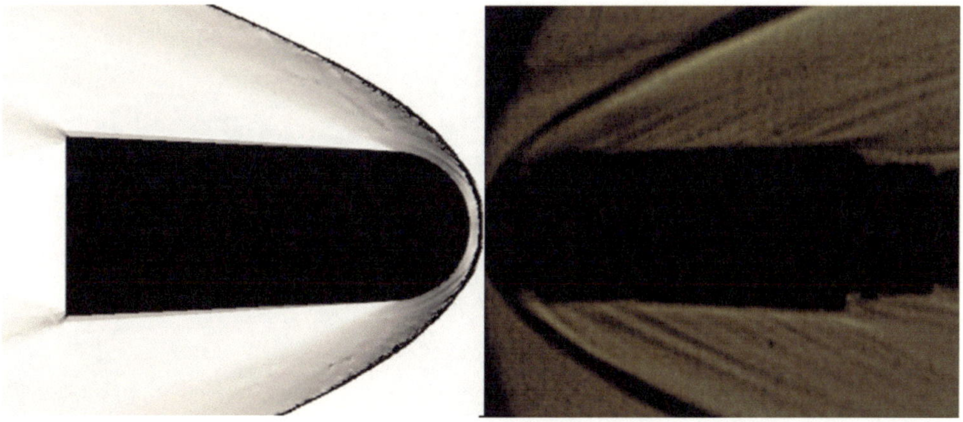

Figure 12.
Numerical Schlieren image versus experimentally observed Schlieren for Mach 4.5 flow over solid nose missile body.

gas equation of state. As discussed below, improved results would likely be observed using STAR-CCM+'s two-specie nonequilibrium gas model.

As a third check on our model, we compare computed shock displacement distances, $\Delta_{bow,model}$, against an order of magnitude estimate, Δ_{est}. As shown in **Table 1**, all three gas models, for both the closed nose and open nose missile bodies, predict $\Delta_{model} \approx 1$ mm. A simple order of magnitude estimate for Δ follows from a theoretical expression for shock standoff distance adjacent blunt bodies [10–12]

$$\frac{\Delta}{R_{nose}} = 0.82\frac{\rho_1}{\rho_2} \tag{7}$$

where ρ_1 and ρ_2 are densities immediately upstream and downstream of the (locally normal) shock, and R_{nose} is the radius of the missile body nose. Assuming isentropic flow between the nozzle plenum and test section,

$$\rho_1 = \frac{P_{01}}{RT_{01}}\left[1+\frac{k-1}{2}M_1^2\right]^{\frac{-1}{k-1}} \tag{8}$$

while ρ_2 is obtained using the normal shock relations, given ρ_1 and M_1.

Geometry	Equation of state	Shock detachment distance (mm)
No cavity	Ideal gas	0.9
	Real gas	1.0
	Thermal non-equilibrium	1.1
Cavity	Ideal gas	1.1
	Real gas	1.2
	Thermal non-equilibrium	1.5

Table 1.
Shock standoff distance as predicted by each equation of state.

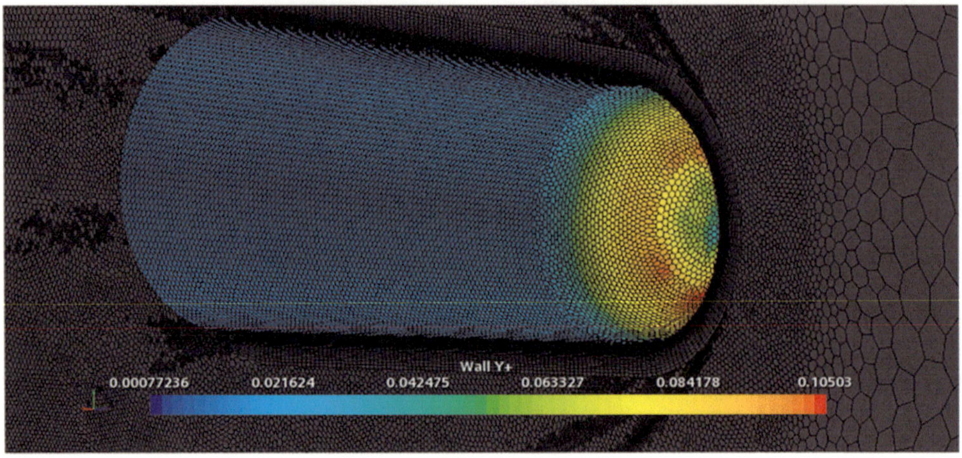

Figure 13.
Distribution of at-surface viscous sublayer Reynolds numbers, as given by y_o^+—solid nose missile body.

Using experimental/model parameters: $P_{01} = 1.63$ Pa, $T_{01} = 291$ K, and $M_1 = 4.5$ (for the closed nose body), we obtain

$$\Delta_{est} \sim 1 \text{ mm} \tag{9}$$

which, as shown in **Table 1**, is comparable to Δ magnitudes predicted by all three gas models.

Finally, a consistency check on the model's resolution of the body-adjacent turbulent boundary layer is obtained via plots of the dimensionless thickness, y_o^+, of the first mesh layer adjacent the body. The parameter $y_o^+ = y_o/\delta_{sublayer}$, represents the Reynolds number associated with the first mesh layer (located well within the viscous sublayer), where y_o is the dimensional thickness of the first cell, $\delta_{sublayer} = \nu/u^*$, is the characteristic sublayer thickness, and $u^* = \sqrt{\tau_{wall}/\rho}$ is the friction velocity. Proper resolution of the viscous sublayer, and by implication, the entire turbulent boundary layer, is indicated by y_o^+ magnitudes smaller than 1. A typical plot of surface y_o^+ magnitudes, showing $y_o^+ < \sim 0.1$ at all locations, is shown in **Figure 13**.

5. Computational experiments

Experimental test section Mach numbers range from 3.5 to 4.5. Thus, according to [5, 13], molecular vibration modes are collisionally excited in N_2 and O_2 molecules, both within the test section and downstream of the bow shock. In this section, we first present scaling arguments to: (a) explain the Mach number-vibration nonequilibrium criterion in [5], (b) estimate the time scale, τ_{rlxn}, for relaxation of excited vibration modes, and (c) show that τ_{rlxn}, for N_2 is orders of magnitude longer than the (bulk) flow time scale, τ_{flow}, downstream of the bow shock.

Importantly, the scaling arguments indicate that the two-specie (N_2 and O_2) nonequilibrium gas model in STAR CCM+ is the appropriate equation of state—as opposed to available ideal gas and real gas models—for investigating flow and heat transfer about our closed and open nose missile bodies.

5.1 Scaling arguments: physical origin of near-nose high temperature gas region, vibrational excitation of molecular N_2 and O_2, and estimated (long) relaxation times

5.1.1 Physical origin of near-nose high temperature gas region

Anderson [13] provides a useful graph indicating temperature ranges over which vibrational excitation, dissociation ($\mathrm{N_2} \rightarrow 2\mathrm{N}$ and $\mathrm{O_2} \rightarrow 2\mathrm{O}$), and ionization.

($\mathrm{N} \rightarrow \mathrm{N^+} + \mathrm{e^-}$ and $\mathrm{O} \rightarrow \mathrm{O^+} + \mathrm{e^-}$) (in air at 1 atm) take place. Theoretically, minimum, pressure-dependent temperatures marking initiation of these processes can be worked out using molecular collision theory; see, e.g. [14].

As shown in **Figure 14**, for Mach 4.5 flow over the solid missile body, all three gas models predict similar temperature distributions, with most of the thin gas layer between the shock and body exhibiting temperatures well in excess of 800 K. Physically, there are three possible sources generating this high temperature gas layer:

a. viscous heating within the bow shock, produced by a large reduction in gas velocity—of order 10^2 m/s, as estimated via simple normal shock theory—over a shock thickness of order of 0.1 μm [15];

b. viscous heating within the gas layer; and

c. pressure heating, taking place between the upstream face of the bow shock, where $P_1 \approx 0.58$ atm, and the missile body nose, where $P = P_{02} \approx 15$ atm.

Focusing on the energy equation and introducing appropriate length, time, and velocity scales, estimates for temperature increases produced by the first two mechanisms are obtained by balancing the temporal change in fluid enthalpy, against the dominant (streamwise) term in the viscous dissipation function:

$$\rho C_p \frac{\partial T}{\partial t} \sim \mu \frac{\partial^2 u}{\partial x^2} \tag{10}$$

Estimating the time scales for mechanisms (a) and (b), respectively, as $\tau_{shock} \sim \delta_{shock}/u_1$, and $\tau_{gaslayer} \sim \Delta/u_2$, we obtain associated estimated temperature increases as follows:

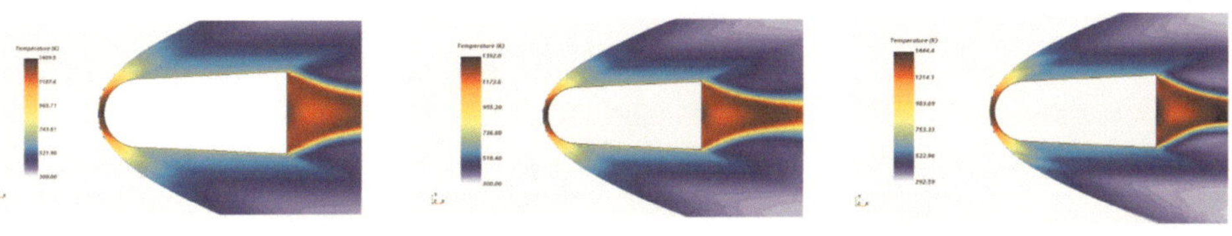

(a) Ideal gas (b) Real gas (c) Two Species Model

Figure 14.
Computed temperature distributions for Mach 4.5 flow over solid nose missile body.

$$\Delta T_{viscshock} \sim \frac{\mu_1}{\rho_1 C_{p1}} \frac{u_1}{\delta_{shock}} \sim 10 \text{ K} \tag{11}$$

$$\Delta T_{viscgaslayer} \sim \frac{\mu_2}{\rho_2 C_{p2}} \frac{u_2}{\Delta} \sim 0.01 \text{ K} \tag{12}$$

Clearly, viscous dissipation, both within the shock and within the near-body gas layer, plays a minimal role in generating the high temperatures observed between the shock and nose.

By contrast, balancing the temporal enthalpy change (of a fluid particle) against the temporal pressure heating term,

$$\rho C_p \frac{\partial T}{\partial t} \sim \frac{\partial P}{\partial t} \tag{13}$$

and again estimating the gas layer (flow) time scale as $\tau_{gas\ layer} \sim \Delta/u_2$ (which cancels), we obtain:

$$\Delta T_{pressure} \sim \frac{P_{02} - P_2}{\rho_2 C_{p2}} \sim 10^2 \text{ K} \tag{14}$$

where again, from normal shock theory, $P_{02} \approx 15$ atm and $P_1 \approx 0.56$ atm. Since $\Delta T = T_{surf} - T_2$, where $T_2 \approx 300$ K, then the estimated maximum temperature, T_{surf}, observed at the missile body surface, is approximately 400 K. While this is approximately one third to one fourth of the maximum temperature magnitudes observed in our numerical experiments, given the approximate nature of our estimate, we can plausibly argue that pressure heating plays the dominant role in generating near nose, elevated gas temperature and missile surface heat transfer.

A similar analysis helps explain the high temperature gas layer observed in flow over the open nose missile body. See **Figure 15**.

5.1.2 Slow relaxation of collision-induced molecular vibration modes

Within the near-nose gas layer, all three gas models predict temperatures well in excess of the approximate (1 atm) 800 K threshold [5, 13] where vibration modes in N_2 and O_2 become significantly excited. Two questions immediately arise:

a. What is the physical origin of the criterion in [5, 13] concerning the temperature-dependent threshold for significant vibrational excitation?

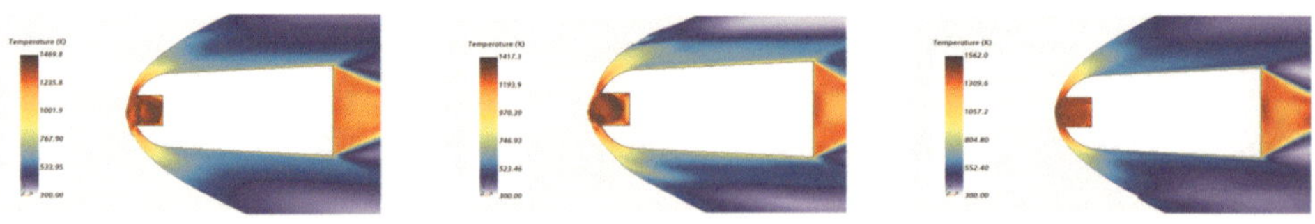

(a) Ideal gas (b) Real gas (c) Two Species Model

Figure 15.
Computed temperature distributions for Mach 4.5 flow over open nose missile body.

b. Since the mean pressure in the layer is on the order of 10 atm ($\sim [P_{01} + P_1]/2$), how does pressure affect the 800 K excitation threshold?

Since collisions induce vibrational excitation, then considering the equilibrium kinetic energy,

$$e_{kinetic} = 3k_B T/2 \tag{15}$$

and equilibrium vibrational energy [16]

$$e_{vib} = \frac{h\nu}{2} + \frac{h\nu}{\exp\left(h\nu/k_B T\right) - 1} \tag{16}$$

of individual N_2 and O_2 molecules at temperature T, then for $T = 800$ K,

$$e_{kinetic}(800\ \text{K}) \approx e_{vib}(800\ \text{K})/2 \tag{17}$$

where k_B, h, and ν are, respectively, Boltzmann's constant, Planck's constant, and the ground state vibration frequency. Thus, in an order of magnitude sense, the 800 K threshold corresponds to the temperature at which kinetic energy is high enough to excite significant vibration.

Considering the second question, since $e_{kinetic}$ and e_{vib} only depend on T, while the collision frequency is proportional to pressure, $\nu_{col} \propto P/\sqrt{T}$ [5], we see that the 800 K excitation threshold does not significantly shift in the high temperature/elevated pressure gas layer.

5.1.3 Slow vibrational relaxation of N_2 and O_2 in the near-nose region

Considering relaxation of vibrationally excited N_2 and O_2 within and downstream of the high temperature near-nose gas region, the Landau-Teller model of vibrational relaxation [17], provides a relatively straightforward description. In order to estimate the time scale, τ_{rlxn}, on which excited molecules relax to a new, elevated equilibrium temperature, the following empirical relationship can be used [13]:

$$\tau_{rlxn} = \frac{C_1}{P} \exp\left(C_2/T\right)^{1/3} \tag{18}$$

where, for N_2 and O_2, the empirical constants are given by: $C_{1,N_2} = 7.12 \times 10^{-3}$ atm $-\ \mu$s, $C_{1,O_2} = 5.42 \times 10^{-5}$ atm $-\ \mu$s, $C_{2,N_2} = 1.91 \times 10^{6}$ K, and $C_{2,O_2} = 2.95 \times 10^{6}$ K. Using $P \sim P_{02} \approx 15$ atm, $T \sim 10^3$ K (based on numerically observed near-nose temperatures), approximate relaxation times are:

$$\tau_{rlxn,N_2} \sim 10^2\ \mu\text{s} \quad \tau_{rlxn,O_2} \sim 10\ \mu\text{s} \tag{19}$$

Importantly, in the high temperature, near-nose gas region, the relaxation time for N_2, comprising 78% (mole fraction) of air, is an order of magnitude longer than the bulk flow time scale, $\tau_{flow} = \tau_{gas\ \ layer} \sim \Delta/u_2$, where $u_2 \sim u_1\rho_1/\rho_2 \approx 1.2\ \Delta/R_{nose}$ Eq. (7), or

$$\tau_{flow} \sim 10\ \mu\text{s} \tag{20}$$

Physically, while O_2 relaxes relatively quickly – $\tau_{rlxn,O_2} \sim \tau_{flow}$ – N_2 relaxes much more slowly, $\tau_{rlxn,N_2} \sim 10 \times \tau_{flow}$, so that significant vibrational energy, again excited by pressure heating between the bow shock and nose, is transported downstream of the nose, consistent with the predicted gas temperature distributions in **Figure 14**.

Finally, and importantly, the close qualitative and quantitative agreement observed between predicted temperature distributions in **Figures 14** and **15**, provides strong evidence that the empirical relations introduced in STAR CCM+'s ideal and real gas models [5] are physically reasonable.

5.2 Boundary layer transition and a near-nose 'ring of fire'

Our numerical experiments indicate that within approximately 50 μm, of the solid nose, a rapid laminar to turbulent boundary layer transition takes place. Evidence of transition, predicted by all three gas models, is captured in **Figures 16** and **17**. Focusing on the nonequilibrium model prediction in **Figure 16**, a rapid, micron-scale drop in near-nose surface heat flux is followed by a slower, millimeter-scale rise. Similar qualitative behavior, although less apparent, is also predicted by the ideal and real gas models. As shown in **Figure 17**, transition produces a nominally circular, localized region of elevated surface heat fluxes. The nonequilibrium and real gas models predict comparable maximum heat flux magnitudes within this zone, while the maximum flux predicted by the ideal gas model is approximately 70 to 75% of these. Interestingly, and likely reflecting the presence of nonequilibrium N_2, the nonequilibrium gas

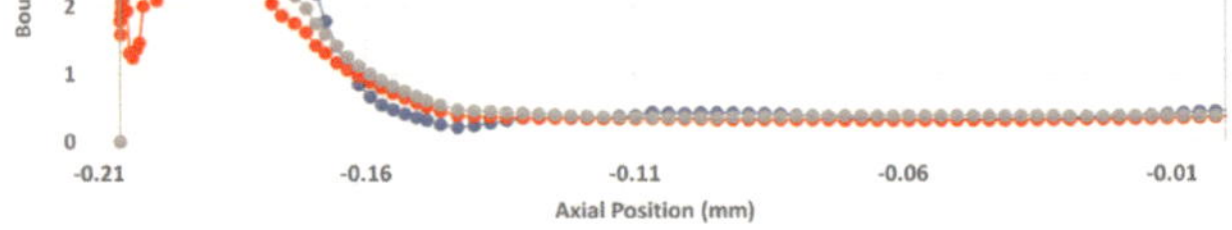

Figure 16.
Boundary heat flux distribution for each gas equation of state—Mach 4.5 flow over solid missile body.

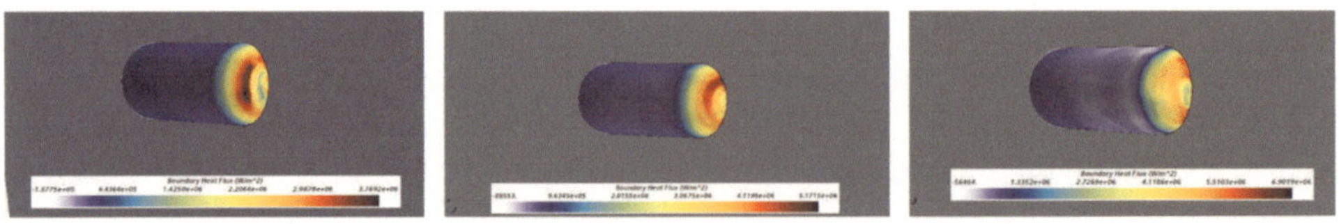

(a) Ideal Gas (b) Real Gas (c) Two Species Model

Figure 17.
Boundary heat flux distribution for each gas equation of state—Mach 4.5 flow over solid missile body—three-dimensional view.

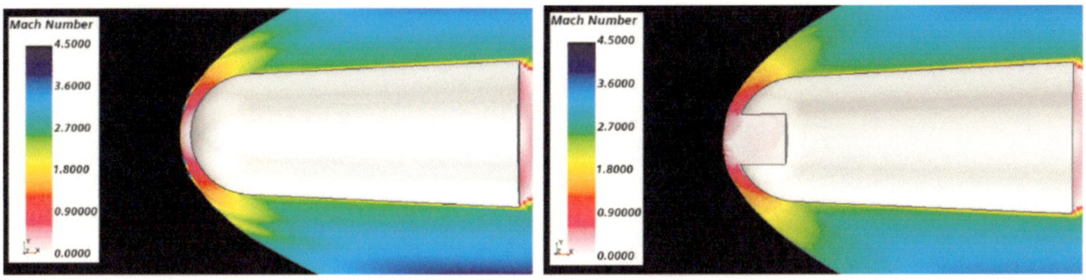

Figure 18.
Computed Mach number fields for Mach 4.5 flow over closed and open nose missile shapes.

model exposes a highly localized, on-nose, heat flux approximately 50% larger than off-nose q_{max} magnitudes (predicted by the nonequilibrium and real gas models). At locations downstream of the (hemispherical) nose shoulder, all three models predict similar, relatively low magnitude, nominally fixed heat flux distributions. For reference, near nose predicted surface fluxes are comparable to those measured during atmospheric reentry of lunar mission Apollo capsules during the 1970s [18].

Interestingly, experimentally observed heat flux distributions on solid nose projectile bodies in [19]—having similar shapes as those used in the present study—were too (spatially) coarse to resolve both the near-nose turbulent boundary layer transition and the resulting ring of intense heat transfer observed here.

The physical origin of the rapid transition to the near-nose turbulent boundary layer, taking place on a ten micron length scale, remains an open question. Based on observed subsonic conditions within the hemispherical gas layer of radius $\sim R$, and thickness $\sim \Delta$, between the bow shock and nose—see **Figure 18**—we postulate that an acoustic feedback mechanism underlies the fast transition. Specifically, upstream-going acoustic disturbances, generated by the near-nose turbulent boundary layer, generate random, small amplitude oscillations in the bow shock. The latter, in turn, generate down-stream going pressure and density waves that drive the fast transition. We note that unsteady bow shocks in hypersonic flow over axisymmetric cone-tipped cylindrical bodies have recently been observed [20]. Intriguingly, in that study, shock oscillations required an unsteady separation bubble and an unstable shear layer passing over the bubble. In our numerical experiments, the unsteady subsonic hemispherical layer may play an analogous role in driving presumed oscillatory shock dynamics.

5.3 Heat sink effect of forward facing cavities

Our numerical experiments reveal that along the cylindrical walls, as well as the circular base of our forward facing cavity, surface heat fluxes exceed the maximum flux observed on the solid nose body; compare **Figures 16** and **19**. The free stream Mach number in both cases is 4.5; since the ideal gas and real gas models appear to under-predict in-cavity heat transfer—see [8]—the results shown in **Figure 19** are obtained using the two-specie non-equilibrium gas model.

We surmise that concentration of thermal energy within the cavity is produced by relatively low in-cavity velocities—see **Figure 20**—which extend residence/flow time scales, thus enhancing deposition of decaying vibrational energy. As discussed in [8], and in contrast to the quantitatively consistent flux predictions for flow over the cavity-free body—see **Figure 16**—STAR CCM+'s ideal and real gas models apparently under-predict this effect when the ratio of flow to (vibration) relaxation time scale becomes too large.

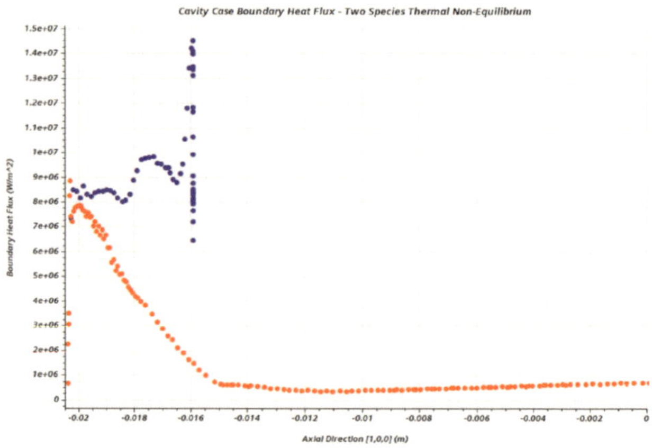

Figure 19.
Computed boundary heat flux distribution for Mach 4.5 flow over open nose body; two-species thermal non-equilibrium gas model.

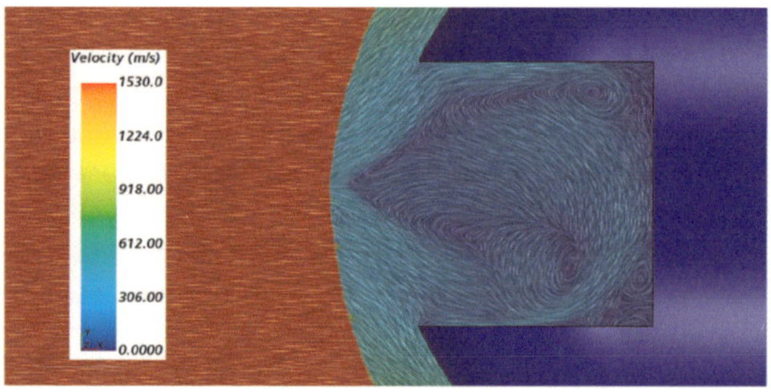

Figure 20.
Computed instantaneous in-cavity and near-nose velocity fields for Mach 4.5 flow over open nose body; two specie, non-equilibrium gas model.

As shown in the numerical experiments in [8], and consistent, for example, with the numerical simulations in [21], introduction of a forward facing cavity reduces heat transfer to the aerodynamic body at all points downstream of the cavity. Physically, this appears to reflect the cavity functioning as a heat sink, transferring gas enthalpy to the walls of the cavity. Clearly, this heat sink effect must be accommodated for in designing open nose hypersonic aerodynamic bodies.

6. Conclusions

1. Schlieren images of Mach 4.5 and 3.5 flow, about closed and open nose missile bodies, respectively, were obtained in the hypersonic wind tunnel at UNC Charlotte.

2. Comparisons between numerically predicted and experimentally observed as well as between numerically calculated and physically estimated drag forces and shock standoff distances suggest that the STAR CCM+ software package provides

high quality predictions of hypersonic aerodynamic flows, consistent with the results reported in [5].

3. A new technique, based on static background removal and SVD decomposition of raw digital schlieren images, provides significantly enhanced detail of experimentally imaged high speed flow fields.

4. Numerical experiments reveal that in hypersonic (Mach 4) flow over solid nose bodies, turbulent boundary layer transition takes place on tens of micron length scales. The physical origin of this extremely rapid transition remains an open question. Near-nose transition produces a ring-shaped zone of enhanced heat transfer immediately downstream of the missile nose.

5. Numerical experiments also reveal that forward facing cavities function as heat sinks, absorbing elevated near-nose gas enthalpy and reducing heat transfer to the body downstream of the cavity.

Author details

Tyler Watkins, Jesse Redford, Franklin Green, Jerry Dahlberg, Peter Tkacik and Russell Keanini*
Department of Mechanical Engineering and Engineering Science, University of North Carolina at Charlotte, Charlotte, North Carolina, USA

*Address all correspondence to: rkeanini@uncc.edu

References

[1] Anon. Skyhaven Systems wins award from the Missile Defense Agency to develop a hypersonic leading edge thermal cooling system. Skyhaven Systems. 2020. Available from: http://skyhavensystems.com/2019/11/skyhaven-systems-wins-award-from-the-missile-defense-agency-to-develop-a-hypersonic-leading-edge-thermal-cooling-system/ [Accessed: May 17, 2022]

[2] Solomon JM. The Design, Calibration, and Commissioning of a Benchmark Hypersonic Wind Tunnel. Charlotte, NC, USA: The University of North Carolina at Charlotte; 2021

[3] White M. Improvement of Optical and Pressure Sensing Capabilities of the UNC Charlotte Hypersonic Wind Tunnel. Charlotte, NC, USA: The University of North Carolina at Charlotte; 2021

[4] Solomon J, Tkacik PT, Dahlberg J. Design and build of a Benchtop hypersonic wind tunnel as a Senior Design Project. In: AIAA Propulsion and Energy 2021 Forum. 2021

[5] Cross PG, West MR. Simulation of Hypersonic Flowfields Using STAR-CCM+. China Lake, CA, USA: Naval Air Warfare Center Weapons Division; 2019

[6] Redford J, Li X. Exploration of SVD for image compression and time series processing. In: 2020 IEEE MIT Undergraduate Research Technology Conference (URTC). 2020. pp. 1-4

[7] Lopez J. Aerodynamic forces and heat transfer of sphere and sharp cone in hypersonic flow [doctoral dissertation, MS thesis]. San Jose, CA, USA: Dept. Aero. Eng., San Jose State Univ.; 2014

[8] Watkins TH. Experimental and Numerical Validation of a Forward Facing Cavity as a Passive Thermal Protection System of a Hypersonic Body. Charlotte, NC, USA: The University of North Carolina at Charlotte; 2022

[9] Andrews H, Patterson C. Singular value decompositions and digital image processing. IEEE Transactions on Acoustics, Speech, and Signal Processing. 1976;**24**(1):26-53. DOI: 10.1109/TASSP.1976.1162766

[10] Ambrosio A, Wortman A. Stagnation-point shock-detachment distance for flow around spheres and cylinders in air. Journal of the Aerospace Sciences. 1962;**29**(7): 875-875

[11] Van Dyke MD. The supersonic blunt-body problem-review and extension. Journal of the Aerospace Sciences. 1958;**25**(8):485-496

[12] Olivier H. A theoretical model for the shock stand-off distance in frozen and equilibrium flows. Journal of Fluid Mechanics. 2000;**413**:345-353

[13] Anderson JD. Hypersonic and High Temperature Gas Dynamics. Reston, VA, USA: AIAA; 2000

[14] Hirschfelder JO, Curtiss CF, Bird RB. Molecular Theory of Gases and Liquids. New York: Wiley; 1964

[15] Robben F, Talbot L. Measurement of shock wave thickness by the electron beam fluorescence method. The Physics of Fluids. AIP. MD, USA: College Park. 1966;**9**(4):633-643

[16] Pathria RK. Statistical Mechanics. Amsterdam, NL: Elsevier; 2016

[17] Landau L. Theory of sound dispersion. Physikalische zeitschrift der Sowjetunion. 1936;**10**:34-43

[18] Jenniskens P, Wercinski P, Olejniczak J, Paiche G, Kontinos D, Allen G, et al. Preparing for Hyperseed MAC: An observing campaign to monitor the entry of the Genesis Sample Return Capsule. In: Modern Meteor Science an Interdisciplinary View. Dordrecht: Springer; 2005. pp. 339-360

[19] Sudarshan B, Saravanan S. Heat flux characteristics within and outside a forward facing cavity in a hypersonic flow. Experimental Thermal and Fluid Science. 2018;**97**:59-69

[20] Sasidharan V, Duvvuri S. Large-and small-amplitude shock-wave oscillations over axisymmetric bodies in high-speed flow. Journal of Fluid Mechanics. 2021; **913**:R7-1-R7-12

[21] Lu H, Liu W. Nummerical simulation in influence of forward-facing cavity on aerodynamic heating of hypersonic vehicle. Procedia Engineering. 2012;**29**:4096-4100

Chapter 9

The Effects of Blade Configurations on Performance of a Tidal Vertical Axis Turbine

Sepideh Amiri Tavasoli, Seyed Jalal Hemmati, Saeed Niazi and Ali Jalali

Abstract

Hydrokinetic energy contains the major uncontrolled source of renewable marine energy. The highest level of converter technology readiness offered in the last three decades is TRL8–9, which is related to the first-generation horizontal axis converters. In low-depth calm waters, one of the best options to harvest tidal energy is vertical axis turbines. About 16% of the conceptual designs presented in the last 30 years apply this type of converter, which does not have a high level of technological readiness. In this study, a laboratory-designed vertical axis turbine has been introduced in which the effects of the number of blades, the blade profile, and attack angle on the performance of the turbine were analyzed. A 3D incompressible viscous turbulent computational finite volume approach is applied, with the spatial second-order and temporal first-order accuracies. The turbulent model k-ω SST was used to obtain the flow inside the turbine. Rotors include two, three, and six blades with three different profiles, including NACA2421, NACA16021, and NACA0020. Computational results reveal that the turbine with three blades and an angle of attack of +8 using the NACA2421 profile has a maximum generation capacity of about 4 kW, with a strength factor of 0.4 and a power factor of about 20%. The capacity, however, was lower for a higher number of blades.

Keywords: renewable tidal energy, in-stream tidal turbine, CFD, vertical axis turbine

1. Introduction

Tidal energy technology is a cluster of four technologies, namely tidal dams, tidal lagoons, tidal current converters (TISECs), and dynamic tidal power (DTP) that use tidal cycles to generate electricity [1]. The greatest unrestrained potential of tidal energy is from the tidal current source. This unlimited source of clean energy can be estimated with high accuracy and reliability. Practically, to extract the tidal energy, the average flow velocity in the spring tide of neighborhoods should be faster than 1.5–2.5 m/s. Nevertheless, in nearly 10 regions in the world, the flow velocity is relatively high.

Zeiner-Gundersen [2] proposed a simple and cost-effective vertical axis turbine with flexible foils designed based on inspiration from hydrodynamic thrust characteristics of aquatic creatures. These novel aspects resulted in a high-performance turbine that attained up to a 0.37 power coefficient in a confined channel and could self-start at low inflow water velocities.

Pongduang et al. [3] studied the helical tidal current turbine and reported its performance and characteristics, aiming to develop the free water flow electric turbine. The scaled model of the tidal turbine was built as 0.5 and 0.6 m in diameters and 1.25 m in length. The turbine cross-section blade was symmetric NACA0020 with a 0.07 m chord length and three blades with the helical angle of 120°, 135°, and 150°. The model was tested in a towing tank.

Priegue and Stoesser [4] investigated the influence of blade roughness on the performance of a vertical axis tidal turbine. In this design, the vertical axis of turbines undergoes stall at some periods, that is, the flow separates from the blade during each revolution. In this study, it is hypothesized that roughening turbine blades causes a delay of the flow separation (similar to flows over rough bluff bodies), which in turn diminishes the turbine stall, and ultimately results in an increase in the performance of the turbine.

Based on an experimental study, Harries et al. [5] analyzed the performance and optimization of a prototype novel drag-driven vertical axis a tidal stream turbine. Hamidi et al. [6] also investigated the power and momentum coefficients of a three-blade vertical tidal turbine using a laboratory model. Bouzaher et al. [7] investigated a vertical axis tidal turbine with flexible blades, with the focus on analyzing the effect of flexible airfoil types and the blade flexibility on the net output power of the turbine. Delafin et al. [8] used the variable pitch in order to control the angle of the attack of blades continuously during their rotation. The same approach is also considered in this study. Based on the 2D blade-resolved unsteady Reynolds-Averaged Navier-Stokes (RANS) simulations, the improvement caused by pitching the blades is evaluated in a three-straight-blade vertical axis tidal turbine. Three pitching laws are defined and tested, aiming to reduce the angle of the attack of the blades in the upstream half of the turbine, but no pitching motion in the downstream half.

Airfoil is a geometric structure to generate mechanical force from the relative motion between the airfoil and the airflow around the airfoil structures. The first wind turbine blades were also designed by airfoils used in aviation applications. However, in the 1980s, due to defects in aeronautical airfoils applied to a wind turbine, special air airfoils were assigned to wind turbines. A wind turbine has been created with the help of special airfoils. The airfoil series for controlled wind turbines, with variable level adjustment by stations, was developed by the National Renewable Energy Laboratory in 1984 [9].

Vertical axis turbines and their hydrodynamic performances have not been well investigated. This type of turbine is very suitable for calm and shallow water conditions, which are very common around the world. In this study, the flow in a microturbine under calm and shallow water conditions is examined based on a simulation. Using the hybrid mesh and finite volume approach simulation methods, the effects of the number and profile and angles of the attack of turbine blades on the performance of vertical axis turbines are studied.

2. Numerical method

The governing equations to solve the 3D incompressible flow are the time-averaged continuity equation and the time-averaged momentum equations:

$$\frac{\partial \overline{u}}{\partial x}+\frac{\partial \overline{v}}{\partial y}+\frac{\partial \overline{w}}{\partial z}=0 \tag{1}$$

$$\frac{\rho D\overrightarrow{\overline{v}}}{Dt}=\rho\vec{g}-\vec{\nabla}\overrightarrow{\overline{p}}+\vec{\nabla}.\overline{\tau'}+\overrightarrow{\overline{f}} \tag{2}$$

where $\vec{g}$ is the force of gravity, $\overrightarrow{\overline{f}}$ is the volumetric force, and $\overline{\tau'}$ is the stress tensor, which is presented as follows:

$$\overline{\tau'}=(\mu+\mathrm{n})\left(\frac{\partial \overline{u}_i}{\partial x_j}+\frac{\partial \overline{u}_j}{\partial x_i}\right) \tag{3}$$

In Eq. (3), both the viscous shear stress and the turbulent or eddy shear stress are included. The time average method is used to simulate turbulence effects. In these models, new terms are introduced to simulate turbulence of the mean flow instead of explicitly examining the turbulence behavior. These are the time-averaged Navier-Stokes equations that are applied for a larger-scale modeling of turbulence problems.

Turbulent modeling is relatively difficult because it requires detailed information about the structure of the turbulence phenomenon in the fluid, which is often not available. Therefore, one of the available turbulence models can be used to determine the magnitude of the stresses. In this study, the k-ω SST[1] model is used as the turbulence model. The k-ω SST model is based on two equations, which is actually a combination of two models—standard k–ω and k–ε. The k-ω SST model effectively uses the high precision of the standard k–ω formulation near walls, as well as the independence of the k–ε model, and for this reason, it has gotten rid of the problem of the k-ω standard model, which is highly sensitive to the turbulent properties of the input current. For this reason, the k-ω SST model has better performance than both k–ε and k–ω models.

The operational performance of marine turbines can be defined as dimensionless coefficients. These coefficients are the tip speed ratio (TSR), thrust coefficient (C_T), and power coefficient (C_P):

$$\mathrm{TSR}=\frac{\omega \mathrm{R}}{\mathrm{V}} \tag{4}$$

$$\mathrm{C_T}=\frac{\mathrm{T}}{\frac{1}{2}\rho\ \mathrm{V2A}} \tag{5}$$

$$\mathrm{C_P}=\mathrm{TSR}\times \mathrm{C_T} \tag{6}$$

where ω is the rotational speed of the turbine, R is the radius of the blade, V is the inlet axial flow velocity, A is the hypothetical surface of the turbine disk, and T denotes the turbine thrust [10]. One of the most important variables, which indicates the effect of the number of blades, is the solidity ratio:

$$\sigma=\frac{NC}{\pi D} \tag{7}$$

where N is the number of blades and C is the length of their airfoil chord.

---

1 Shear stress transport.

In this chapter, a 3D laboratory-designed vertical turbine is modeled, and a finite volume method is used to simulate the hydrodynamic performance of the turbine. The length of the airfoil chord is 120 mm, the spiral angle (the angle between the blade and the horizontal plate) is 60°, the height and the diameter of the turbine are both 300 mm, and the number of blades is two, three, and six (**Figure 1**). Among the available options, profiles of the NACA2421, NACA0020, and NACA16021 are chosen. **Figure 2** illustrates the cross-sectional shape of each of the three selected profiles.

The outer part of the model is considered similar to that used by Revuz et al. [11]. As shown in **Figure 3**, the computational domain has a length of 4500 mm, width of 3000 mm, and height of 1500 mm, which includes a rotating part with a diameter of 400 mm and a height of 320 mm. The diameter of the internal section is 200 mm and its height is 320 mm.

Figures 4–6 show grids, the turbine blade, and total solution domain meshes, respectively. As mentioned before, the moving mesh was employed, and the adjacent areas to the object have smaller mesh size than the other areas. The mesh sensitivity analysis was also performed; the results of which are presented in **Figure 7**. Optimally, the number of computational cells was about 3.5 million. The Ansys fluent software is used. A high-performance (HP) server with five nodes, each with the ability to connect to 56 clients and 256 Gbyte ram, and a total of 5 terabyte hard disk space were used. Each performance took between 2 and 8 days.

Simulations are performed for different TSRs and the results are compared. A complete cycle of sea tides lasts for 14 days, during which every 6 hours, the

Figure 1.
3D model of a rotor designed with three blades.

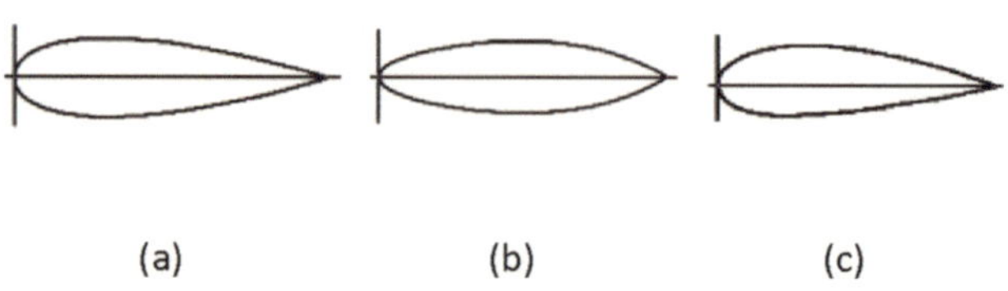

Figure 2.
A top view of the blade section with profiles of (a) NACA0020, (b) NACA16021, and (c) NACA2421.

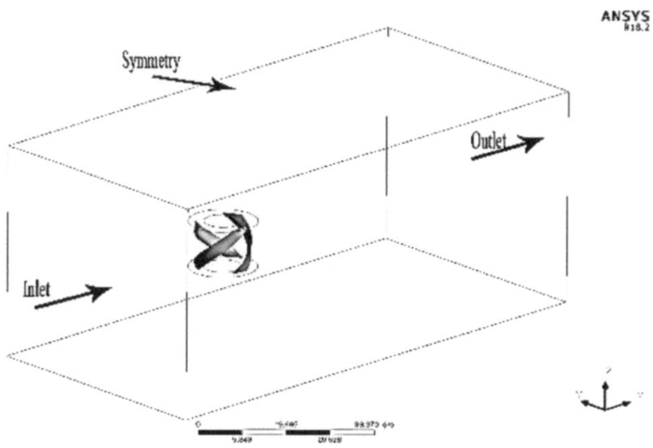

Figure 3.
Computational domain of the simulation of the hydrodynamic turbine.

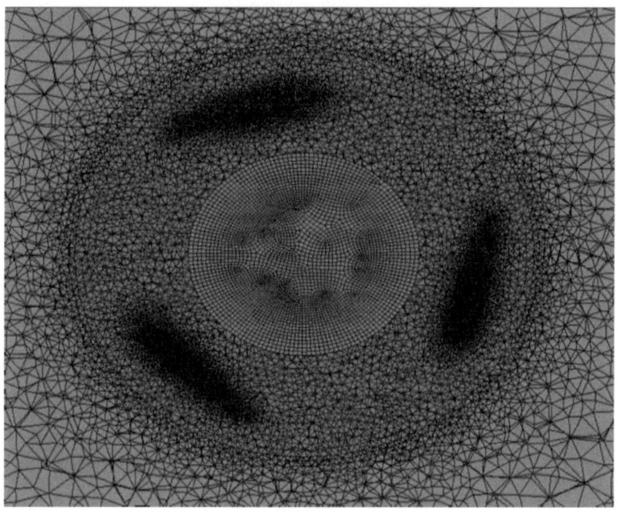

Figure 4.
Internal computational grids.

horizontal velocity of the current rises to the maximum from zero, and then falls back to zero. **Figure 8** depicts an example of speed series on the Qeshm channel in the Persian Gulf [12].

Laboratory results reported by Talukdar et al. [13] were used to validate our results. In their study, the water flow velocity was 0.87 m/s, and the density of seawater was 1.025 kg/m^3. Torque coefficients and computational power obtained in our study are compared against the laboratory data present in a study by Talukdar et al. [13] and presented in **Figures 8** and **9**. As can be seen, for the NACA0020 type, the profiles are similar in both studies. A comparison of analytical and experimental results reveals that the maximum relative error for the torque coefficient and the computational power coefficient was 18% and 20% less than the experimental values, respectively. The way the generator is connected to the blade, as well as uncertainties in experimental data, contributed to some errors, but changes in torque and power coefficients in different TSRs were similar in both cases.

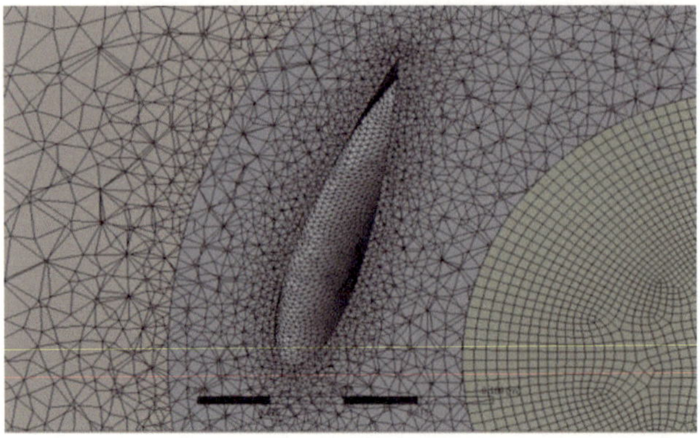

Figure 5.
Grids of the turbine blade.

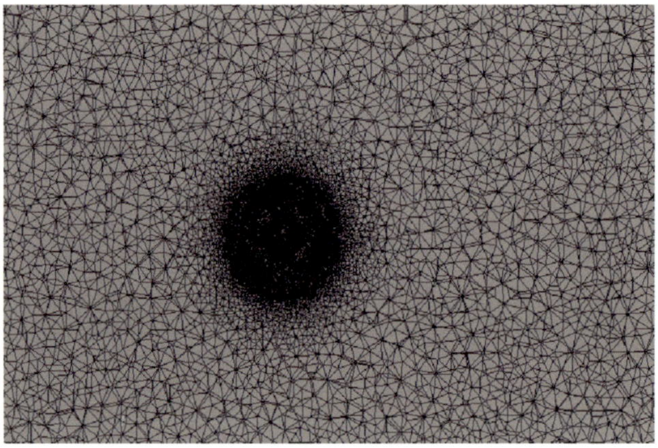

Figure 6.
Total solution of the domain mesh.

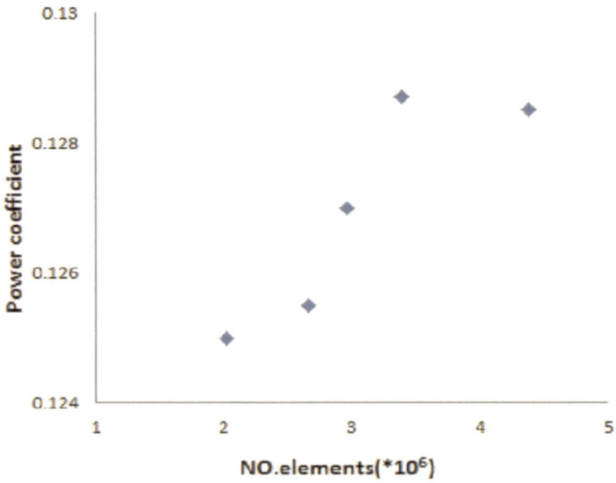

Figure 7.
The grid sensitivity analysis.

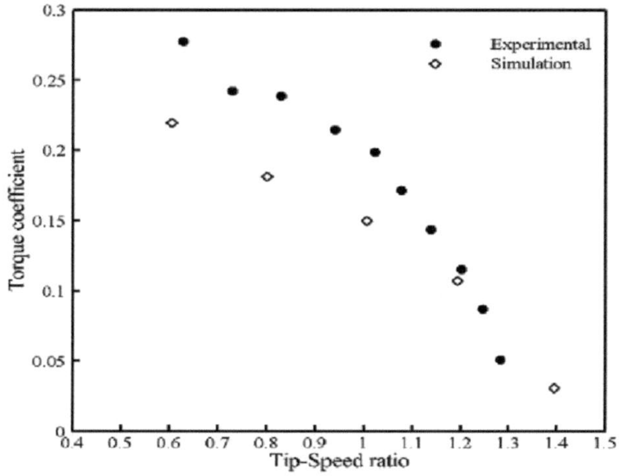

Figure 8.
Comparison of torque coefficient for different TSRs with experimental data [13].

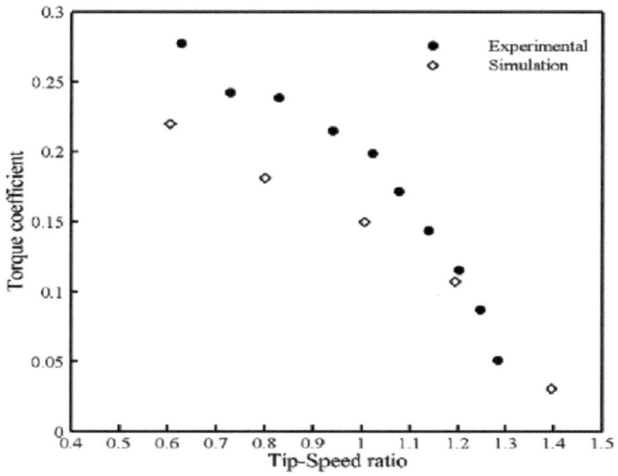

Figure 9.
Comparison of power factor for different TSRs with experimental data [13].

3. Results and discussion

Figures 10–16 depict simulated velocity and pressure and the velocity vector around the blades for different blade numbers and profiles. Only results of the simulations under the maximum velocity (1.8 m/s) are shown. In these figures, it is assumed that the stream enters horizontally from the left side at a constant mean speed and exits from the right side.

Figure 10 shows that the maximum flow velocity inside the turbine when collides with blades is 3.7 m/s for turbines with two blades. This value is 3.6 and 2.8 m/s for turbines with three and six blades, respectively. The most turbulent flow is behind the turbine with six blades, as illustrated in **Figure 10c**.

Analysis of the velocity distributions for the turbines with different blade numbers indicates that the current deviation in the increasing order is for the two-, three-, and,

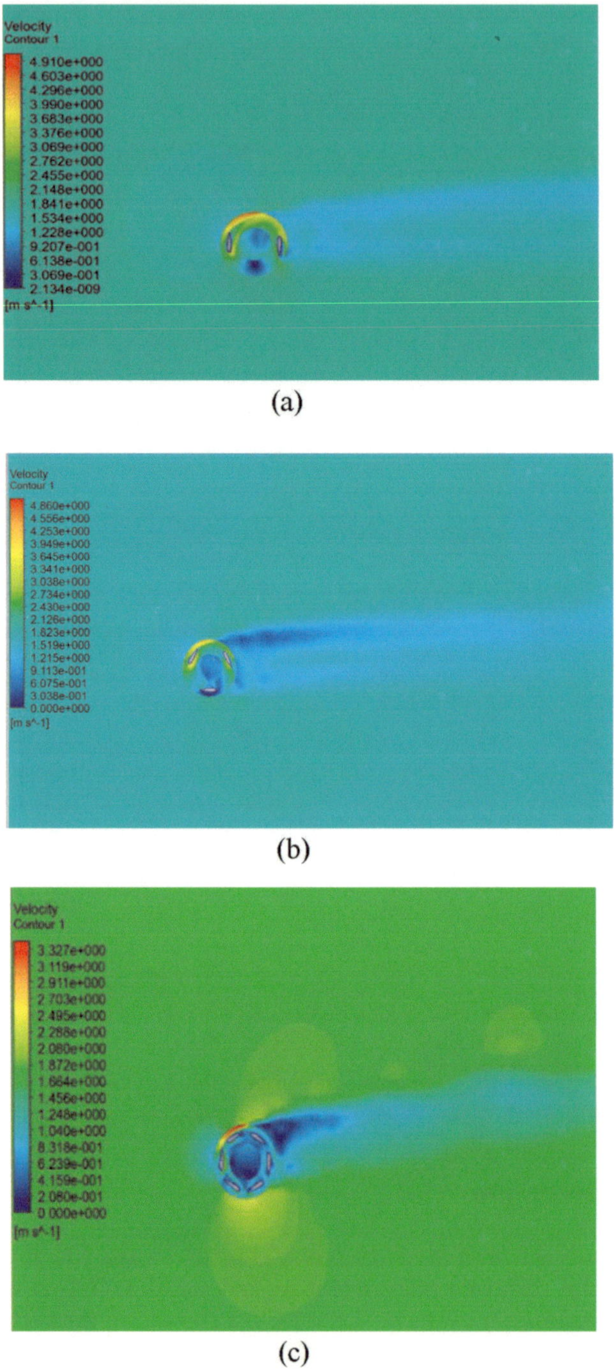

Figure 10.
A top view of the velocity contour for (a) two-blade, (b) three-blade, and (c) six-blade turbines with a speed of 1.8 m.

six-blade turbines, respectively (**Figure 10**). If the turbine has a large flow deviation, more energy should be consumed. Thus, the smaller the flow deviation, the better the turbine in terms of energy production. From the edge of the attack to the tail of the

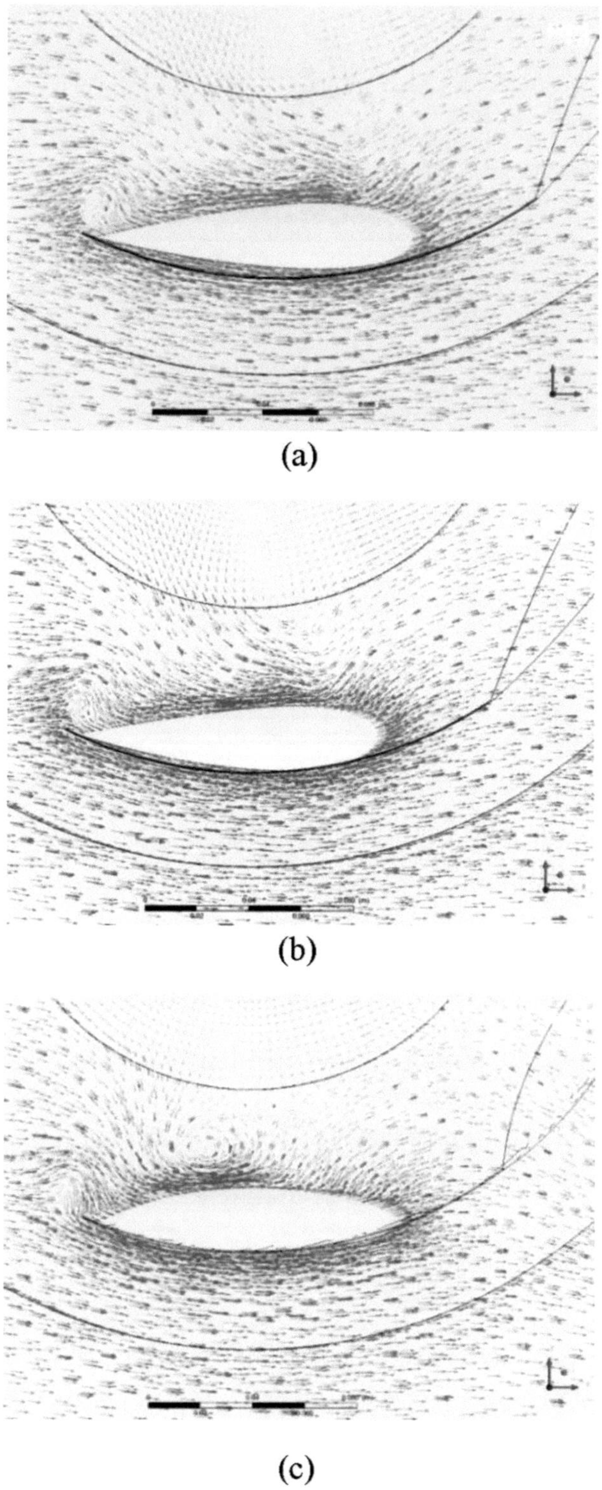

Figure 11.
Speed vector around the blade for the input speed of 1.8 m/s in the profile mode of (a) NACA16021, (b) NACA2421, and (c) NACA16021.

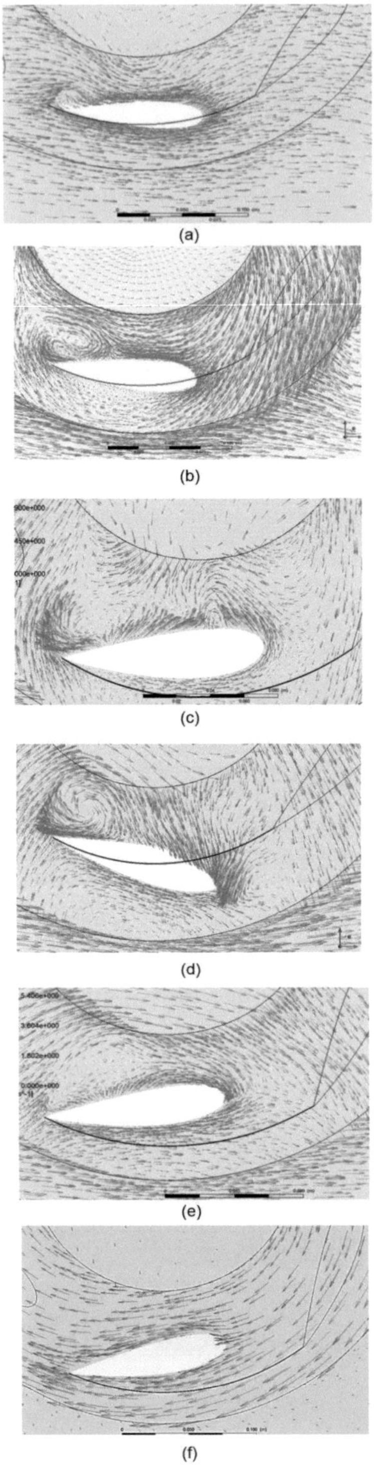

Figure 12.
Speed vector around a blade for the input speed of 1.8 m/s in the angle of the attack mode (a) The angle of +4° (b) The angle of +8° (c) The angle of +12° (d) The angle o—– 4° (e) The angle of −8° (f) The angle of- 12°.

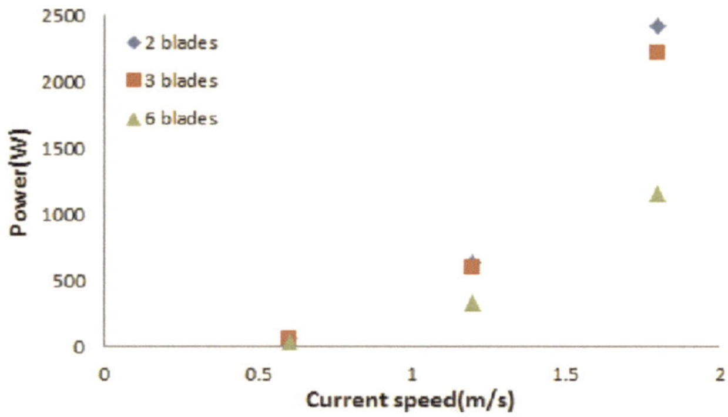

Figure 13.
Tidal energy versus stream velocity for different blade numbers.

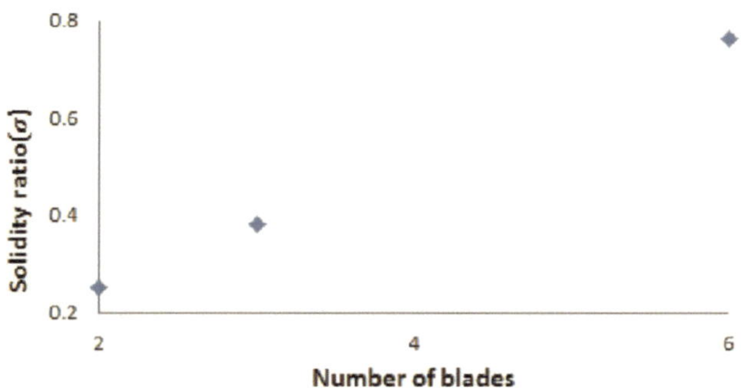

Figure 14.
The solidity coefficient for different number of blades.

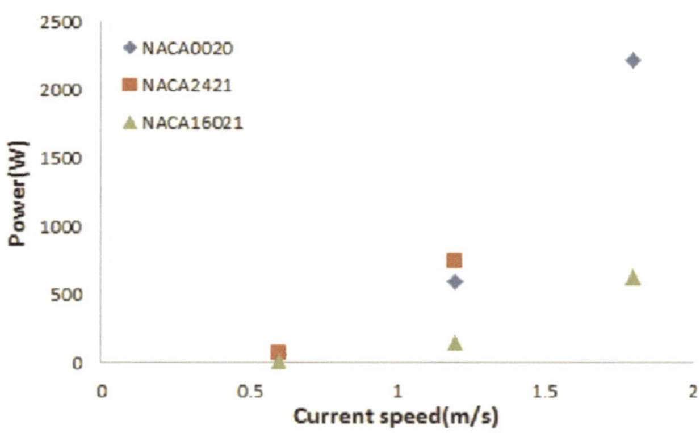

Figure 15.
Tidal energy versus stream velocity for different blade profiles.

airfoil, pressure initially decreases but then begins to increase, and the increase in pressure leads to the separation of the boundary layer, which increases the pressure gradient and may cause problems in the turbine blades.

Figure 11 shows the velocity distribution for turbines with different blade profiles, but the inlet maximum speed of 1.8 m/s. Flow deviation in the turbine with the NACA2421 profile is less than the others, causing the extraction of more energy from

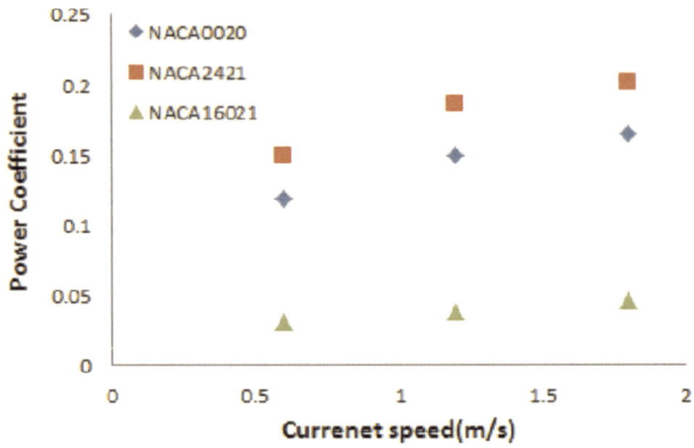

Figure 16.
Power coefficient versus stream speed for different blade profiles.

the tidal water. **Figure 11** also shows that the vorticity value in the turbine with the NACA2421 profile is less than the other two types, which improves the performance of the turbine.

As a turbine with the NACA2421 profile is the most optimal at the flow velocity of 1.8 m/s, it was designed and shaped under different angles.

One of the issues that always complicates the simulation of vertical axis turbines is the continuous interaction of the fluid with blades in each section of the rotor. In fact, a fluid that is affected by a blade in the upstream area should hit the blades in the lower half of the rotor.

In general, by reducing the angle of the attack of both negative and positive angles at upstream, some current separation occurs at the end of the airfoil, which causes a severe pressure difference at the upper and lower levels of the airfoil (**Figure 12**), which leads to an increase in the return time of the current from turbulence to laminar. This pressure difference, along with the destructive phenomenon of dynamic fatigue and the circular currents of the airfoil tip, leads to the destructive drag force. In response to this force, the speed may be reduced and some severe vibrations may be created in the turbine. If this condition continues to increase, the pressure on the blade structure can have far more dangerous effects, such as breaking the blades. These vortices will be increased in the negative angles as far as the reverse pressure is observed. There is also increased in positive angles with increasing the angle of these vortices. At an angle of +8°, which is in line with the flow direction, the end of the blade is covered with vortices, but at an angle of +4°, it can be seen that in addition to the end of the blade, vortices are formed on the curvature of the blade.

Figure 13 presents the amount of tidal energy obtainable for different velocities of flow for turbines with two, three, and six blades. As the speed increases, the output power increases, but the increase in the two-blade mode is greater than that of three or six blades. This result can also be seen in **Figure 10**. In general, with increasing the number of blades, the speed of the rotor rotation decreases, but torque increases. Similarly, as the number of blades increases, the solidity of the turbine increases (**Figure 14**). As in the case of a three-blade turbine, the solidity factor is more than two blades and the power that can be produced does not decrease considerably when the number of blades increases, the pressure distribution decreases. Thus, the fatigue stress with oscillating loads created in turbines with fewer blades causes the blades to break and damage. The two-blade turbine produces more energy than a three-blade turbine, but because of the high pressure on its blades, it causes the turbine to sag. Naturally, this loosening motion causes stability problems for the turbine. It also increased stress on the components of the turbine, such that they may break down over time. The two-blade turbine generally causes environmental problems due to issues such as loss of components and the need for repair and also increases the maintenance costs of the turbine. However, the amount of power produced in a three-blade turbine is not that much different from that in a two-blade turbine. The cost of a three-blade turbine is also lower, and it leads to less environmental impact. On the other hand, when blades are more than three, there is more water resistance and the power generation is slower. Thus, a three-blade turbine is more efficient. For these reasons, a three-blade turbine is ideal because of its high-energy performance, greater stability, and durability. As a result, a three-blade turbine is selected as the most optimal turbine in terms of the blade number.

Airfoil is a geometrically shaped structure for mechanical force generation from the relative movement between the airfoil and the surrounding airflow of the airfoil structures [14]. For wind turbines, the airfoil shape of the blades influences the turbine power production. The lifting efficiency of blades determines the effectiveness of rotor rotation for the productive energy conversion from wind kinetics to rotor rotation, which leads to higher electricity generation from the drive unit.

The boundary layer of the airfoil is exerted by additional pressure generated by the curvature shape of the airfoil compared to the constant pressure on the boundary layer made of the plate with zero incidences. The pressure distribution on the edge of the boundary layer is the same as the pressure distribution on the wall in the plate. However, due to the streamline curvature of the airfoil surface, pressure gradients and compensation for the centrifugal force of the streamline flow are generated inside the boundary layer. Furthermore, the transition point of the boundary layer on the airfoil is determined by the outer flow and its pressure difference generated by the curvature shape of the surface [15].

As each profile is from different NACA series, there are some differences. In the NACA0020, the thickness of the profile was 20%, while in the NACA2421 and NACA16021, it was 21%.

As can be seen in **Figure 15**, the NACA2421, which is thicker than the NACA0020 and is more curved than both the NACA0020 and NACA16021, produced more energy. On the other hand, the NACA0020, which is less thick than the NACA16021, produced more energy. As a result, the amount of energy production decreases with increasing the thickness and increases with increasing the curvature. As a result, the most optimal type of profile among the studied profiles is the NACA2421, which is a four-digit and asymmetric NACA series. Thus, this profile is selected as the optimal profile.

As expected, power generation by different profiles increases when the speed increases, but for the NACA2421 profile, this increase is the most. Hence, in the maximum current velocity of 1.8 m/s, it is predicted that a power of 2.5 kW can be obtained.

Figure 16 shows the tidal energy in terms of stream velocity in turbines with different profiles. A turbine with three blades (NACA2421) generates the highest power at different speeds, with a maximum of 20%.

Figures 17 and **18** show the coefficient of power and extractable power relative to the ratio of different tip velocities for different attack angles, and **Figures 19** and **20** show the extractable power and power factor relative to different attack angles for different velocity ratios.

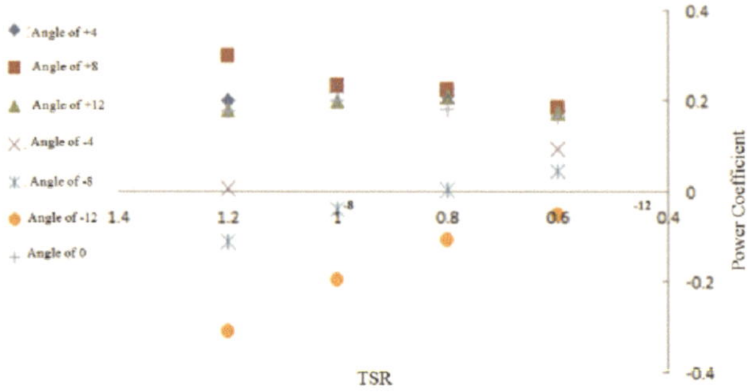

Figure 17.
Power factor diagram of the tip speed ratio for different attack angles.

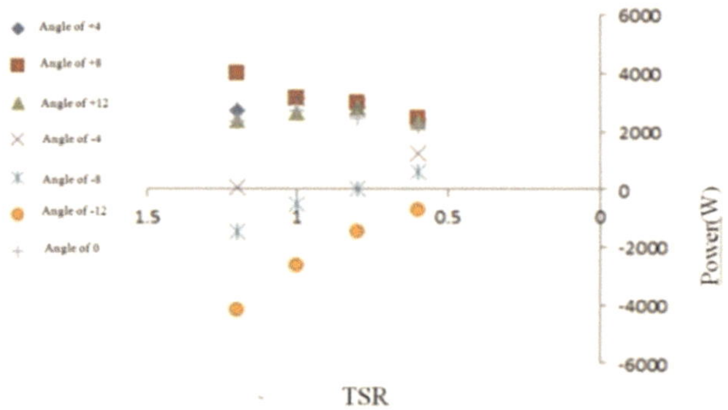

Figure 18.
Power diagram based on the tip speed ratio for different attack angles.

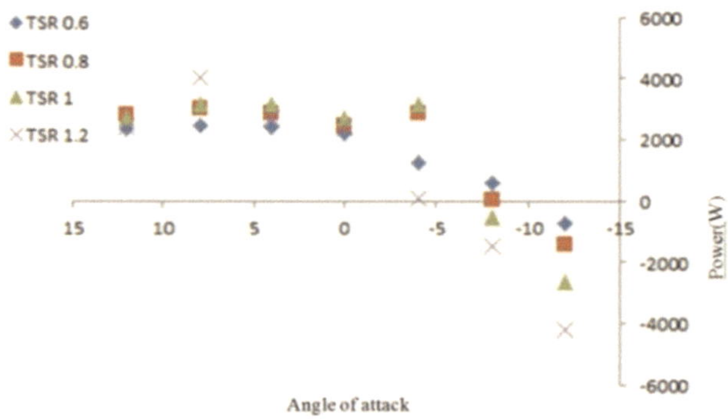

Figure 19.
Power diagram of different attack angles for the tip speed ratio.

As shown in **Figures 17–20**, the turbine has the highest power factor at a certain tip speed ratio, which is the optimal tip speed ratio, and changes depending on the blade angle of the attack. The optimal tip velocity ratio changes with the angle of the attack relative to the reference mode, so that the maximum increase is at the angle of the attack of +8°. At negative angles, it first increases at the angle of the attack of −4°, but then it decreases. As the airfoil angle of the attack increases, the power factor decreases over time.

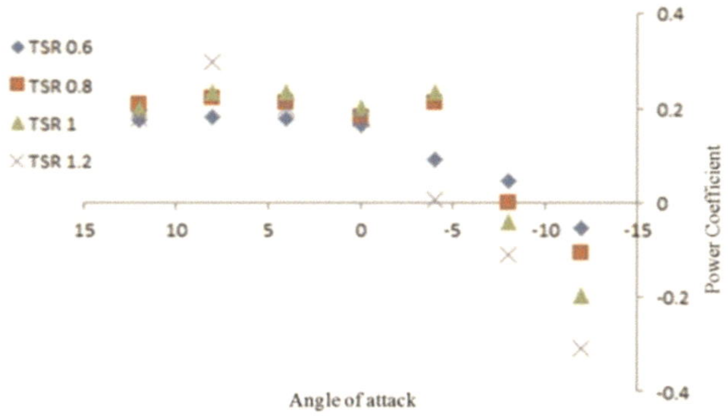

Figure 20.
Power factor diagram of different attack angles for the tip speed ratio.

According to these diagrams, it can be seen that the optimum power output of the turbine in terms of the attack angle is a turbine with an attack angle of +8, which produces a power of about 4000 watts, about 1.5 times more than the power that can be produced by a turbine with the zero-degree angle of the attack. Nevertheless, the pressure on the turbine blades increases by 7% in the turbine with the optimum power output compared to that with the zero-degree angle of the attack, although this increase in pressure is negligible.

4. Conclusion

Figure 21 shows the power factor in two turbine rotation cycles for different speed ratios. It can be seen that for different tip speed coefficients, the amount of power extracted from the three-blade turbine was the most. With increasing the speed of the tip, the amount of power further increased, but decreased for three-blade and two-blade turbines, respectively.

According to **Figure 22**, at different blade tip speeds, the power factor in a turbine with the NACA2421 blade profile is higher than a turbine with the NACA0020 profile. According to **Figure 22**, the optimal tip velocity ratio is equal to TSR1 because it has the highest output power factor.

According to **Figure 23**, the power factor has changed for angles of 0 and +8°, with the maximum value in TSR 1.2 for the angle of +8°. Increasing the angle of the attack from 0 to +8° increases the power factor, but at the angle of +12, it decreases. The maximum power belongs to the turbine with the attack angle of +8° and the tip speed ratio of 1.2 (**Figure 24**).

Our analysis showed that an increase in the number of blades negatively affects the obtainable power. In other words, the solidity coefficient increases with increasing the number of blades. This increase has destructive environmental impacts, including the noise generated by the blades, as well as the pollution during construction, transmission, and installation of the blades [16]. As a result, according to the simulations, the turbine axis with three blades yields better performance.

The simulation results also predict that the curvature and thickness of the blade profile affect the production capacity. The production capacity is directly related to the increase in the curvature of the blade profile and is inversely related to its

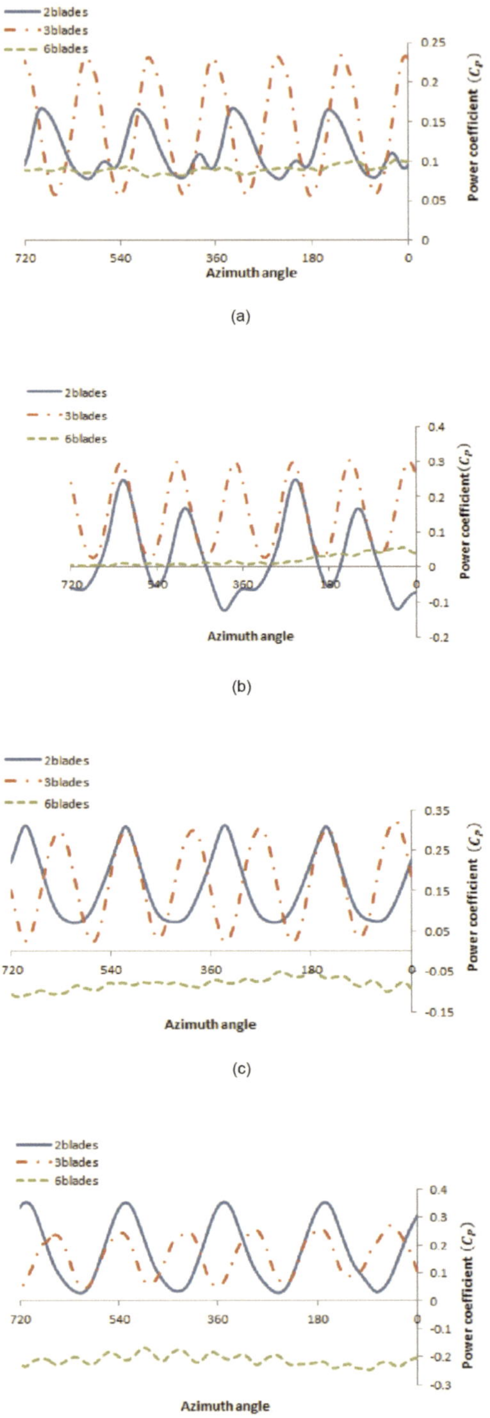

(d)

Figure 21.
Comparison of power factor diagram of a torsion angle at different tip speed ratios for a turbine with two, three, and six blades. (a) TSR 0.6, (b) TSR 0.8 (c) TSR 1 (d) TSR 1.2.

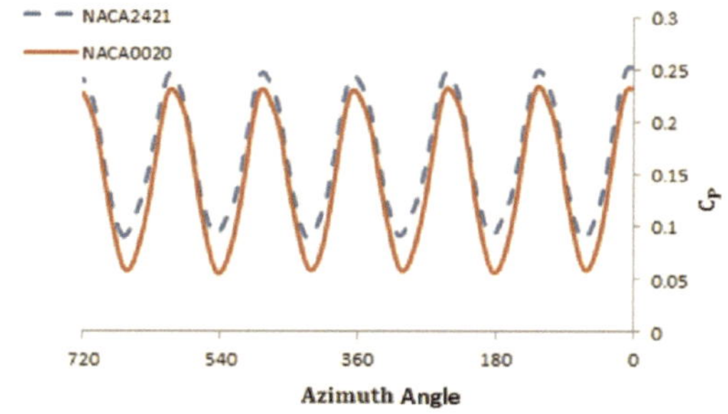

(a)

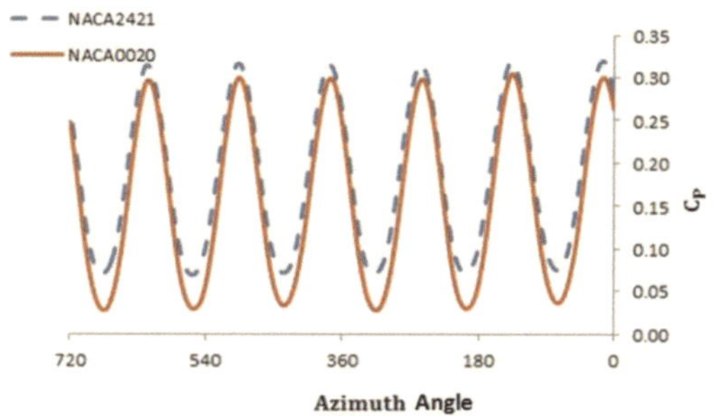

(b)

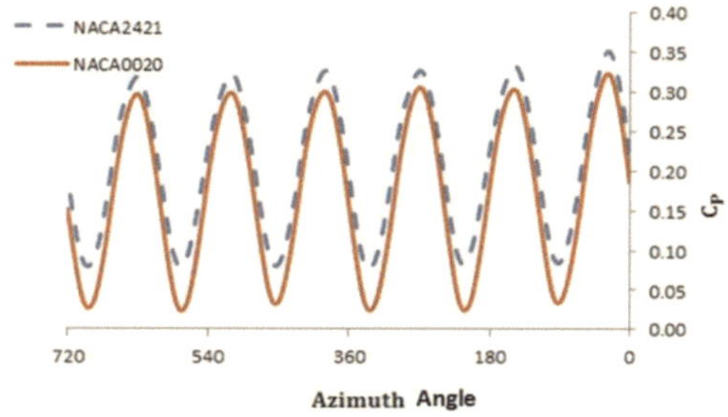

(c)

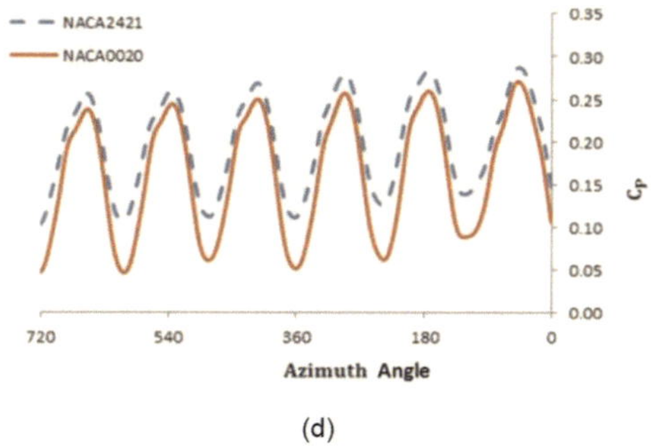

(d)

Figure 22.
Comparison of power factor diagrams in two turbine revolutions for the NACA0020 and NACA2421 profiles at different tip speeds. (a) TSR 0.6, (b) TSR 0.8 (c) TSR 1 (d) TSR 1.2.

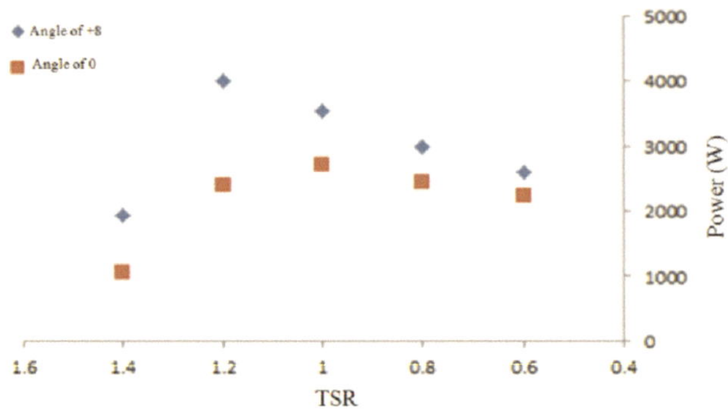

Figure 23.
Comparison of the power graphs of the tip velocity ratios for attack angles of 0 and +8°.

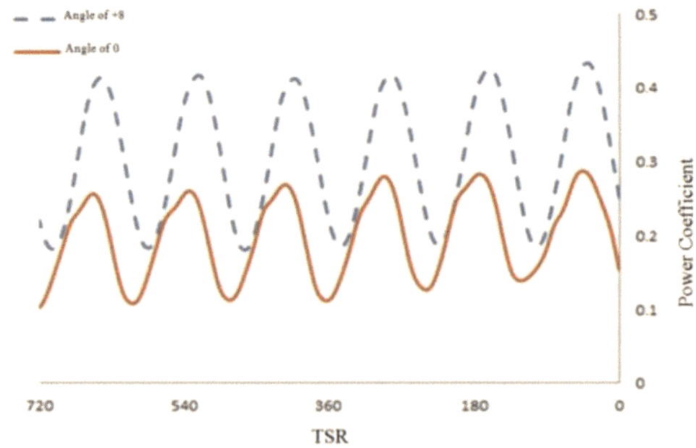

Figure 24.
Comparison of power factor diagrams in two turbine revolutions for the angle of +8° and 0° in TSR1.2.

thickness. The production capacity in this study was estimated to be 4 Kw for a three-blade turbine and the NACA2421 profile.

Although in experimental tests [13], the experimental power has been reported to be higher than the computational value in the current research, trend changes in design indicators and the amount of existing error are acceptable. This method can be used to design suitable microturbines at the entrance of bays, shrimp ponds, and in the water outlet of boilers.

Acknowledgements

This research was carried out with the support of the Electricity Company of Hormozgan province, Iran, as a "plan to measure the production of electricity from marine energy in Hormozgan region in the Persian Gulf." The authors appreciate and thank the esteemed management and members of the company's research committee.

Author details

Sepideh Amiri Tavasoli[1], Seyed Jalal Hemmati[1*], Saeed Niazi[1] and Ali Jalali[2]

1 Faculty of Engineering, University of Hormozgan, Department of Mechanical Engineering, Bandar Abbas, Iran

2 Department of Mechanical Engineering, Mashhad Payam Noor University, Mashhad, Iran

*Address all correspondence to: hemmati@hormozgan.ac.ir

References

[1] Magagna D, Uihlein A. JRC Ocean Energy Status Report: Technology, Market and Economic Aspects of Ocean Energy in Europe. European Union, 2015; 2014

[2] Zeiner-Gundersen DH. A novel flexible foil vertical axis turbine for river, ocean, and tidal applications. Applied Energy. 2015;**151**:60-66

[3] Pongduang S, Kayankannavee CH, Tiaple Y. Experimental investigation of helical tidal turbine characteristics with different twists. Energy Procedia. 2015, 2015;**79**:409-414

[4] Priegue L, Stoesser TH. The influence roughness on the performance of a vertical axis tidal turbine. International Journal of Marine Energy. 2017;**17**: 136-146

[5] Harries T, Kwan A, Brammer J, Falconer R. Physical testing of performance characteristics of a novel drag-driven vertical axis tidal stream turbine; with comparisons to a conventional Savonius. International Journal of Marine Energy. 2016;**14**:215-228

[6] Hamidi R. CFD Simulation of Experimental Vertical Axis Tidal Turbine in Persian Gulf Region. 2019

[7] Bouzaher MT, Guerira B, Hadid M. Performance Analysis of a Vertical Axis Tidal Turbine with Flexible Blades. 2017. DOI: 10.1007/s11804-017-1391-0

[8] Delafin P, Deniset F, Astolfi J, Hauville A. Performance improvement of a Darrieus Tidal turbine with active variable pitch. Energies. 2021;**14**:667

[9] Tangler JL, Somers DM. NREL airfoil families for HAWT's. In: Wind Power. Washington; 1995. pp. 117-123

[10] Shamsi R, Ghasemi H. Hydrodynamic performance analysis of horizontal marine current turbines using computational fluid dynamics. In: 15th Conference on Marine Industries, MIC2013 (in Prersian). 2013

[11] Revuz J, Hargreaves DM, Owen JS. On the domain size for the steady-state CFD modeling of a tall building. Wind and Structures. 2012;**15**(4):313-329

[12] Monitoring and Modeling Studies of Iranian Coasts. Ports and Maritime Organization, Private Communication; 2019

[13] Talukdar PK, Kulkarni V, Saha UK. Field-testing of model helical bladed hydrokinetic turbines for small scale power generation. Renewable Energy Journal. 2018

[14] Manwell J, McGowan F, Rogers JG, A. L. Wind Energy Explained: Theory, Design and Application. 2nd ed. UK: Wiley; 2009. p. 704

[15] Schlichting H, Gersten K. Boundary-Layer Theory. Berlin, Heidelberg: Springer; 2017. p. 805

[16] Mohammadi A, Farahat S, Mohammadi M, Mohammadi MR. Investigating the effect of using different airfoils inOutput power of horizontal axis wind turbine blade. In: Fourth International Conference on New Approaches to Maintenance Energy

Section 4

Atmospheric Turbulence and Numerical Models

Chapter 10

Air Pollution Dispersion Using Coupled AERMOD-WRF Modeling System and Generation of Gridded Emission Inventory of NO_X over Nagpur

Rahul Boadh, Yogendra Kumar Rajoria and Vallampati Ramachandra Prasad

Abstract

This study, a work used to distinguish the dispersion patterns of air pollutants, is designs particularly NO_X because of vehicular and modern sources over a quickly creating metropolitan city, Nagpur (21.15° N, 79.09° E), India, during winter and summer representative months, that is, January and April respectively in 2009. Utilizing the emission variables of industry and various vehicles, 1 km x 1 km high-resolution Gridded Emission Inventory (GEI) has produced over Nagpur city. Here, two PBL schemes form WRF model *viz.*, local (MYJ) and non-local (YSU) schemes considered for obtaining the PBL as well as meteorological parameters. Both the schemes MYJ and YSU representing the boundary layer parameters are reasonably well and measured for generating the input parameter for AERMOD model. During the winter and summer month's fundamentally unique scattering patterns of NO_X have been observed. After carefully examining the NO_X concentration from the AERMOD model, the reasonable well-by integrated YSU schemes of WRF were found to be superior to the other schemes. The current review advocates the utility of the created GEI of NO_X with coupled AERMOD-WRF offline model for air quality appraisal over Nagpur.

Keywords: gridded emission inventory, air quality, AERMOD, PBL schemes, WRF model

1. Introduction

Air pollution is a not unserious risk to the climate influencing human's health [1–3], and horticulture and normal biological systems [4]. There are many wellsprings of air pollution, *viz.*, usual sources, transportation, horticultural, and generation of power and industrial industries. According to World Health Organization (WHO) standards, among many Indian urban communities with populace of over a million, air contamination levels surpassed the edge. It is revealed by WHO [5] that nearly 500,000 premature deaths are in India due to only indoor air pollution. Mesoscale atmospheric flows play significant character in dispersion of air pollutants, local air quality problems, and transference [6–12]. Emission estimates and trends during 1990 to 2000 and its implications over Delhi have been reported [13]. AERMOD [14–16], a Gaussian plume air dispersion model with explicit representation of PBL processes, is widely being used for the assessment of air pollution dispersion of various pollutants and estimation of ground-level concentrations (GLC) as well. WRF-AERMOD coupling is employed for the methodology to study the dispersion patterns of PM_{10} over Pune [17]. Gurjar et al. [18] have evaluated emission and air quality in selected mega-cities of different countries. Mohan et al. [19] implemented AERMOD dispersion model in standalone mode for the exposure assessment of particulate matter pollution over Delhi. Goyal et al. [20] have developed grid-based mobile source emission inventory over Delhi using IVE modeling tool. It is appraised the enactment of WRF, for the situation usage as the meteorological pre-processor aimed at investigative modeling of air quality using AERMOD over Cuba [21]. A sensitivity study of different PBL schemes of parameterization schemes of WRF model coupled with AERMOD NO_X dispersion over a shoreline city of Visakhapatnam [22] and over Gurugram [23] was conducted.

The good activist resolution of various parameters such as upper air meteorological and surface parameters in standalone mode is required by AERMOD model. In maximum Indian cities, this type of the PBL data is absent and over Nagpur such data are not available. The high resolution (3 km) has been utilized for integrating the meteorological variables from WRF model, which is required in air quality simulation. The result obtained is used with offline coupler [22]. The presented coupler in the present work originates the PBL and surface meteorological parameters over Nagpur and generates required input variables for the integration of AERMOD. It is obviously true that PBL processes assume a significant part in the scattering of air toxins and is crucial for utilizing reasonable PBL plot in WRF-ARW model. It has been reported that YSU followed by MYJ has shown better simulation PBL processes after validating it with available observations over Nagpur region [22, 23].

In this study, AERMOD [24, 25] and a Mesoscale weather prediction model WRF-ARW [26, 27] with YSU and MYJ schemes are employed for calculation of NO_X dispersion and estimation of ground-level concentrations (GLC), due to vehicular traffic during two contrast seasons such as winter and summer during the year 2009 over Nagpur city. January has been chosen as a representative month of winter and April as summer month following the criteria of India Meteorological Department, India. In this study, GEI with high resolution (1 km $\times$ 1 km) using emission factors of different types of vehicular traffic over the study area has been developed. An additional NO_X emission from a thermal power plant (TPP) located in the north direction which is 11 km away from Nagpur city has also been considered. The model-simulated NO_X concentrations are validated using the monitored observations by CPCB to assess the presentation of the displaying framework with two distinct PBL plans during summer and winter.

2. Study area

Nagpur (21.15° N, 79.09° E) is the biggest city in focal of India and the second capital of the territory of Maharashtra. Nagpur is the quickly developing city and third most crowded city after Mumbai and Pune in Maharashtra. Additionally, it is the Center for advancement, industrialization, urbanization, and business action (**Figure 1**). The main sources of air pollution are vehicular, and industrial and domestic sources. Over Nagpur city, according to the National Ambient Air Quality Standards (NAAQS)—2008, suspended particulate matter (SPM), and respirable suspended particulate matter (RSPM) fundamentally high, the degree of nitrogen dioxide (NO_2) is moderate [28]. NO_X is of specific worry as it is radiated from industrial and vehicle cycles that consume non-renewable energy sources. NO_X draws consideration as it responds with volatile organic compounds (VOCs) to shape ground-level ozone and it likewise responds with different constituents in air to frame nitrates and corrosive vapor sprayers. Nitrogen dioxide disturbs the nose and throat, and it seems to build vulnerability to respiratory diseases separated from nearby wellsprings of air contamination. Territorial and metropolitan air quality peculiarities are affected by a few different factors like the local meteorology.

According to climate classification by Köppen, the dry climate and tropical wet climate with dry conditions found for throughout year over Nagpur. Nagpur is situated at a height of 312.42 meters above sea level. Over Nagpur region, the winter season lies from late November to January and the temperature can goes below 10°C. Summer lies from late March to June and it is extremely hot and May being the hottest month over Nagpur region. On May 22, 2013, highest temperature (47.9°C) was recorded in Nagpur city.

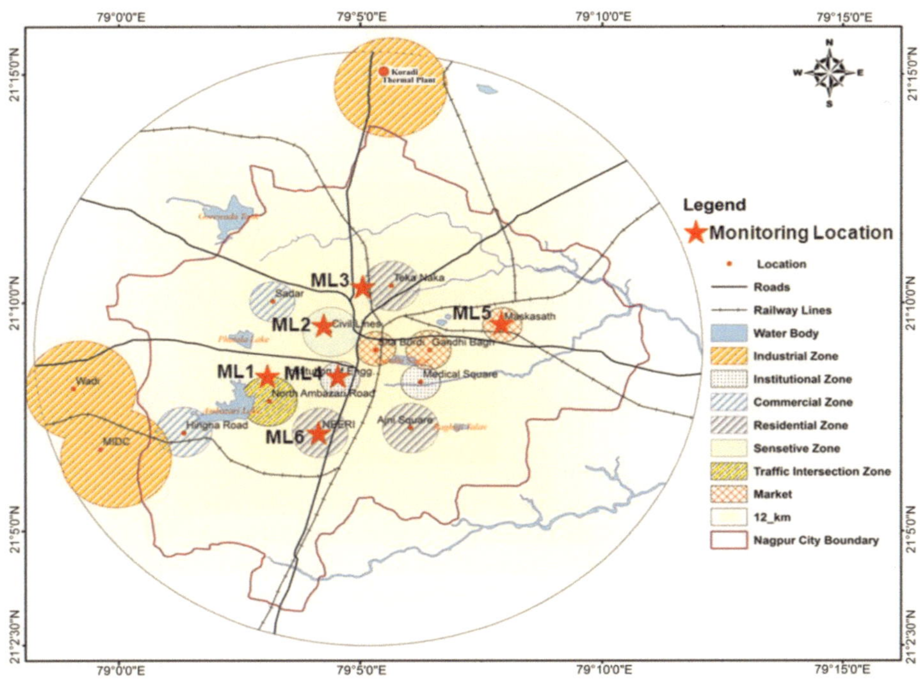

Figure 1.
Study area (30 km × 30 km) over Nagpur [monitoring locations (ML1 = .MIDC industrial area, ML2 = MIDC office, ML3 = Govt. polytechnic college, ML4 = Institute of Engineering, ML5 = Maskasath, and ML6 = NEERI lab)].

3. Data and methodology

The data required for running WRF-ARMOD coupled modeling system and methodology of emission inventory and prediction of pollutant concentrations over Nagpur city using integrated WRF-ARMOD coupler are described in the following sections.

3.1 Data

In the present section, the mandatory data for offline coupled modeling system, gridded data of emission inventory from area and point sources over Nagpur, and ambient air quality monitored by CPCB are described in detail.

3.1.1 Data required for WRF-ARW model

In the current review, here accessible 6-h global final analysis (FNL) utilized with $1.0° \times 1.0°$ grid resolution produced by the National Center for Environmental Prediction (NCEP's) global forecast system (GFS) as initial conditions, during study period in January and April 2009.

3.1.2 Emission data

In this study, two types of emission data have been used. One is the vehicular data along with exhaust emissions considered as an area emission and the other is thermal power plant (TPP) data considered as point source over the Nagpur city. The CPCB has six monitored stations over Nagpur city. They are MIDC industrial area, MIDC office, Government Polytechnic College, Institute of Engineering, Maskasath, and NEERI Lab. The NO_X-monitored data are accessible as 24 h average from CPCB, designed for all above six monitoring locations.

The data are not uniform w. r. to time at all monitoring locations; that is, the monitored data at each location are available on different days. Hence, in the current study, the comparison of observed and predicted concentrations has been done at those monitoring locations where data are available for the days, the concentrations are estimated.

3.1.3 Estimating number and types of vehicles pursuing over Nagpur

In this study, a number of vehicles plying in Nagpur city are calculated using RTO (Regional Transport Office) report for Maharashtra state [29]. Based on this report, there is an increase in the number of vehicles registered at the rate of 7.21% per year. Accordingly, we have estimated the number of vehicles at Nagpur city from the base year 2011. The yearly vehicular growth during the last 16 years is shown in **Figure 2**. According RTO, vehicles registered before 15 years from the current year will not be allowed to play on the road.

We considered this as assumption that the vehicles registered in 1994 were not plying on the roads in 2009. For year 2009 (base year), diverse types of vehicles with pollutant emission factor for NO_X obtained from Automotive Research Association of India (ARAI) are shown in **Table 1** [30]. The TPP located at Koradi area of Nagpur is 11 km away from Nagpur city zero miles, which is considered as point source.

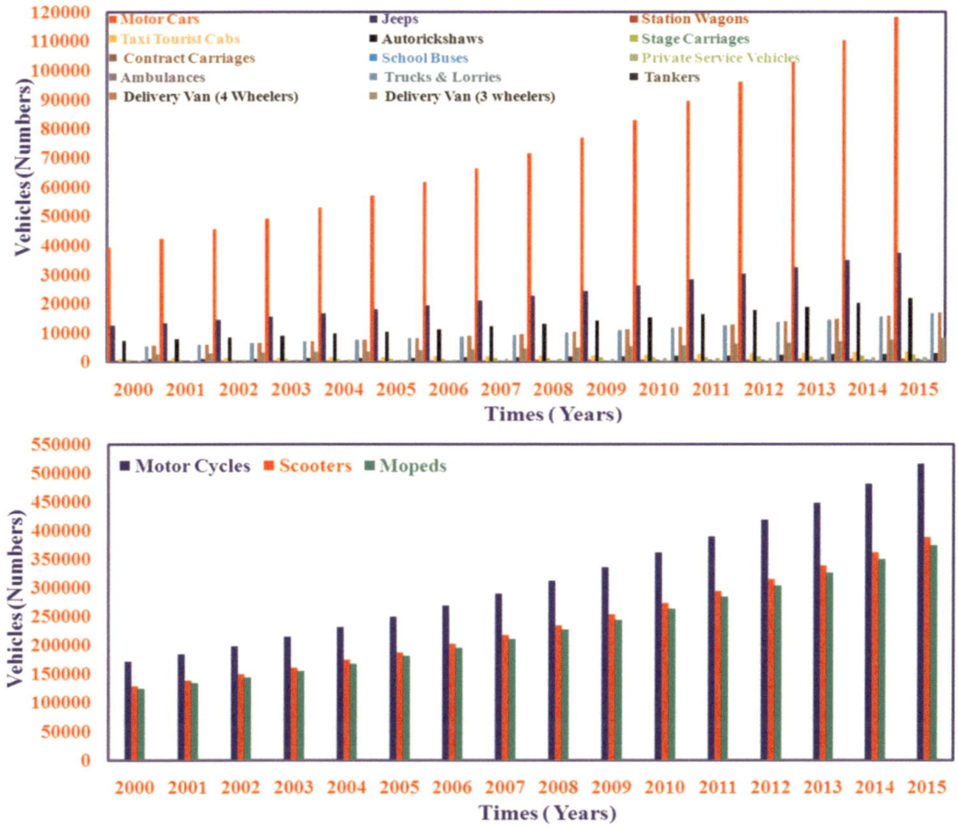

Figure 2.
Yearly vehicle growth for the past 16 years over Nagpur city.

For the calculation of NO_X emissions from the area sources, the Nagpur city (9 km × 8 km) area of 72 km^2 has been considered with 1 km × 1 km grid resolution. The selected area covers all six monitoring stations of CPCB, where urban activities take place due to different sources, and predominantly vehicular exhaust is the main source of air pollution in Nagpur city. Considering this criterion, emissions from 64 grid cells were used as input into the model.

The calculation of emissions from vehicles is based on the emission factors for different types of vehicles according to ARAI 2007, vehicles explicit kind [29], the distance went by a specific vehicle, and their appropriation in light of the sort of the fuel utilized. The methodology to obtain emission from vehicular estimates was based on an earlier work [13, 19, 31];

$$E_i = \sum \left(Veh_j \times D_j\right) \times E_{i;j;km}$$

The emission of the compound (i) is denoted by E_i, per type of vehicles numbers denoted by Veh_j and different vehicle type (j) traveled distance defined by D_j in a grid box. This study has assumed based on the previous study according to ARAI, Ei;j; km the emission factor (g km^{-1}) of different vehicles of compound (i) and vehicle type (j) per driven kilometer. The GEI of NO_X over Nagpur city is shown in **Figure 3**. Based on the previous study, this study assumed only 50% vehicles are moving on

Vehicle type	Number of vehicles	Fraction	Emission factor (gkm^{-1})
Motor Cycles (Petrol)	226,558	113,279	0.15
Scooters (Petrol)	170,702	85,351	0.14
Mopeds (Petrol)	164,827	82,413	0.02
Personal Cars (Petrol)	52,266	26,133	0.09
Multi Utility Vehicles (Diesel)	16,892	8446	0.67
Personal Cars (Diesel)	1545	773	0.28
Three Wheelers (Petrol)	9534	4767	0.16
Three Wheelers (Diesel)	3578	1789	0.51
Busses (Diesel)	1772	886	6.53
Heavy Commercial Vehicles (Diesel)	8672	4336	9.3
Light Commercial Vehicles (Diesel)	8238	4119	2.12

Table 1.
Different types of vehicles based on ARAI-2007 and emission factor over Nagpur city in 2009.

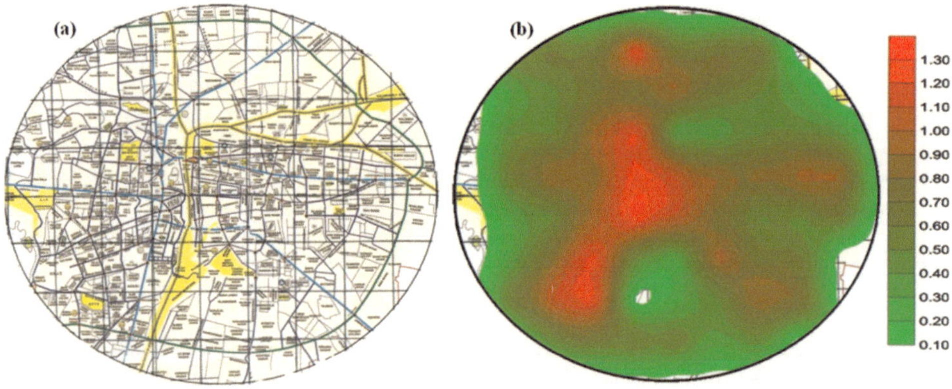

Figure 3.
(a) Nagpur city map (9 km × 8 km) and (b) emission inventory (gs^{-1} m^{-2}) over Nagpur city.

road in per grid box. According of ARAI emission factors, vehicular emissions of NO_X are designed for different types of the vehicles in each 1 km × 1 km grid box.

3.2 Methodology

3.2.1 Air pollution dispersion model

AERMOD [14] is a consistent state Gaussian plume model valuable for the calculation of contamination scattering relevant to metropolitan and provincial region and several sources (volume, point and area) of emissions [14–16].

AERMOD expects the likelihood circulation work for the toxin concentration to be Gaussian in the both horizontal and vertical steady boundary layers. Then, at that point, in present instance of the convective boundary layer, while the flat scattering is Gaussian, the upward dissemination is addressed by a bi-Gaussian likelihood conveyance function.

In light of perceptions of meteorological predictions using a similarity connection, the rising profiles of the essential components, such as wind and temperature, are constructed.

AERMOD represents the vertical in-homogeneity of the PBL in its scattering estimation by averaging the boundaries of real PBL into successful boundaries of an identical homogeneous PBL. The AERMOD meteorological preprocessor, recognized as AERMET, calculates the PBL boundaries, *viz.*, friction velocity, temperature scale, Monin-Obukhov length, surface heat flux, convective speed scale, blending tallness by involving nearby surface attributes as unpleasantness of surface, and Bowen proportion in mix with the standard meteorological perceptions [22, 32]. AERMOD is an analytic model and uses limit layer boundaries alongside routine meteorological information as contribution for assessing the contamination focuses. The overproduction of concentrations of pollutants is because of the lower reversal/blending stature under solid stable circumstances anticipated by WRF.

For predicting 24 hourly normal NO_X concentration utilizing metropolitan scattering choice, the model has been utilized in this study. In this work, two different domains were chosen: Domain-1 with 30 km × 30 km with 1-km grid resolution considering the entire city including TPP and Domain-2 a finer 12 km × 12 km with 1-km grid resolution without considering TPP emissions. This exercise has been conducted so as to assess whether TPP emission does have any influence in the concentration of NO_X over Nagpur. The consequences of the integrated concentration of NO_X are approved against CPCB information on 24-h noticed concentration of NO_X over Nagpur.

3.2.2 Weather research and forecast model

WRF version 3.6 mesoscale model established by the National Center for Atmospheric Research (NCAR) is utilized in the existing study [26, 27, 33]. This model contains the bother amounts of tension, scalars (cloud water, water vapor mixing ratio etc.), turbulent kinetic energy, surface pressure, geo-potential, three-dimensional wind, and potential temperature and includes the prognostic variables. WRF is also completely squeezable with non-hydrostatic equations. From the literature, one can say that PBL cycles and land surface affect the incorporation of winds, choppiness, and other state factors in the lower atmosphere.

The two PBL schemes YSU and MYJ both have shown better performance than the other schemes of WRF-ARW chosen based on the works over the Nagpur region and three nested domains used in study (**Figure 4**) [22, 25]. The complete model over view consists of model domain; different parameterizations used are presented in **Table 2**.

3.2.3 Offline coupler of WRF and AERMOD models

Hourly meteorological observations of surface and upper air remain mandatory to AERMOD. The model acknowledges a solitary station information and expects that the condition of the weather is evenly homogenous over the whole study region. In any case, the sort of meteorological information, which is fundamental to process the necessary boundary layer variables that can fill in as contribution to AERMOD, is not sweeping the study region. Subsequently, in the current work, we have utilized the offline coupler to incorporate the weather research forecast and AERMOD models.

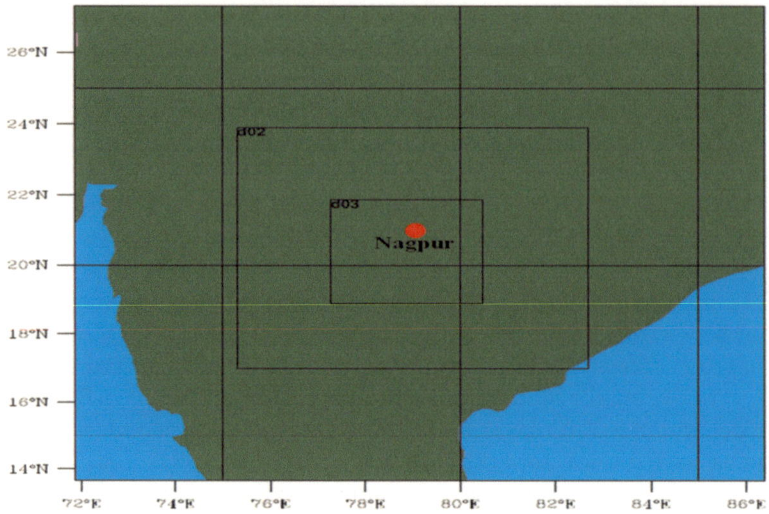

Figure 4.
WRF-ARW model domain used over Nagpur.

PBL scheme	1) YSU [34] 2) MYJ [35, 36]
Vertical coordinates	Terrain-following hydrostatic pressure vertical coordinate with 27 vertical levels
Covered area	13.8°-27.4° N and 71.9°-86.2° E
Interval	6 hrs
Spatial differencing scheme	6th order center differencing
Microphysics	Eta microphysics [37]
Horizontal grid system	Arakawa-C grid
Map projection	Mercator
Surface layer parameterization	Noah land Surface Scheme
Integration time step	90 sec
Data	NCEP FNL
Time integration scheme	3rd-order Runga-Kutta Scheme
Long-wave radiation	RRTM scheme radiation [38]
Short-wave radiation	Dudhia scheme [39]
Dynamics	Non-hydrostatic
Resolution	Domain1: 27 km × 27 km Domain2: 9 km × 9 km Domain3: 3 km × 3 km
Grid size	Domain1: (60× 60) × 27 Domain2: (91× 91) × 27 Domain3: (112× 112) × 27
Cumulus parameterization	Kain-Fritsch scheme [40]

Table 2.
Overview of the WRF-ARW modeling system.

Presented coupler directly generates meteorological input files mandatory in AERMOD model [22] from the surface and PBL parameters over Nagpur from the output of WRF model.

4. Results and discussion

The current section approval of GLCs and scattering patterns of NO_X acquired by utilizing the coordinated coupler of WRF-AERMOD are examined.

4.1 Simulation and dispersion of NO_X

In segment, the coupled WRF-AERMOD displaying framework reenacted dispersion patterns and GLCs of NO_X utilizing the different information (presented in Section 3) and Emission Inventory (EI); approval practices utilizing the accessible checked perceptions from factual investigation are introduced. The meteorological information obtained from WRF-ARW by utilizing PBL plans YSU (consequently alluded to as Model-1) and MYJ (hereafter alluded to as Model-2) are utilized in the coupled modeling framework for foreseeing NO_X concentrations over the review area. The exhibition assessments of integrated NO_X concentrations by Model-1 and Model-2 are learned through the accessible perceptions. The investigation and approval of the obtained GLC of NOX subsequent to coordinating the modeling framework more than two chosen domains (Domain-1 and Domain-2) of Nagpur city are presented in Section 3.2.1 with the accessible observations (given in Section 3.1.3).

4.2 Validation and GLCs of NO_X

For the model combination, we utilized EI of area sources over Nagpur (without TPP in Domain-2) for integrating the concentrations of NO_X pollutant. Concentrations of NOX pollutant in January and April 2009 anticipated by Model-1 and Model-2 with the accessible CPCB checked information are portrayed in **Figures 5** and **6**, individually.

The convergence of NO_X anticipated by Model-1 shifts 52 to 110 μgm^{-3}, though by Model-2 fluctuates between 53 and 180 μgm^{-3}, while the noticed focus differs from 28

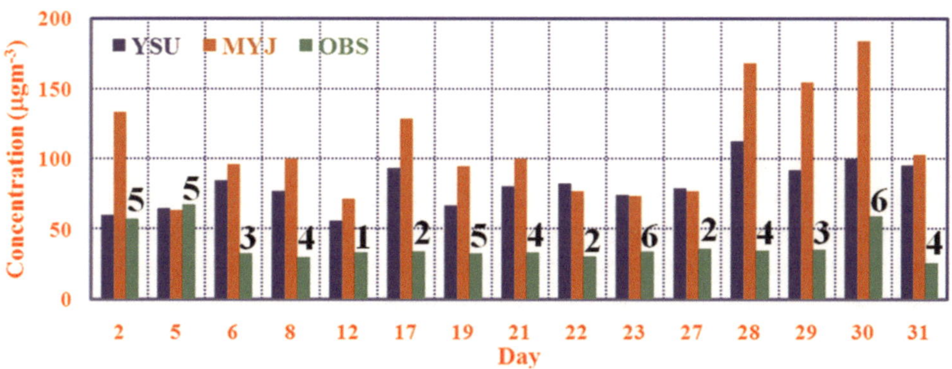

Figure 5.
Validation of model predicted 24-hour NO_X concentration (μgm^{-3}) with CPCB observation in January 2009 over Nagpur (with the monitoring location number given in ***Figure 1****).*

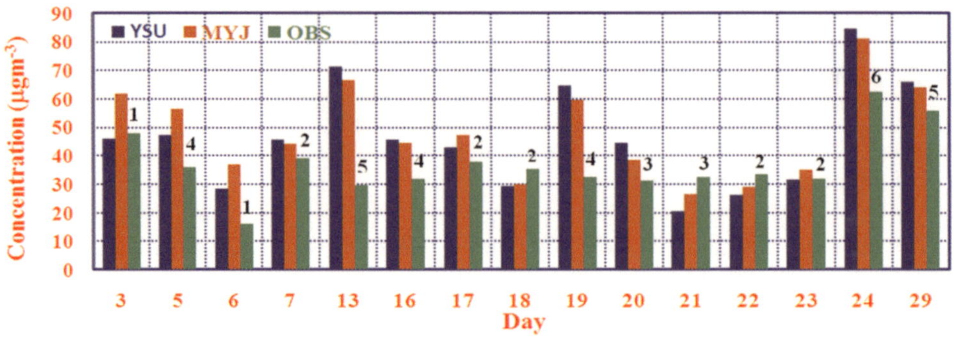

Figure 6.
Validation of model predicted 24-hour NO_X concentration (μgm-3) with CPCB observation in April 2009 over Nagpur (with the monitoring location number given in ***Figure 1****).*

to 54 μgm^{-3} during winter. On the correlation with the observed qualities, over-forecast of re-enacted convergence of NO_X by both Model-1 and Model-2 is seen (**Figure 5**). This might be credited to that reality the more grounded stable soundness conditions and low breezes reproduced by the WRF-ARW model prompting more catching of contamination interface of NO_X.

During summer month, the anticipated fixation by Model-1 and Model-2 shifts between 20 to 85 μgm^{-3} and 28 to 80 μgm^{-3} separately. During same month, CPCB perception of NO_X fixations fluctuates somewhere in the range of 18 and 61 μgm^{-3}. On the nearby assessment of the outcomes presented in **Figure 6** uncovers, the scope of fixations by the Model-1 and Model-2 is practically identical with the perceptions, with generally better execution by Model-1. The primary explanation of the over expectation of the NO_X focus could be because of the exclusion of metropolitan geology in the current review since this sort of information getting is extremely intense in India. The metropolitan regions would adjust the thermo-dynamical design of the barometrical limit layer and changes the strength definition. Because of the avoidance of metropolitan elements, there are potential outcomes of over assessment of stable circumstances by the models and prompts more catching of NO_X results in over forecast more. To find out the best execution of the models, factual examination is directed as given in the accompanying section.

4.3 Statistical analysis

The model presentation was assessed utilizing classical statistical [41] parameters such as geometric variance (VG), normalized root mean square error (NMSE), fractional bias (FB), mean absolute error (MAE), correlation coefficient (CC), root mean square error (RMSE), and fraction of two (FAC2) (prediction and observations). The results for the performance of statistical indicators have been shown in **Table 3**.

It is found from these results that the RMSE and MAE of Model-1 are less as compared to Model-2 during winter month and summer month, as reported in previous studies [19, 23, 42, 43]. The statistical parameters, for example, FB and NMSE would give the necessary dependability rules in the model execution assessment. Chang and Hanna [43] examined the correlation of execution north of five field tests for three models specifically AERMOD, AMDS, and ISC-3. Hanna et al. [44] contemplated and observed and getting error VG and FAC2 separately 2.9 and 0.46 from

Parameter (ideal value)	January 2009 (Winter)		April 2009 (Summer)	
	YSU scheme	MYJ scheme	YSU scheme	MYJ scheme
MAE (0)	48.61	77.38	18.52	21.85
RMSE (0)	54.34	87.90	24.73	28.92
CC (1)	0.12	0.36	0.55	0.50
NMSE (0)	0.99	1.94	0.35	0.44
FB (0)	0.80	1.03	0.34	0.42
FAC2 (1)	0.43	0.32	0.71	0.66
VG(1)	2.02	3.23	1.07	1.22

Table 3.
Statistical analysis of NO_X concentrations during winter and summer months over Nagpur city.

AERMOD. This is comparable to model-1 error for both of the month as compared to model-2.

The performance of AERMOD model is examined over Delhi [19]. They observed that RMSE error is higher for each of the seven stations, and the scope of the error lies between 90.05 and 245.66. In the current work, RMSE error is sensibly well in contrast to the past review and this is given affirmation that model-1 is given better performance in contrast to the model-2.

Less values of FB and NMSE indicate that the overall deviation is less by Model-1 compared to Model-2. These boundaries uncover that exhibition of Model-1 is superior to the Model-2 during winter and summer months. FB demonstrates how well the calculation creates the normal qualities around the normal upsides of observed factors. The ideal value is zero of FB, but it can range from −2 to 2. An absolute value of 0.67 corresponds to a prediction inside an influence of two of observation [45]. Interestingly, correlation coefficient of Model-2 is to found better as compared to Model-1 for during winter and summer months. Yet, in light of the in general factual assessment, the presentation of expectation of NO_X concentrations utilizing the meteorological factors integrated from Model-1 is superior to Model-2. Large values of MAE, RMSE, NMSE, FB, and VG1 indicate overprediction of concentrations by both models due to low wind conditions.

4.4 Spatial dispersion of GLC NO_X concentrations

In the current segment, spatial scattering examples of GLC NO_X concentrations are obtained from WRF-AERMOD modeling system utilizing two definition plans; specifically, YSU and MYJ are introduced over Domain-1 and Domain-2 over Nagpur city (as presented in Section 3.2.1). The coupled demonstrating framework has been incorporated for the entire long periods of winter and summer months over the picked spaces at whatever point the CPCB monitoring of GLC perceptions is accessible during winter- and summer-represented months. In this chapter, the aftereffects of 1 day each from these 2 months, January 8 and April 21, 2009 are just introduced.

The dispersion of 24-hourly averaged NO_X concentrations for 30 km × 30 km domain on January 8 by utilizing PBL schemes of YSU and MYJ is represented in **Figure 7a–e** along with wind roses. On January 8, 2009, the most extreme GLCs are anticipated by MYJ plot (**Figure 7b**) contrasted with the YSU (**Figure 7a**), in light of the fact that MYJ (**Figure 7d**) re-enacted low breezes than that of YSU as displayed in **Figure 7c**. The extent of winds obtained from MYJ is lesser as compared with YSU.

This clearly shows that the winds stay improved arrested by YSU upon an evaluation by the observations (**Figure 7e**) with MYJ. Maximum concentrations of NO_X by using both schemes (YSU and MYJ) inside the city stay only due to vehicular emission. From the concentration patterns, it is clearly seen that nearby is nope influence of TPP-generated NOX over Nagpur as seen in **Figure 7a** and **b**.

The NO_X concentrations of April 21, 2009 are depicted in **Figure 8a–e** along with wind roses. The similar patterns of the wind roses are shown by both schemes YSU and MYJ, respectively (as seen in **Figure 8c** and **d**) as shown in observations (**Figure 8e**). The wind simulations of MYJ scheme are showing the low winds in comparison with YSU scheme but are in better arrangement with the observation. The GLCs of NO_X predicted by both schemes YSU and MYJ are presented in **Figure 8a** and **b**, respectively. MYJ predicted higher concentration compared to YSU, because MYJ simulated low wind compared with YSU. This is on the grounds that the impact of wind speed is contrarily corresponding to the centralizations of the pollutants [46, 47].

In comparison to the winter day, there is significantly less NO_X emission from TPP over Nagpur during this summer day. Comparative sort of results is seen during every one of the times of the both study months. Henceforth, we can infer that the impact of the power plant emissions of NO_X concentrations is negligible over Nagpur city for both winter and summer months. Based on this result, a fine resolution of 12 km × 12 km domain was selected for predicting the GLCs of NO_X over the Nagpur city for these two contrasting months and results during January 8, 2009 and April 21, 2009 are presented.

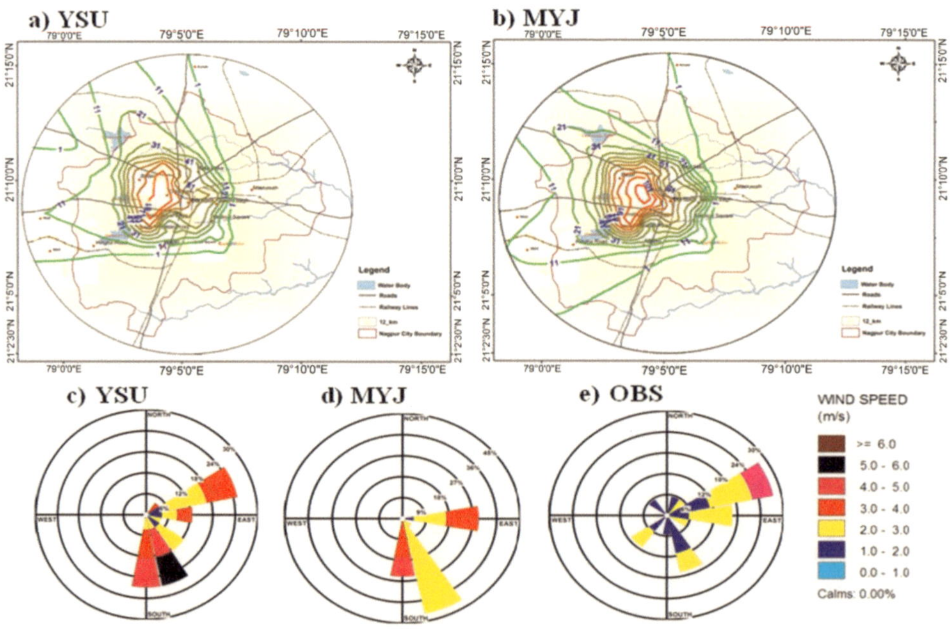

Figure 7.
Spatial distribution of 24-hour predicted NO_X GLCs (μgm^{-3}) (a) YSU, (b) MYJ (c) wind roses of YSU (d) wind roses of MYJ and (e) OBS wind roses on January 8, 2009, over Nagpur by using 30 km ×30 km domain.

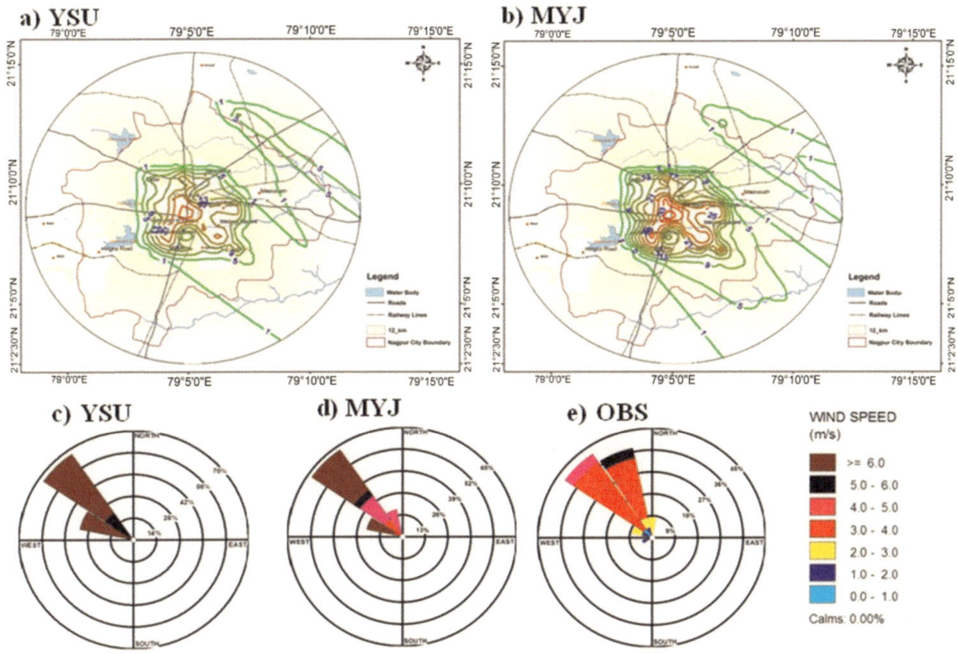

Figure 8.
Spatial distribution of 24-hour predicted NO_X GLCs (μgm^{-3}) (a) YSU, (b) MYJ (c) wind roses of YSU (d) wind roses of MYJ, and (e) OBS wind roses on April 21, 2009 over Nagpur by using 30 km ×30 km domain.

The forecasted concentration on January 8, 2009 from both (YSU and MYJ) schemes along with wind roses is presented in **Figure 9**. The determined NO_X concentration forecasted by PBL schemes YSU and MYJ (**Figure 9a** and **b** respectively) on January 8 is mostly in western part of the city, because the predominant wind directions are east northeasterly and southeasterly (**Figure 9c** and **d**). The predicted concentration of NO_X by both (YSU scheme and MYJ scheme) the schemes on April 21, 2009 is presented in **Figure 10**. The predominant wind direction is northwesterly (**Figure 10c** and **d**) and the results (the maximum concentration of the NO_X) predicted by YSU (**Figure 10a**) and MYJ (**Figure 10b**) schemes are in southeastern quadrant of the Nagpur city.

Similar kind of signatures of maximum concentrations is noticed during study period of these two contrasting months. These varieties of scattering of concentrations of NO_X are critical during these two differentiating months' outcomes in particular unique air quality over Nagpur. In light of the model inferred most extreme NO_X concentrations, not many spots of the Nagpur city are distinguished as area of interest areas meaning decayed air quality and are displayed in **Table 4**.

The recognized hotspot sites are Sita Buldi, Central bazaar road, North Ambazari road, Anjumna square, Liberty square, Low College square, Katol road, police commissioner office, and Sankar Niger square located in the western part to Nagpur city. These locations have high concentrations in January month and low concentrations in April month. Other locations such as medical chowk, Baidwanath chowk, Great nag road, NMC office Mahal, Sakkardara road, and Rajendra nagar are in north and southeastern part of the Nagpur city. These locations have higher concentrations of NO_X in January and low concentrations in April 2009. These changes in concentrations in the different locations of the Nagpur city during these two representative

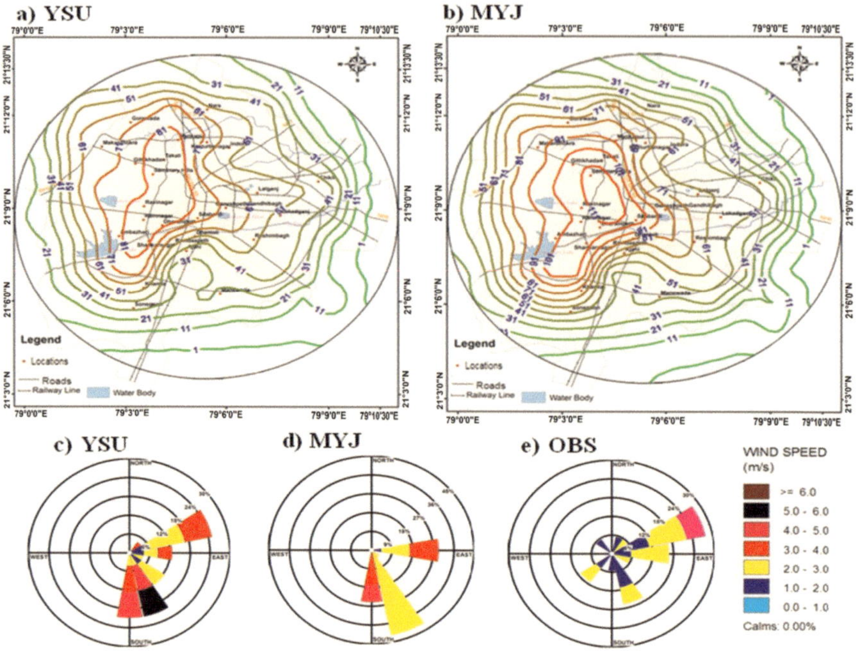

Figure 9.
Spatial distribution of 24-hour predicted NO_X GLCs (μgm^{-3}) (a) YSU, (b) MYJ (c) wind roses of YSU (d) wind roses of MYJ and (e) OBS wind roses on January 8, 2009 over Nagpur by using 12 km × 12 km domain.

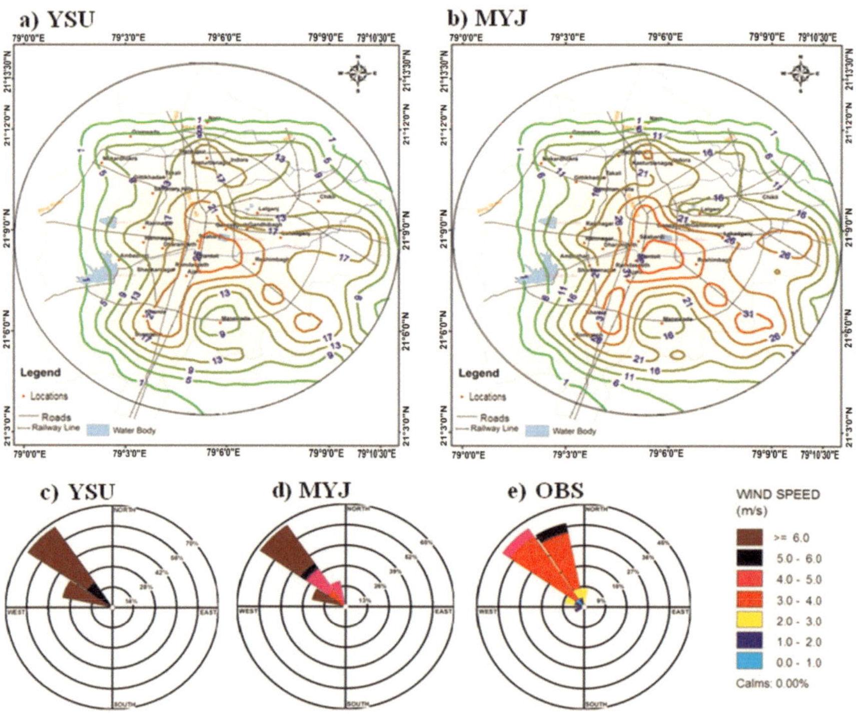

Figure 10.
Spatial distribution of 24-hour predicted NO_X GLCs (μgm^{-3}) (a) YSU, (b) MYJ (c) wind roses of YSU (d) wind roses of MYJ and (e) OBS wind roses on April 21, 2009 over Nagpur by using 12 km ×12 km domain.

Hotspot locations over Nagpur City based on higher NO_X concentrations	Average NO_X Concentration (μgm^{-3}) of January 2009 (Winter)		Average NO_X Concentration (μgm^{-3}) of April 2009 (Summer)	
	YSU	MYJ	YSU	MYJ
Sita Buldi	92	105	41	44
Central Bazar Road	80	90	38	25
North Ambazari Road	65	87	32	28
Anjuman Square	78	96	30	24
Liberty Square	62	84	29	26
Sankar Nagar Square	78	98	34	28
Low College Square	75	84	37	23
Katol Road	65	75	28	22
Police Com. Office	73	96	35	26
Medical Chowk	48	52	34	36
Baidwanath Chowk	35	48	29	34
Great Nag Road	54	62	33	35
NMC Office Mahal	62	67	38	29
Sakkardara Chowk	38	43	30	31
Rajendra Nagar	29	35	21	29

Table 4.
Identified deteriorated air quality (hotspot) locations over Nagpur City.

months of summer and winter season have discovered the monthly variability of deteriorated air quality locations in relations to extreme concentrations of NO_X.

5. Conclusions

WRF-ARW model is combined with AERMOD for processing GLCs of NO_X over Nagpur city. Scattering of NO_X by utilizing YSU and MYJ planetary limit layer plans of WRF-ARW model over Nagpur has been assessed.

Accurate assessment of the air pollution dispersion as well as air quality issues requires high-resolution emission inventory and air pollution dispersion models with better representation of the PBL parameters. One of the contributions emanated from this study is the preparation of gridded emission inventory using the vehicular traffic data over Nagpur city with 1-km resolution. The offline coupled modeling system of WRF-AERMOD reasonably simulated GLCs concentrations of NO_X during both representative months of winter and summer seasons over Nagpur city.

The coarse resolution of the (30 km × 30 km) has shown negligible impact of NO_X emission released from Thermal Power Plant in both the contrasting. The GLCs obtained from the finer resolution (12 km × 12 km) over of the Nagpur city has revealed significant dispersion patterns during the 2 months. Deteriorated air quality locations referred to as hotspot locations, based on the maximum NO_X concentrations, are identified for Nagpur. Significant variations in the concentrations are noticed during these two seasons. This study advocates the better representation of air pollution dispersion has shown by offline coupled modeling system of WRF-AERMOD and given confidence for its utility in the air quality assessment.

Author details

Rahul Boadh[1], Yogendra Kumar Rajoria[1] and Vallampati Ramachandra Prasad[2]*

1 Department of the Mathematics, School of Basic and Applied Science, K.R. Mangalam University, Gurugram, Haryana, India

2 Department of Mathematics, School of Advanced Sciences, Vellore Institute of Technology, Vellore, Tamil Nadu, India

*Address all correspondence to: v.ramachandraprasad@vit.ac.in; rcpmaths@gmail.com

References

[1] Dockery DW, Pope CA, Xu X, Spengler JD, Ware JH, Fay ME, et al. An association between air pollution and mortality in six U.S. cities. The New England Journal of Medicine. 1993; **329**(24):1753-1759

[2] Solomon S, editor. Intergovernmental Panel on Climate Change, Summary for policymakers, in Climate Change. In: The Physical Science Basis. Contribution of Working Group I to the Fourth Assessment Report of the Intergovernmental Panel on Climate Change. New York: Cambridge University Press; 2007. pp. 1-18

[3] Ramanathan V, Feng Y. Air pollution greenhouse gases and climate change: Global and regional perspective. Atmospheric Environment. 2009;**43**: 37-50

[4] Grimm NB, Faeth SH, Golubiewski NE, Redman CL, Wu J, Bai X, et al. Global change and the ecology of cities. Science. 2008; **319**(5864):756-760

[5] World health organization. Quantifying Selected Major Risks to Health, Chapter 4 from the World Health Report, World Health Organization, Geneva. 2002

[6] Srinivas CV, Venkatesan R, Baskaran R, Rajagopal V, Venkatraman B. Regional scale atmospheric dispersion simulation of accidental release of radionuclides from Fukushima Dai-ichi reactor. Atmospheric Environment. 2012; **61**:66-84

[7] Angevine WM, Eddington L, Durkee K, Fairall C, Bianco L, Brioude J. Meteorological model evaluation for CalNex 2010. Monthly Weather Review. 2012;**140**:3885-3906

[8] Fast JD, Gustafson WI, Berg LK, Shaw WJ, Pekour M, Shrivastava M, et al. Transport and mixing patterns over Central California during the carbonaceous aerosol and radiative effects study (CARES). Atmospheric Chemistry and Physics. 2012;**12**: 1759-1783

[9] Baker KR, Misenis C, Obland MD, Ferrare RA, Scarino AJ, Kelly JT. Evaluation of surface and upper air fine scale ARW meteorological modeling of the may and June 2010 CalNex period in California. Atmospheric Environment. 2013;**80**:299-309

[10] Hu XM, Nielsen GJW, Zhang F. Evaluation of three planetary boundary layer schemes in the WRF model. Journal of Applied Meteorology and Climatology. 2010;**49**:1831-1844

[11] Hu XM, Klein PM, Xue M. Evaluation of the updated YSU planetary boundary layer scheme with in WRF for wind resource and air quality assessments. Journal of Geophysical Research – Atmospheres. 2013;**118**: 10490-10505

[12] Srikanth M, Satyanarayana ANV, Srinivas CV, Kumar M. Mesoscale atmospheric flow-field simulations for air quality modelling over complex terrain region of Ranchi in eastern India using WRF. Atmospheric Environment. 2015;**107**:315-328

[13] Gurjar BR, Aardenne JA, Lelieveld J, Mohan M. Emission estimates and trends (1990–2000) for mega city Delhi and implications. Atmospheric Environment. 2004;**38**:5663-5681

[14] Cimorelli AJ, Perry SG, Venkatram A, Weil JC, Paine RJ, Wilson RB, Lee RF, Peters WD,

Brode RW, Paumier JO. AERMOD: Description of model formulation. US Environmental Protection Agency, EPA Report No. 454/R-03-002d, 2004; p 85.

[15] Cimorelli AJ, Perry GS, Venkatram A, Weil JC, Paine RJ, Wilson RB, et al. AERMOD: A dispersion model for industrial source applications. Part 1: General model formulation and boundary layer characterization. Journal of Applied Meteorology. 2005;**44**: 682-693

[16] Perry SG, Cimorelli AJ, Paine RJ, Brode RW, Weil JC, Venkatram A, et al. AERMOD: A dispersion model for industrial source applications. Part II: Model performance against 17 field study databases. Journal of Applied Meteorology. 2005;**44**:694-708

[17] Kesarkar AP, Dalvi M, Kaginalkar A, Ojha A. Coupling of the weather research and forecasting model with AERMOD for pollutant dispersion modeling, a case study for PM10 dispersion over Pune, India. Atmospheric Environment. 2007; **41**:1976-1988

[18] Gurjar BR, Butler TM, Lawrence MG, Lelieveld J. Evaluation of emissions and air quality in megacities. Atmospheric Environment. 2008;**42**: 1593-1606

[19] Mohan M, Bhati S, Rao A. Application of air dispersion modeling for exposure assessment from particulate matter pollution in mega City Delhi. Asia-Pacific Journal of Chemical Engineering. 2011;**6**:85-94

[20] Goyal P, Mishra D, Kumar A. Vehicular emission inventory of criteria pollutants in Delhi. Springer Plus. 2013; **2**:216

[21] Carbonell LT, Mastrapa GC, Rodriguez YF, Escudero LA, Gacita MS, Morlot AB, et al. Assessment of the weather research and forecasting model implementation in Cuba addressed to diagnostic air quality modeling. Atmospheric Pollution Research. 2013;**4**: 64-74

[22] Boadh R, Satyanarayana A NV, Rama Krishna TVBPS, Madala S. Sensitivity of PBL parameterization schemes of Weather Research Forecasting Model and coupling with AERMOD in the dispersion of NO_X over Visakhapatnam (India). Asia-Pacific Journal of Chemical Engineering. 2015. DOI: 10.1002/apj.1876

[23] Sharma N, Singh R, Rojoria YK, Rajendra P, Boadh R. Coupled WRF-AERMOD modeling system by using dispersion of air pollutant and generation of gridded emission inventory of NO_X over Faridabad and Gurugram. International Journal of Scientific Research in Mathematical and Statistical Sciences. 2021;**8**(1):01-12

[24] Boadh R, Satyanarayana ANV, Rama Krishna TVBPS, Madala S. Sensitivity of PBL schemes of WRF-ARW model in simulating boundary layer flow parameters for its application to air pollution dispersion modeling over a Tropical Station. Atmosfera. 2016;**29**(1): 61-81

[25] Boadh R, Satyanarayana ANV, Rama Krishna TVBPS, Subba Rao A, Rajendra P. Numerical simulation of boundary layer flow parameters by using wrf-arw model over a tropical region. Journal of Industrial Pollution Control. 2017;**33**(1):1148-1154

[26] Skamarock WC. Positive-Definite and Monotonic Limiters for Unrestricted-Time-Step Transport Schemes. Monthly Weather Review. 2006;**134**: 2241-2250

[27] Skamarock WC, Weisman ML. The impact of positive-definite moisture transport on NWP precipitation forecasts. Monthly Weather Review. 2009;**137**:488-494

[28] CPCB (Central Pollution Control Board), India. National Ambient Air Quality Standards (NAAQS). Gazette notification, New Delhi. 2009.

[29] MVS. Motor transport statistics of Maharashtra. 2011. http://www.mahatranscom.in/pdf/STATISTICAL.BOOK.10-11.pdf

[30] ARAI. Draft report on "Emission Factor development for Indian Vehicles" as a part of Ambient Air Quality Monitoring and Emission Source Apportionment Studies. 2007. p. 1-94

[31] Kandlikar M, Ramachandran G. The causes and consequences of particulate air pollution in urban India: A synthesis of the science. Annual Review of Energy and the Environment. 2000;**25**:629-684

[32] Boadh R, Satyanarayana ANV, Rama Krishna TVBPS. Assessment of dispersion of oxide of nitrogen using AERMOD model over a tropical industrial region. International Journal of Computer Applications. 2014;**90**(11): 43-50

[33] NCEP-DSS1. NCEP Global Tropospheric Analyses, (1 × 1 deg), from Data Support Division (DSS). NCAR. 2005; Dataset # 083.2. /http://dss.ucar.edu/datasets/ds083.2/ [Accessed June 2005]

[34] Hong SY, Noh Y, Dudhia J. A new vertical diffusion package with explicit treatment of entrainment processes. Monthly Weather Review. 2006;**134**: 2318-2341

[35] Janjic ZA. The step-mountain eta coordinate model, further developments of the convection, viscous sub layer and turbulence closure schemes. Monthly Weather Review. 1994;**122**: 927-945

[36] Janjic ZI. Non-singular Implementation of the Mellor–Yamada Level 2.5 Scheme in the NCEP. Meso Model. 2002;**437**:61

[37] Ferrier BS, Lin Y, Black T, Rogers E, DiMego G. Implementation of a new grid-scale cloud and precipitation scheme in the NCEP Eta model. American Meteorological Society: Conference Proceeding; 2002. pp. 280-283

[38] Mlawer EJ, Taubman SJ, Brown PD, Iacono MJ, Clough SA. Radiative transfer for inhomogeneous atmosphere RRTM, a validated correlated-k model for the longwave. Journal of Geophysical Research. 1997;**102**(D14):16663-16682

[39] Dudhia J. Numerical study of convection observed during the winter monsoon experiment using a mesoscale two-dimensional model. Journal of the Atmospheric Sciences. 1989;**46**: 3077-3107

[40] Kain JS. The Kain–Fritsch convective parameterization an update. Journal of Applied Meteorology. 2004; **43**(1):170-181

[41] Wilks DS. Statistical Methods in the Atmospheric Science. Second ed. U.S.A: Elsevier Academic Press; 2008. p. 627

[42] Kumar A, Luo J, Bennett GF. Statistical evaluation of lower flammability distance (LFD) using four hazardous release models. Process Safety Progress. 1993;**12**(1):1-11

[43] Chang JC, Hanna SR. Air quality model performance evaluation. Meteorology and Atmospheric Physics. 2004;**87**:167-196

[44] Hanna S, Yang R. Evaluations of mesoscale models' simulations of near-surface winds, temperature gradients, and mixing depths. Journal of Applied Meteorology. 2001;**40**:1095-1104

[45] Arya SPS. Air Pollution Meteorology and Dispersion. New Yrok: Oxford University Press; 1999. p. 310

[46] Harrison RM. An Introduction to Pollution Science. RSC Publication; 2006. p. 29. DOI: 10.1039/9781847555410

[47] Rao MN, Rao HVN. Air pollution. 1st ed, New Delhi: Tata McGraw-Hill; 1990. p. 15

Section 5

Laminar and Turbulent Flows Past Bodies of Different Shapes

Chapter 11

Thermo-Rheological Effect on Weak Nonlinear Rayleigh-Benard Convection under Rotation Speed Modulation

S.H. Manjula and Palle Kiran

Abstract

The effects of rotation speed modulation and temperature-dependent viscosity on Rayleigh-Benard convection were investigated using a non-autonomous Ginzburg-Landau equation. The rotating temperature-dependent viscous fluid layer has been considered. The momentum equation with the Coriolis term has been used to describe finite-amplitude convective flow. The system is considered to be rotating about its vertical axis with a non-uniform rotation speed. In particular, we assume that the rotation speed is varying sinusoidally with time. Nusselt number is obtained in terms of the system parameters and graphically evaluated their effects. The effect of the modulated system diminishes the heat transfer more than the un-modulated system. Further, thermo-rheological parameter V_T is found to destabilize the system.

Keywords: rotation speed modulation, Coriolis force, thermo-rheological parameter, weak nonlinear theory

1. Introduction

For the past few decades, researchers are trying to explore various possibilities of controlling convective phenomena in a horizontal fluid layer by imposing external physical constraints. Some of them are temperature modulation, gravity modulation, magnetic field modulation and rotation speed modulation, etc. Motivated by the experiments of Donnelly [1], about the effect of rotation speed modulation on the onset of instability in fluid flow between two concentric cylinders, Venezian [2] was the first to perform a linear stability analysis of Rayleigh-Benard convection under temperature modulation by considering free-free surfaces. Later on, Rosenblat and Tanaka [3] studied the same problem of temperature modulation for rigid-rigid boundaries, using the Galerkin method. Bhadauria and Kiran [4] investigated thermal modulation on an-isotropic porous media. Their investigation reveals that an-isotropy of the medium and modulation has been used to regulate heat transfer in the system. It was concluded that an-isotropy play a role in instability of the porous media.

The study of rotation and thermal modulation on onset convection in a rotating fluid layer was investigated by Rauscher and Kelly [5]. Their results conclude that, for small values of Pr an oscillatory mode exists. However, they found that deviation of instability occurs at Pr =1. Liu and Ecke [6] analyzed heat transport of turbulent Benard convection under rotational effect. Malashetty and Swamy [7], investigated the thermal instability of a heated fluid layer subjected to both boundary temperature modulation and rotation. Kloosterziel and Carnevale [8] investigated the thermal modulation effect on the stability analysis of the modulated system. They have analytically determined critical points on the marginal stability boundary above which an increase of either viscosity or diffusivity is destabilizing. They also show that, if the fluid has zero viscosity the system is always unstable, in contradiction to Chandrasekhar's conclusion. Bhadauria [9] studied the effect of rotation on thermal instability in a fluid-saturated porous medium under temperature modulation. Further, Bhadauria [10] studied the double-diffusive convection in a rotating porous layer with temperature modulation of the boundaries. Also, Bhadauria [11] studied the magneto fluid convection in a rotating porous layer under modulated temperature on the boundaries. Malashetty and Swamy [12] studied the effect of thermal modulation on the onset of instability in a rotating fluid layer. Bhadauria [13] investigated the rotational influence on Darcy convection and found that both rotation and permeability suppress the onset of thermal instability. Bhadauria et al. [14] investigated the nonlinear thermal instability in a rotating viscous fluid layer under temperature/gravity modulation. They have concluded that both modulations may use alternatively to regulate heat transfer in the system.

The idea of rotation speed modulation was the originating idea of thermal as well as gravity modulation, but research work in this field is scarce. There are many studies that reported only linear thermal instability than nonlinear thermal instability. The effect of temperature modulation and the effect of rotation speed modulation on thermal instability has been investigated in detail both theoretically and experimentally by Niemela and Donnelly [15], Kumar et al. [16], Meyer et al. [17], Walsh and Donnelly [18]. In general, the effect of thermal modulation is supposed to stabilize conduction state. Also, it breaks the reflection symmetry about the mid-plane and forms hexagons, rather than cylinders. Then it constitutes the convection plan form immediately above the threshold. However, the above problem may be avoided if the rotation speed is modulated. Amongst the literature, the study due to Bhattacharjee [19] is of great importance, in which he studied the effect of rotation speed modulation on Rayleigh-Benard convection in an ordinary fluid layer. He found that the effect of modulation is to stabilize for most of the configurations. Bhadauria and Suthar [20] investigated the effect of the rotation speed modulation on the onset of free convection in a rotating porous layer placed farther away from the axis of rotation. Further, Om et al. [20] investigated the effect of rotation speed modulation on the onset of free Darcy convection. It is found that by applying modulation of proper frequency to the rotation speed, it is possible to delay or advance the onset of convection.

It is known that viscosity [4] plays a significant role in the study of fluid flows. It is mainly affected by temperature fluctuation due to its nature. From the above literature, the fluid viscosity is considered to be constant. However, in certain situations, fluid viscosity is not constant. This may vary with temperature, pressure, and distance. In most of the applications related to thermal transportation problems the temperature distribution is not uniform, i.e., the fluid viscosity may change noticeably if large temperature differences exist. That is why one needs to consider temperature-dependent viscosity in the energy equation.

The top and bottom structures of the fluid layer are different when the fluid viscosity varies with temperature. Kafoussius and Williams [21], investigated the effect of variable viscosity on the free convective boundary layer flow along with a vertical isothermal plate. It is concluded that when the viscosity of a working fluid is sensitive to the variation of temperature, care must be taken to include this effect, otherwise, considerable error can result in the heat transfer processes. Kafoussius et al. [22] examined the effect of temperature-dependent viscosity on the mixed convection laminar boundary layer flow along with a vertical isothermal plate. Molla et al. [23], studied the natural convection flow from an isothermal circular cylinder with temperature-dependent viscosity. Pal and Mondal [24] examined the influence of temperature-dependent viscosity and thermal radiation on MHD-forced convection over a non-isothermal wedge. Ching and Cheng [25], investigated temperature viscosity effects on natural convection for boundary layer flow. Nadeem and Akbar [25, 26], studied the effects of temperature-dependent viscosity on the peristaltic flow of a Jeffrey-six constant fluid in a uniform vertical tube. However, most of these studies are done with a steady temperature gradient across the fluid layer. Nield [27] investigated the effect of temperature-dependent viscosity on the onset of convection in a saturated porous medium. There is a stabilizing effect when the Rayleigh number is a function of mean viscosity.

Bhadauria and Kiran [28] investigated weak nonlinear analysis of magneto-convection [29] under magnetic field modulation. Using Landau mode, the effect of rotational speed modulation on heat transport in a fluid layer with temperature-dependent viscosity and internal heat source has been studied by Bhadauria and Kiran [29]. Centrifugally driven convection in a nanofluid saturated rotating porous medium with modulation has been investigated by Kiran and Narasimhulu [30]. The effect of gravity modulation on different flow models has been presented in [31–37]. These studies present results of gravity modulation on thermal instability in fluid or porous layer. Also, the effect of thermal modulation on convective instability has been investigated in [38–56]. The effect of thermal modulation on chaotic convection [38], throughflow effects in the presence of thermal modulation [39, 41, 42, 49, 56], thermal modulation with internal heating [40], thermal modulation on magneto-convection [43, 45, 48, 54, 55] have been investigated. The effect of rotation on nonlinear fluid convection under temperature modulation has been reported by Kiran and Bhadauria [57].

In most of the previous studies, only linear theory has been considered. This linear theory is not supportive to quantify heat/mass transfer. Also, other studies in the literature provide gravity modulation and thermal modulation on convective flows than rotation modulation. To the best of the authors' knowledge, no nonlinear study is available in the literature in which the effect of rotation speed modulation has been considered in a temperature-dependent viscous fluid layer. This situation motivated us to study a rotational speed modulation on temperature-dependent viscous fluid layer.

2. Governing equations

An infinitely and horizontally extended viscous-incompressible fluid layer has been considered. It is confined between two parallel planes which are at $z = 0$, lower plane and $z = d$, upper plane. The lower surface is heated and the upper surface is cooled to maintain an adverse temperature gradient across the fluid layer. We assume that the system is rotating with variable rotation speed $\vec{\Omega}(t) = (0, 0, \Omega(\mathrm{t}))$, about the z-axis as shown in **Figure 1**. The effect of rotation speed is restricted to the Coriolis term; thus, we neglect the centrifugal force term. The effect of density variation is given by the Boussinesq approximation. With these assumptions the basic governing equations are:

$$\nabla.\vec{\mathrm{q}} = 0 \tag{1}$$

$$\frac{\partial \vec{q}}{\partial t} + \left(\vec{q}.\nabla\right)\vec{q} + 2\left(\vec{\Omega}(t) \times \vec{q}\right) = -\frac{1}{\rho_0}\nabla p - \frac{\rho}{\rho_0}\vec{g} + \frac{\mu(T)}{\rho_0}\nabla^2\vec{q} \tag{2}$$

$$\frac{\partial T}{\partial t} + \left(\vec{q}.\nabla\right)T = \kappa_T \nabla^2 T \tag{3}$$

$$\rho = \rho_0(1 - \alpha_\mathrm{T}(T - T_0) \tag{4}$$

$$\mu(T) = \frac{\mu_0}{1 + \varepsilon^2\delta_0(T - T_0)} \tag{5}$$

Eq. 5 has been considered by many authors [4, 25–27, 29]. The rotation speed is assumed to vary sinusoidally with respect t to time, as

$$\vec{\Omega}(t) = \Omega_0\left(1 + \varepsilon^2\delta\ \cos\left(\omega\ \mathrm{t}\right)\right)\vec{k} \tag{6}$$

where $\vec{q}$ is velocity, ν is a kinematic viscosity, T is temperature, p is reduced pressure, $\kappa_T$is the thermal diffusivity, α_T is thermal expansion coefficient, ρ is the density, ρ_0 and T_0 are the reference density and temperature, $\vec{g}$ =*(0,0,-g)* is gravitational acceleration, δ, ω are the small amplitude and frequency of modulation, ε^2 is a quantity that indicates the smallness in order of magnitude of modulation and t is the time. The gravity is being considered in the downward direction.

The constants and variables used in the above Eqs. (1)–(5) have their usual meanings and are given in the nomenclature. The thermo-rheological relationship in Eq. (5) is guided by Nield [27] and Bhadauria and Kiran [29]. The considered thermal boundary conditions are:

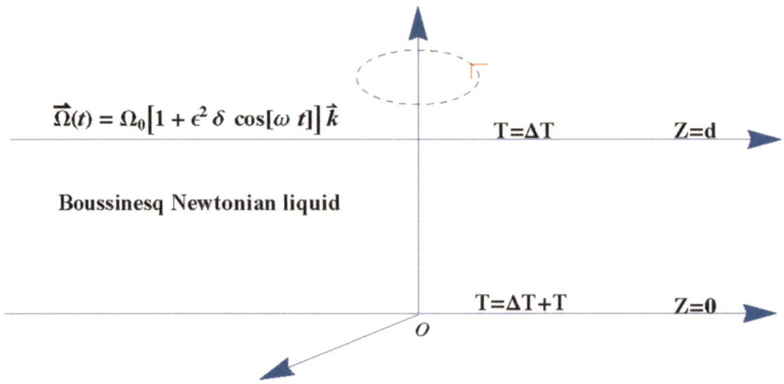

Figure 1.
Physical configuration of the problem.

We consider the following steady thermal boundary conditions are:

$$\begin{aligned} T &= T_0 + \nabla T \qquad \text{at z} = 0 \\ T &= T_0 \qquad \text{at z} = \text{d} \end{aligned} \tag{7}$$

The conduction state is assumed to be quiescent i.e. $q_b = (0,0,0)$. The other quantities of the basic state are:

$$\rho = \rho_b(z), p = p_0(z) \text{ and } T = T_b(z) \tag{8}$$

Substituting the Eq. (8) into Eqs. (1)–(5), we get the basic state equations, respectively for pressure and temperature, as

$$\frac{\partial p_b}{\partial z} = -\rho_b g \tag{9}$$

$$\frac{d^2 T_b}{dz^2} = 0 \tag{10}$$

$$\rho_b = \rho_0(1 - \alpha_\text{T}(T_b - T_0) \tag{11}$$

where "b" refers the basic state. The Eq. (10) is solved for $T_b(z)$ subject to the boundary condition Eq. (7), we get

$$T_b = T_0 + \Delta T\left(1 - \frac{z}{d}\right) \tag{12}$$

The perturbations on the basic state are superposed in the form

$$\vec{q} = q_b + \vec{q}^{/}, T = T_b + T^{/}, \rho = \rho_b + \rho^{/} \text{ and } p = p_b + p^{/} \tag{13}$$

where the perturbations are of finite-amplitude. Substituting Eq. (13) i in Eqs. (1)–(5) and using the basic state results Eqs. (9)-(12), we obtain

$$\nabla.\vec{\text{q}}^{/} = 0 \tag{14}$$

$$\frac{\partial \vec{\text{q}}^{/}}{\partial t} + \left(\vec{\text{q}}^{/}.\nabla\right)\vec{\text{q}}^{/} + 2\left(\vec{\Omega}(t) \times \vec{\text{q}}^{/}\right)k = -\frac{1}{\rho_0}\nabla p + \alpha_\text{T} g T^{/} + \frac{\mu(T)}{\rho_0}\nabla^2 \vec{\text{q}}^{/} \tag{15}$$

$$\frac{\partial T^{/}}{\partial t} + w\frac{dT_b}{dz} + \left(\vec{q}.\nabla\right)T^{/} = \kappa_T \nabla^2 T^{/} \tag{16}$$

We consider, two-dimensional convection in our study, introducing, stream function ψ as $u = \frac{\partial\psi}{\partial z}$ and $w = -\frac{\partial\psi}{\partial x}$ .Non-dimensionalizing the physical variables $(x, y, z) = (x^*, y^*, z^*)$ $t = \frac{d^2}{k_t} t^*$, $\vec{q} = \frac{\kappa_T}{d} \vec{q}^*$, $\psi = \kappa_T \psi^*$, $T^{/} = \Delta T\, T^*$ and $\omega = \frac{\kappa_T}{d^2}\omega^*$ by eliminating the pressure term and finally dropping the asterisk, we obtain:

$$\begin{aligned} \frac{1}{Pr}\frac{\partial}{\partial t}\nabla^2\psi - \frac{1}{Pr}\frac{\partial(\psi, \nabla^2\psi)}{\partial(x,z)} = -R_{aT}\frac{\partial T}{\partial x} + \overline{\mu}(T)\nabla^4\psi + \sqrt{Ta}\left(1 + \varepsilon^2\delta\cos(\omega\,\text{t})\right)\frac{\partial V}{\partial z} \\ + \frac{d\overline{\mu}(T)}{dz}\frac{\partial}{\partial z}\nabla^2\psi \end{aligned} \tag{17}$$

$$-\frac{dT_b}{dz}\frac{\partial\psi}{\partial x} - \nabla^2 T = -\frac{\partial T}{\partial t} - \frac{\partial(\psi,T)}{\partial(x,z)} \tag{18}$$

$$\frac{1}{Pr}\frac{\partial V}{\partial t} - \overline{\mu}(T)\nabla^2 V = -\sqrt{Ta}\left(1+\varepsilon^2\delta\cos(\omega\,t)\right)\frac{\partial\psi}{\partial z} + \frac{1}{Pr}\frac{\partial(\psi,V)}{\partial(x,z)} \tag{19}$$

The non-dimensional parameters which are obtained in the above equations are $Pr = \frac{\nu}{\kappa_T}$ is the Prandtl number, $R_{aT} = \frac{\alpha_T g \Delta T d^3}{\kappa_T \nu}$ is thermal Rayleigh number, $Ta = \frac{4\Omega_0^4 d^4}{\nu^2}$ is the Taylor number, $\nu = \frac{\mu}{\rho_0}$ is the kinematic viscosity. We assume small variations of time and re-scaling it as $t = \varepsilon^2\tau$ to study the stationary convection of the system. It is to be noted that in our study we are not considering overstable solutions of the system. To solve Eqs. (17)–(19) we considering stress-free and isothermal boundary conditions as given bellow

$$\psi = \frac{\partial^2\psi}{\partial z^2} = \frac{\partial V}{\partial z} = \mathrm{T} = 0 \text{ at } \mathrm{T} = 0 \text{ and } \mathrm{T} = \mathrm{z} \tag{20}$$

The above boundary conditions are free-free and stress-free isothermal, they are used by [28, 29, 32, 45, 54, 55].

3. Heat transport for stationary instability

We introduce the following asymptotic expansions (also used in [31–36]) in the above Eqs. (17)–(19):

$$\begin{aligned}(\mathrm{R_{aT}}, \psi, \mathrm{T}, \mathrm{V}) = \mathrm{R_{0c}} + \varepsilon^2\mathrm{R_2} + \varepsilon^2\mathrm{R_2} + \ldots, \varepsilon\psi_1 + \varepsilon^2\psi_2 + \ldots, \\ \varepsilon\mathrm{T_1} + \varepsilon^2\mathrm{T_2} + \ldots \quad \varepsilon V_1 + \varepsilon^2 V_2 + \ldots\end{aligned} \tag{21}$$

Where R_{0c} is the critical value of the Rayleigh number at which the onset of convection takes place in the absence of modulation. Now we solve the system for different orders of ε. For reference see the studies of [28, 29, 32, 45, 51].

At the lowest order, we get the following system:

$$\begin{pmatrix} -\nabla^4 & R_{0c}\frac{\partial}{\partial x} & -\sqrt{Ta}\frac{\partial}{\partial z} \\ \frac{\partial}{\partial x} & -\nabla^2 & 0 \\ \sqrt{Ta}\frac{\partial}{\partial z} & 0 & -\nabla^2 \end{pmatrix}\begin{pmatrix}\psi_1 \\ T_1 \\ V_1\end{pmatrix} = \begin{pmatrix}0\\0\\0\end{pmatrix} \tag{22}$$

The solution of the lowest order system subject to the boundary conditions Eq. (20), is

$$\psi_1 = A(\tau)\sin(k_c x)\sin(\pi z)$$

$$T_1 = -\frac{kc}{\beta^2}A(\tau)\cos(k_c x)\sin(\pi z)$$

$$V_1 = -\frac{\pi\sqrt{Ta}}{\beta^2}A(\tau)\sin(k_c x)\cos(\pi z) \tag{23}$$

where $\beta^2 = \pi^2 + k_c^2$ is the total wavenumber. The critical value of the Rayleigh number for the onset of stationary convection is calculated numerically and the expression is given by

$$R_{0c} = \frac{\beta^6 + \pi^2 Ta}{k_c^2} \tag{24}$$

For the system without rotation i.e. Ta = 0, we get $R_{0c} = \frac{\beta^6}{k_c^2}$ and it's critical wave number is given as $k_c = \frac{\pi}{\sqrt{2}}$, which is the classical results obtained by Chandrasekhar [58]. At the second order, we have the following terms on RHS (similar to the system Eq. (22)).

$$R_{21} = 0 \tag{25}$$

$$R_{22} = \frac{\partial(\psi_1, T_1)}{\partial(x, z)} \tag{26}$$

$$R_{23} = \frac{\partial(\psi_1, V_1)}{\partial(x, z)} \tag{27}$$

The second order solutions subjected to the boundary conditions Eq. (20), is obtained as follows

$$\psi_2 = 0 \tag{28}$$

$$T_2 = -\frac{k_c^2}{8\pi\beta^2} A(\tau)^2 \sin(2\pi z) \tag{29}$$

$$V_2 = \frac{\pi^2\sqrt{Ta}}{8k_c \, \Pr\beta^2} A(\tau)^2 \sin(2k_c x) \tag{30}$$

The horizontally averaged Nusselt number $\mathrm{Nu}(\tau)$, for the stationary mode of convection is given by:

$$Nu(\tau) = 1 + \left(\frac{\frac{k_c}{2\pi} \int_0^{\frac{k_c}{2\pi}} \frac{dT_2}{dz} dx}{\frac{k_c}{2\pi} \int_0^{\frac{k_c}{2\pi}} \frac{dT_b}{dz} dx} \right)_{z=0} \tag{31}$$

Substituting the expression of T_2, T_b in the Eq. (31) and simplifying we get:

$$Nu(\tau) = 1 + \frac{k_c^2}{4\beta^2} A(\tau)^2 \tag{32}$$

At third order system the expressions in RHS (Eq. (33)) will have many terms and it is difficult to simplify and find expressions for (ψ_3, T_3, V_3).

$$\begin{pmatrix} -\nabla^4 & R_{0c}\dfrac{\partial}{\partial x} & -\sqrt{Ta}\dfrac{\partial}{\partial z} \\ \dfrac{\partial}{\partial x} & -\nabla^2 & 0 \\ \sqrt{Ta}\dfrac{\partial}{\partial z} & 0 & -\nabla^2 \end{pmatrix} \begin{pmatrix} \psi_3 \\ T_3 \\ V_3 \end{pmatrix} = \begin{pmatrix} R_{31} \\ R_{32} \\ R_{33} \end{pmatrix} \tag{33}$$

The expression given in the Eq. (33) may be obtained at third order system as:

$$R_{31} = \frac{1}{Pr}\frac{\partial}{\partial t}\nabla^2\psi_1 - R_2\frac{\partial T_1}{\partial x} + \sqrt{Ta}(1 + \delta\cos(\omega\, t))\frac{\partial V_1}{\partial z} + V_T T_b \nabla^4\psi_1 \tag{34}$$

$$R_{32} = -\frac{\partial T_1}{\partial t} + \frac{\partial \psi_1}{\partial x}\frac{\partial T_2}{\partial z} \tag{35}$$

$$R_{33} = -\frac{1}{Pr}\frac{\partial V_1}{\partial t} - \sqrt{Ta}(\mathrm{V_T T_b} + \delta\cos(\omega\, t))\frac{\partial \psi_1}{\partial z} + \frac{1}{Pr}\frac{\partial \psi_1}{\partial z}\frac{\partial V_2}{\partial x} \tag{36}$$

Substituting the first and second order solutions into Eqs. (34)–(36), we can easily simplify the expressions R_{31}, R_{32} and R_{33}. Now, by applying the solvability condition for the existence of third order solution, we get the Ginzburg-Landau equation for stationary mode [28–36, 38–45, 52–56] of convection, with time-periodic coefficients, in the form:

$$Q_1\frac{\partial A(\tau)}{\partial \tau} - Q_2 A(\tau) + Q_3 A(\tau)^3 = 0 \tag{37}$$

Where $Q_1 = \frac{\beta^2}{Pr} + \frac{R_{0c}k_c^2}{\beta^4} - \frac{\pi^2 Ta}{Pr\beta^4}$, $Q_2 = \frac{R_{0c}k_c^2}{\beta^2} - \frac{2\pi Ta}{\beta^2}\delta\cos(\omega\, t) - H_1$, and $Q_1 = \frac{R_{0c}k_c^4}{8\beta^4} + \frac{\pi^4 Ta}{\genfrac{}{}{0pt}{}{2}{4\mathrm{Pr}\beta^4}}$, $H_1 = \frac{R_{0c}V_T k_c^2}{\beta^2} - \frac{3V_T\pi^2 Ta}{\beta^2} + \frac{\left(2V_T\pi\beta^2 Ta - V_T\beta^4\right)}{2.}$

The equations given by Eq. (37) is known as the Bernoulli equation. This equation is solved numerically using NDSolve of Mathematica 8. A suitable initial condition $A(0) = a_0$ maybe chosen to solve the amplitude equation. In our calculations we take $R_2 = R_{0c}$ to analyze instability near to critical Rayleigh number.

4. Analytical solution for un-modulated case

In the case of un-modulated system δ = 0, we obtain the following analytical expression for $A_u(\tau)$

$$A_u(\tau) = \frac{1}{\sqrt{\frac{A_3}{2A_2} + C_1 e^{\frac{-2A_2}{A_1}}}} \tag{38}$$

where C_1 is an integrating constant to be calculated for a given initial condition. The horizontal Nusselt number is obtained from the Eq. (32) substituting $A_u(\tau)$ in place of $A(\tau)$.

5. Results and discussion

In this problem, we address a weakly nonlinear Rayleigh-Benard convection with a variable viscous liquid under rotational speed modulation. Here, we have presented the results of weakly nonlinear stability analysis to investigate the effect of rotational speed modulation and thermo-rheological parameters on heat transport. The modulation of the Rayleigh-Benard system has been assumed to be of order $O(\varepsilon^2)$, which shows, that we consider the only small amplitude of rotation speed modulation (see the article of [29–36, 38–45] for different modulations at the third order of Landau equation). This assumption will help us in obtaining the amplitude equation in a simple manner, and much easier than the Lorenz model. The present work is important to study nonlinear convection and quantify heat transport. External regulation of convection is important in the study of thermal instability in a fluid layer, therefore, in this paper, we have considered rotational speed modulation for either enhancing or inhibiting convective heat transport as is required by a real-life application.

The results of rotational speed modulation on heat transport have been depicted in the **Figures 2–10**. Here, we have drawn the figures Nu versus slow time. We observe the nature of the graphs where Nu is a function of time and vary with respect to different parameters. To see the results corresponding to the parameters Pr, V_T, Ta, δ, and ω, we have drawn figures with variations in slow time. The mentioned parameters describe the convective flow of heat transport. The first two parameters are related to the fluid layer and the next three parameters concern the external mechanism of controlling convection. The fluid layer is not considered to be highly viscous, therefore, only moderate values of Pr are taken for calculations. Because of small amplitude modulation, the values of δ are considered around 0.5. Further, modulation considers low frequency, due to maximum heat transport. The effect of frequencies of modulation influences the onset of convection as well as on heat transport. The thermo-rheological parameter V_T, has taken to be small values.

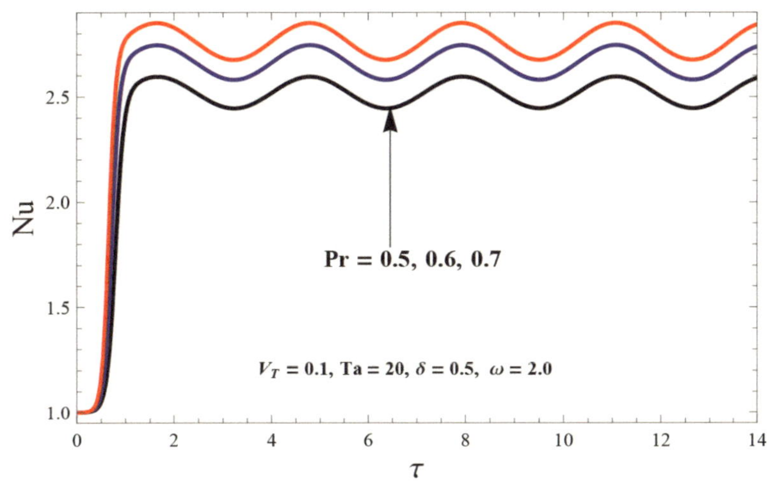

Figure 2.
Nu verses τ for different values of Pr.

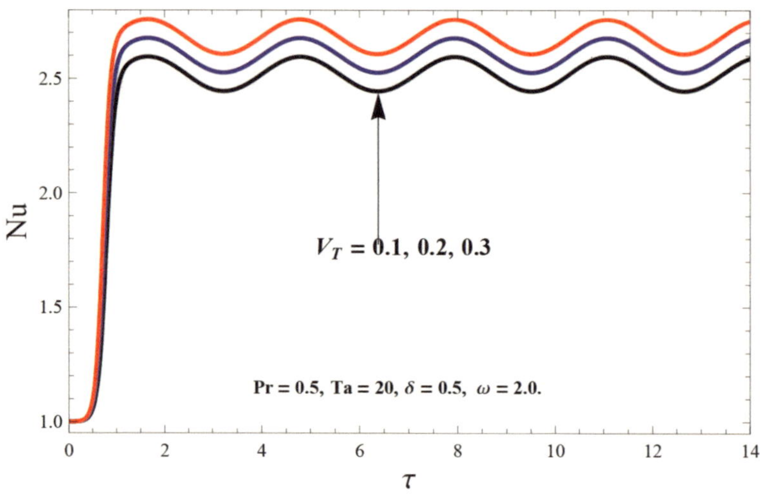

Figure 3.
Nu verses τ for different values of V_T.

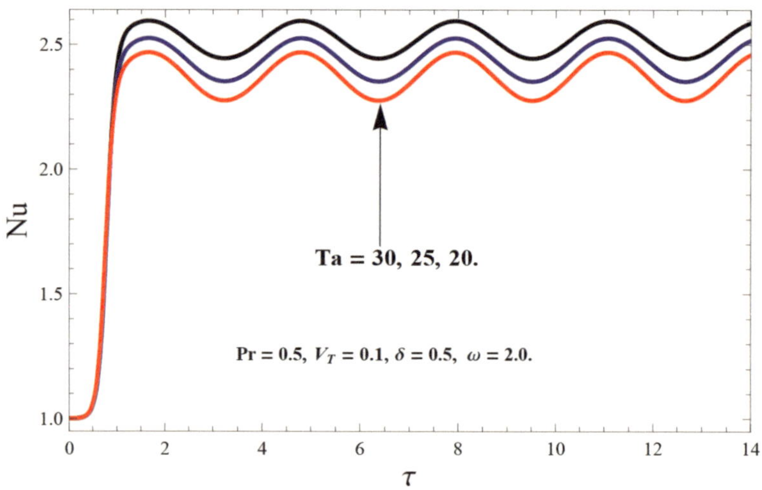

Figure 4.
Nu verses τ for different values of Ta.

Before writing the discussion of the results, we mention some features of the following aspects of the problem.

1. The importance of nonlinear stability analysis to study heat transfer.

2. The relation of the problem to the real application (Temperature, viscosity).

3. The selection of all the dimensional parameters utilized in computations.

4. Numerical computation.

In **Figures 2–8**, we have plotted the Nusselt number $Nu(\tau)$ with slow time τ for the case of rotational speed modulation. From the figures, we find that for small-time τ,

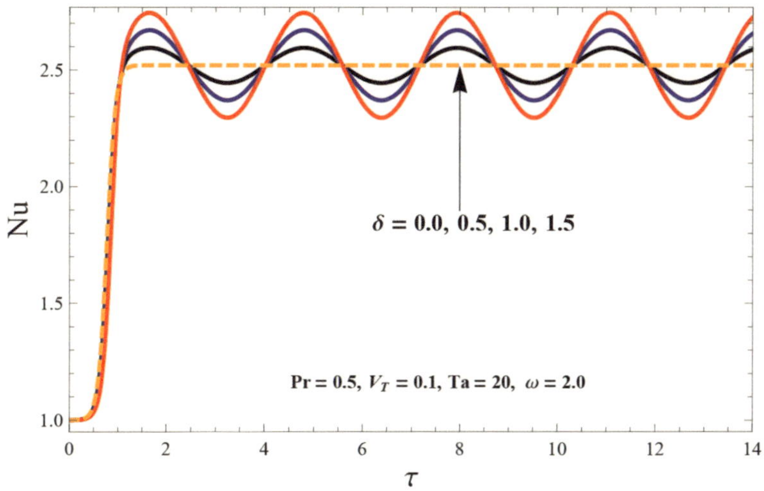

Figure 5.
Nu verses τ for different values of δ.

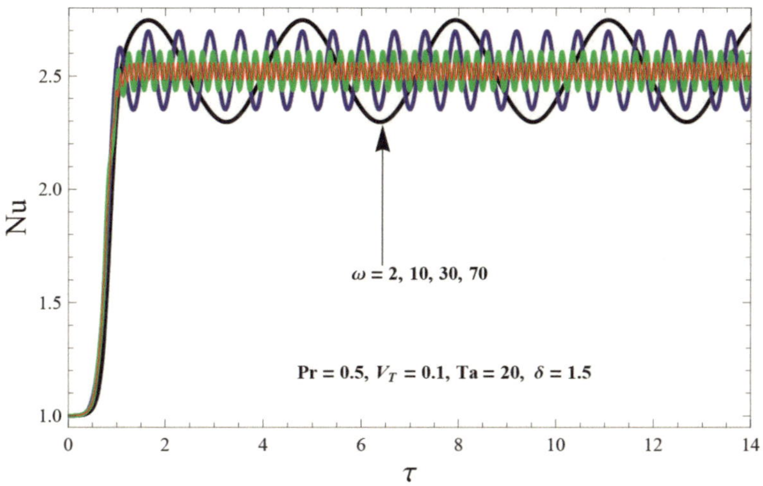

Figure 6.
Nu verses τ for different values of ω.

the value of $Nu(\tau)$ does not alter and remains almost constant, then it increases on increasing τ, and finally becomes oscillatory on further increasing τ. It is clear from the figure that $Nu(\tau)$ starts with one showing the conduction state, initially.

In **Figure 2**, we see that $Nu(\tau)$ increases upon increasing Pr for fixed values of other parameters (see the studies of [46–48]). This may happen due to the dominating role of thermal diffusivity κ_T over kinematic viscosity ?. As Prandtl number Pr increases, then for no change in kinematic viscosity, probably there is a large decrement in thermal diffusivity, and this makes the sudden increase in the temperature gradient. So, convection takes place early, and there is an enhancement in heat transfer. Thus, the effect of an increment in Prandtl number Pr is to advance the convection. There are few studies, which relate the effect of Pr given by [28, 29, 32, 43,

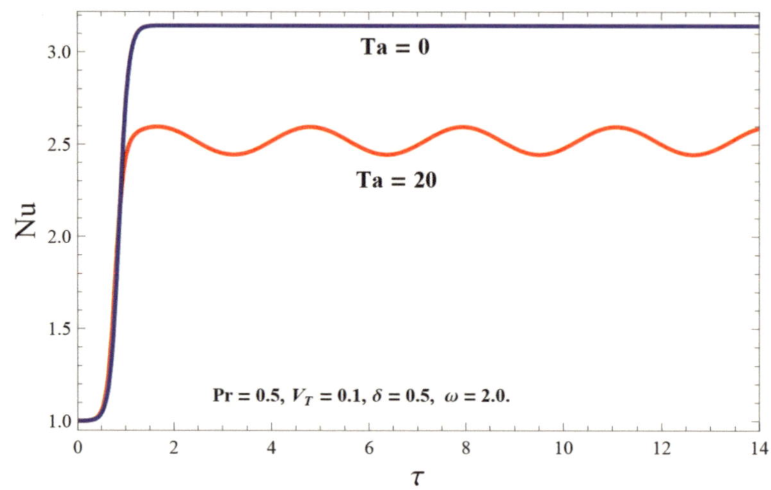

Figure 7.
Nu verses τ with or without Ta.

45, 48, 52]. A similar effect may be observed for V_T given in **Figure 3**. The reader may notify that the heat transport is more in the presence of variable viscosity. We have

$$Nu_{Pr=0.5} < Nu_{Pr=0.6} < Nu_{Pr=0.7}$$

$$Nu_{V_T=0.1} < Nu_{V_T=0.2} < Nu_{V_T=0.3}$$

From the **Figure 4**, we depict the effect of the Taylor number on $Nu(\tau)$ for fixed values of other parameters. An increment in Ta increases the value of critical Rayleigh number $\boldsymbol{R_{0c}}$, and it delays the onset of convection, hence, heat transport decreases.

Figure 4 shows that as ***Ta*** increases, the amplitude of modulation enhances. From the Eq. (37) we observe that the rotation is multiple of amplitude modulation. This means that the amplitude is dependent on rotation. Generally, if there is no rotation (***Ta*** = 0), it is meaningless to talk about rotation speed modulation. Further, for no rotation ***Ta*** = 0, the effect of frequency diminishes, so the effect of frequency of modulation can be seen when rotation is not there. In our study consider Ta non-zero otherwise modulation effect disappears.

$$Nu_{Ta=30} < Nu_{Ta=25} < Nu_{Ta=20}$$

For reference, the reader may observe the studies of [29, 41, 50, 51]. In **Figure 5**, we depict the effect of amplitude of modulation for moderate values of ***Ta*** and for the fixed values of other parameters. On increasing the value of δ, the value of $Nu(\tau)$ increases, hence advancing the convection so the heat transport. This means that an increasing amplitude of modulation increases heat transfer.

In the case of un-modulated δ = 0 system shows no influence on heat transport for larger values of time τ. Similar results can be obtained analytically for an unmodulated system given by Eq. (38).

$$Nu_{\delta=0.0} < Nu_{\delta=0.5} < Nu_{\delta=1.0} < Nu_{\delta=1.5}$$

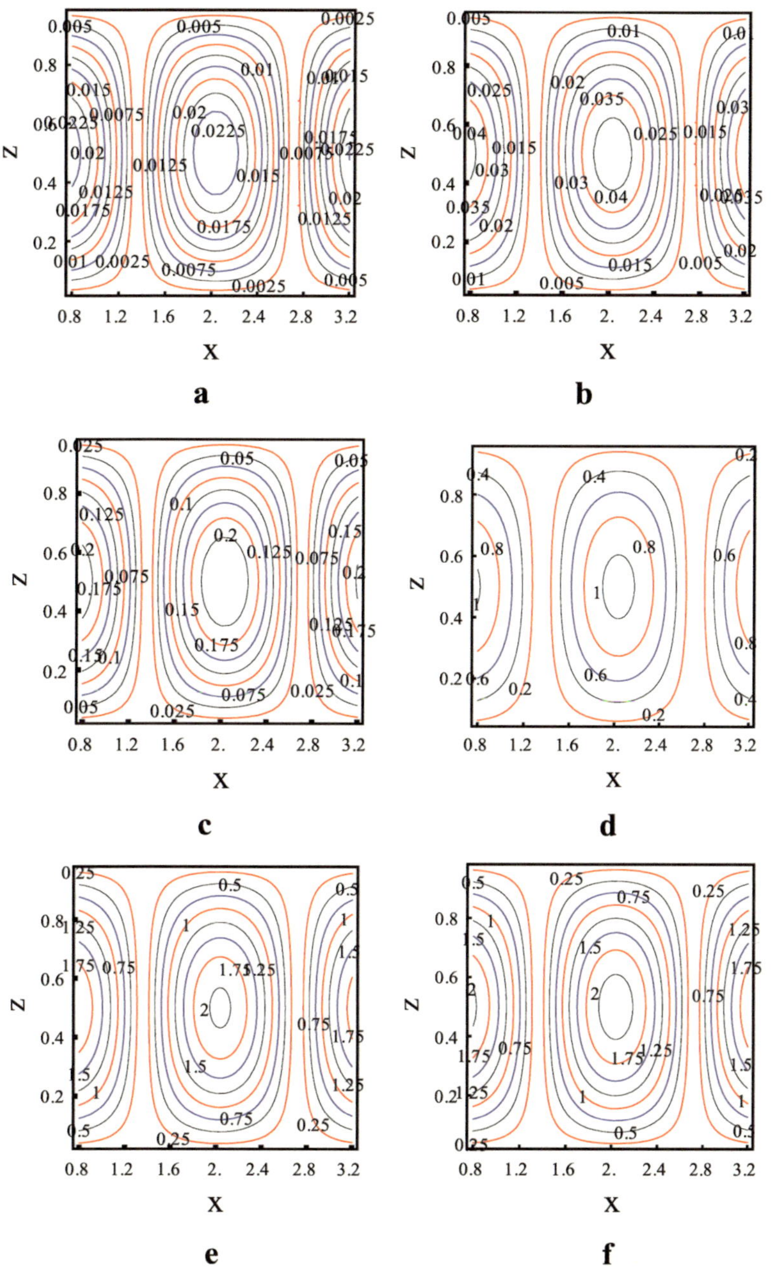

Figure 8.
Streamlines at (a) τ =0.0 (b) τ =0.1 (c) τ =0.4 (d) τ =0.7(e) τ =1.0 (f) τ =1.5.

Figure 6, shows that the effect of frequency on Nu, for small values of ω heat transports is more. An increment in the value of ω decreases the magnitude of heat transfer and shortens the wavelength. Upon frequency increasing from 2 to 100, $Nu(\tau)$ decreases. Further value of ω, the effect of modulation disappears altogether. Hence, the effect of ω is to stabilize the system. It is clear from the studies of [29–36, 38–43] that frequency of modulation can reduce heat transfer. We have the following inequality

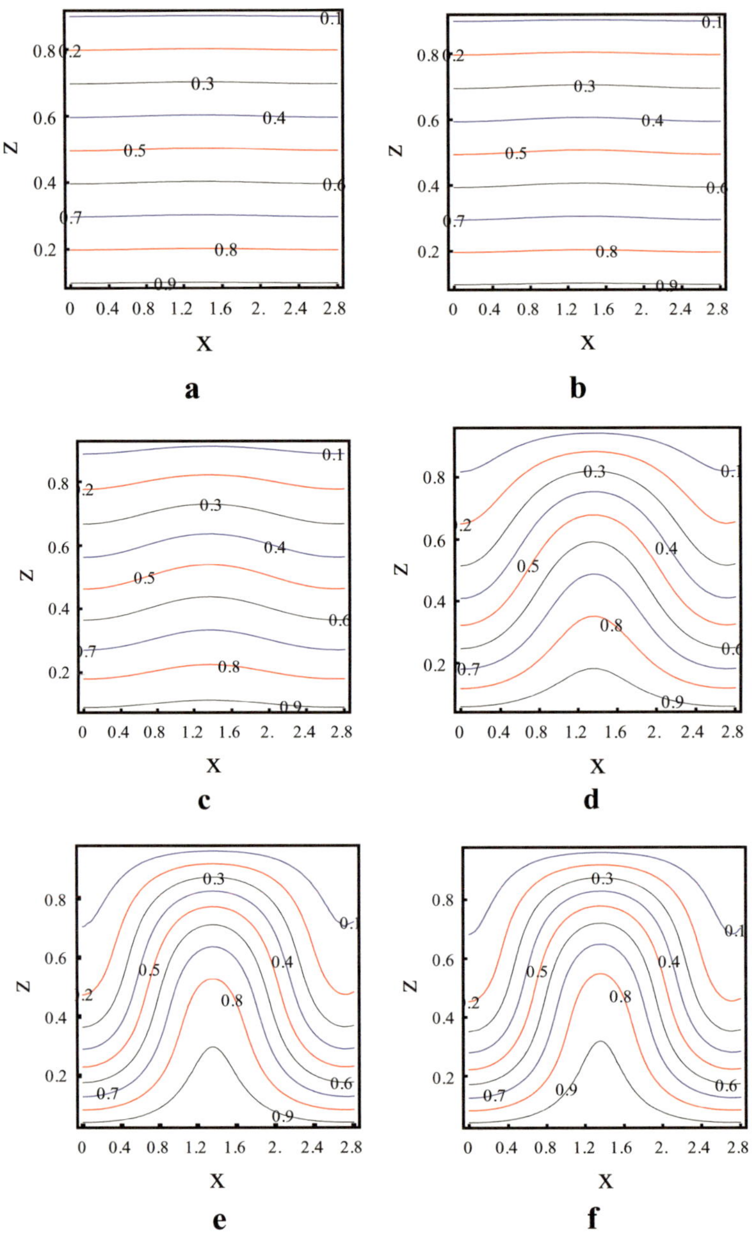

Figure 9.
Isotherms (a) τ *=0.0 (b)* τ *=0.1 (c)* τ *=0.4 (d)* τ *=0.7(e)* τ *=1.0 (f)* τ *=1.5.*

$$Nu_{\omega=70} < Nu_{\omega=30} < Nu_{\omega=10} < Nu_{\omega=2}$$

The results corresponding to δ and ω one can see the recent results obtained by the studies of [49–52]. **Figure 7** shows the heat transport is more when there is no rotation ***Ta*** = 0 (which means no modulation) than in the presence of rotation and modulation. Hence, rotation strongly stabilizes the system. The results of rotational effects on convection may compare with the results obtained in [29, 44, 57].

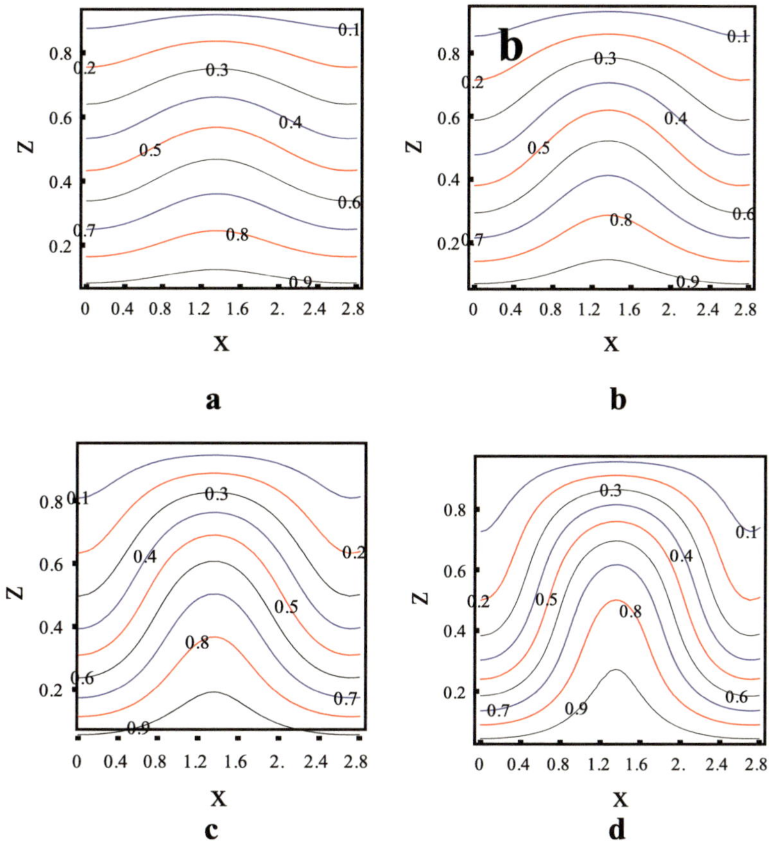

Figure 10.
Isotherms (a) V_T = 0.2 (b) V_T = 0.6 (c) V_T = 1.0(d) V_T = 1.4.

$$Nu_{Ta\neq 0} < Nu_{Ta=0,\omega\neq 0}$$

We have drawn streamlines and isotherms in **Figures 8–10**. The fixed values of τ = 0.0, 0.1,0.4,0.7,1.0 and 1.5 for Pr = 0.5, Ta = 20.0, δ = 0.5 and ω=2.0 has been consider. We found that initially, the magnitude of streamlines is small (**Figure 8a, b**), and isotherms are straight **Figure 9a, b** showing the conduction state. However, as time increases, the magnitude of streamlines increases, and the isotherms lose their evenness. This shows that convection is taking place in the system. Convection becomes faster by further increasing the value of time τ. However, steady-state achieves beyond τ =1.0 as there is no change in the **Figures 8d–f** and **9d–f. Figure 10** shows that as V_T increases the isotherms lose their evenness which shows heat transfer increasing as V_T. This nature of streamlines and isotherms is quite natural for convective flow models. Similar results can be observed from the studies of [31–36].

6. Conclusions

Heat transfer results are obtained in terms of Nusselt number, which is derived from Ginzburg-Landau equation and the effect of various parameters depicted graphically. The Prandtl number $\boldsymbol{Pr}$ and variable viscosity $\boldsymbol{V_T}$ is to increase heat transfer. The modulation loses its effect at sufficiently large values of frequency of modulation. Overall, the effect of rotation speed modulation is highly significant and can be used to reduce heat transfer. As the time τ passes the magnitude of streamlines increases, and isotherms lose their evenness, showing that convection is taking place. At τ =1.0 the system achieves an equilibrium state. Further, It was observed that amplitude and frequency of modulation have no effect on un-modulated system but, in the case of the modulated system shows sinusoidal behavior.

Acknowledgements

The authors are acknowledging the support of their institutions for doing research. Also, the author MJ is grateful to the dept. of Mathematics, VFSTR for support and encouragement. The author PK acknowledges the support of the Department of Mathematics, CBIT for providing research specialties.

Author details

S.H. Manjula[1] and Palle Kiran[2*]

1 Department of Mathematics (S & H), Vignan's Foundation for Science, Technology and Research (VFSTR), Guntur, Andhra Pradesh, India

2 Department of Mathematics, Chaitanya Bharathi Institute of Technology, Hyderabad, Telangana, India

*Address all correspondence to: pallekiran_maths@cbit.ac.in

References

[1] Donnelly RJ. Experiments on the stability of viscous flow between rotating cylinders III: Enhancement of hydrodynamic stability by modulation. Proceedings of the Royal Society of London Series A. 1964;**281**:130-139

[2] Venezian G. Effect of modulation on the onset of thermal convection. Journal of Fluid Mechanics. 1969;**35**: 243-254

[3] Rosenblat S, Tanaka GA. Modulation of thermal convection instability. Physics of Fluids. 1971;**14**:1319-1322

[4] Bhadauria BS, Kiran P. Heat transport in an anisotropic porous medium saturated with variable viscosity liquid under temperature modulation. Transport in Porous Media. 2013;**100**: 279-295

[5] Rauscher JW, Kelly RE. Effect of modulation on the onset of thermal convection in a rotating fluid. International Journal of Heat and Mass Transfer. 1975;**18**:1216-1217

[6] Liu Y, Ecke RE. Heat transport scaling in turbulent Rayleigh-Benard convection: Effects of rotation and Prandtl number. Physical Review Letters. 1997;**79**:2257-2260

[7] Malashetty MS, Swamy M. Combined effect of thermal modulation and rotation on the onset of stationary convection in porous layer. Transport in Porous Media. 2007;**69**: 313-330

[8] Kloosterziel RC, Carnevale GF. Closed-form linear stability conditions for rotating Rayleigh Benard convection with rigid stress-free upper and lower boundaries. Journal of Fluid Mechanics. 2003;**480**:25-42

[9] Bhadauria BS. Fluid convection in a rotating porous layer under modulated temperature on the boundaries. Transport in Porous Media. 2007;**67**(2): 297-315

[10] Bhadauria BS. Magneto fluid convection in a rotating porous layer under modulated temperature on the boundaries. ASME Journal of Heat Transfer. 2007;**129**:835-843

[11] Bhadauria BS. Double diffusive convection in a rotating porous layer with modulated temperature on the boundaries. Journal Porous in Media. 2007;**10**(6):569-584

[12] Malashetty MS, Swamy M. Effect of thermal modulation on the onset of convection in rotating fluid layer. International Journal Heat and Mass Transport. 2008;**51**: 2814-2823

[13] Bhadauria BS. Effect of temperature modulation on Darcy convection in a rotating porous medium. Journal of Porous Media. 2008;**11**(4):361-375

[14] Bhadauria BS, Siddheshwar PG, Om Suthar P. Non-linear thermal instability in a rotating viscous fluid layer under temperature/gravity modulation. Journal of Heat Transfer. 2012; **34**:102-502

[15] Niemela JJ, Donnelly RJ. Externally modulation of Rayleigh-Benard convection. Physical Review Letters. 1987;**59**:2431-2434

[16] Kumar K, Bhattacharjee JK, Banerjee K. Onset of the first instability in hydrodynamic flows: Effect of parametric modulation. Physical Review A. 1986;**34**:5000-5006

[17] Meyer CW, Cannell DS, Ahlers G, Swift JB, Hohenberg PC. Pattern competition in temporally modulated Rayleigh-Bénard convection. Physical Review Letters. 1988;**61**: 947-950

[18] Walsh TJ, Donnelly RJ. Taylor-Couette flow with periodically corotated and counter rotated cylinders. Physical Review Letters. 1988;**60**:700-703

[19] Bhattacharjee JK. Rotating Rayleigh-Benard convection with modulation. Journal of Physics A Mathematical and General. 1989;**22**:L1135-L1139

[20] Bhadauria BS, Om P. Suthar, Modulated centrifugal convection in a rotating vertical porous layer distant from the axis of rotation. Transport in Porous Media. 2009;**79**(2):255-264

[21] Kafoussius NG, Williams EM. The effect of temperature-dependent viscosity on the free convective laminar boundary layer flow past a vertical isothermal plate. Acta Mechanica. 1995; **110**:123-137

[22] Kafoussius NG, Rees DAS. Numerical study of the combined free and forced convective laminar boundary layer flow past a vertical isothermal flat plate with temperature-dependent viscosity. Acta Mechanica. 1998;**127**: 39-50

[23] Molla MM, Hossain MA, Gorla RSR. Natural convection flow from an isothermal circular cylinder with temperature-dependent viscosity. Heat and Mass Transfer. 2005;**41**:594-598

[24] Pal D, Mondal H. Influence of temperature-dependent viscosity and thermal radiation on MHD-forced convection over a non-isothermal wedge. Applied Mathematics and Computation. 2009;**212**:194-208

[25] Ching YC. Natural convection boundary layer flow of fluid with temperature-dependent viscosity from a horizontal elliptical cylinder with constant surface heat flux. Applied Mathematics and Computation. 2010; **217**:83-91

[26] Nadeem S, Akbar NS. Effects of temperature-dependent viscosity on peristaltic flow of a Jeffrey-six constant fluid in a non-uniform vertical tube. Communications in Nonlinear Science and Numerical Simulation. 2010;**15**:3950

[27] Nield DA. The effect of temperature-dependent viscosity on the onset of convection in a saturated porous medium. ASME Journal of Heat and Transfer. 1996;**118**:803-805

[28] Bhadauria BS, Kiran P. Weak nonlinear analysis of magneto–convection under magnetic field modulation. Physica Scripta. 2014;**89**: 095209

[29] Bhadauria BS, Kiran P. Effect of rotational speed modulation on heat transport in a fluid layer with temperature dependent viscosity and internal heat source. Ain Shams Engineering Journal. 2014;**5**:1287-1115

[30] Kiran P, Narasimhulu Y. Centrifugally driven convection in a nanofluid saturated rotating porous medium with modulation. Journal of Nanofluid. 2017;**6**(1):01-11

[31] Bhadauria BS, Kiran P. Nonlinear thermal Darcy convection in a nanofluid saturated porous medium under gravity modulation. Advanced Science Letters. 2014;**20**:903-910

[32] Bhadauria BS, Kiran P. Weak nonlinear double diffusive magneto-convection in a Newtonian liquid under

gravity modulation. Journal of Applied Fluid Mechanics. 2014;**8**(4):735-746

[33] Kiran P. Throughow and g-jitter effects on binary fluid saturated porous medium. Applied Mathematics and Mechanics. 2015;**36**(10):1285-1304

[34] Kiran P. Nonlinear thermal convection in a viscoelactic nanofluid saturated porous medium under gravity modulation. Ain Shams Engineering Journal. 2015;**7**:639-651

[35] Kirna P, BS Bhadauria., V. Kumar. Thermal convection in a nanofluid saturated porous medium with internal heating and gravity modulation. Journal of Nanofluid. 2016;**5**:1-12

[36] Kirna P. Nonlinear throughflow and internal heating effects on vibrating porous medium. Alexandria Engineering Journal. 2016;**55**(2):757-767

[37] Kiran P. Throughflow and gravity modulation effects on heat transport in a porous medium. Journal of Applied Fluid Mechanics. 2016;**9**(3):1105-1113

[38] Kiran P, Bhadauria BS. Chaotic convection in a porous medium under temperature modulation. Transport in Porous Media. 2015;**107**:745-763

[39] Kiran P, Bhadauria BS. Nonlinear throughflow effects on thermally modulated porous medium. Ain Shams Engineering Journal. 2015;**7**(1): 473-482

[40] Kiran P, Narasimhulu Y. Internal heating and thermal modulation effects on chaotic convection in a porous medium. Journal of Nanofluids. 2018; **7**(3):544-555

[41] Kiran P, Bhadauria BS, Narasimhulu Y. Nonlinear throughflow effects on thermally modulated rotating porous medium. Journal of Applied Nonlinear Dynamics. 2017;**6**:27-44

[42] Kiran P. Throughflow and non-uniform heating effects on double diffusive oscillatory convection in a porous medium. Ain Shams Engineering Journal. 2016;**7**:453-462

[43] Kiran P, Bhadauria BS, Narasimhulu Y. Weakly nonlinear and nonlinear magneto-convection under thermal modulation. Journal of Applied Nonlinear Dynamics. 2017;**6**(4):487-508

[44] Bhadauria BS, Kiran P. Time periodic thermal boundary and rotation effects on heat transport in a temperature dependent viscosity liquid. International Journal of Applied Mathematics and Mechanics. 2014; **10**(9):61-75

[45] Bhadauria BS, Kiran P. Weak nonlinear double-diffusive Magnetoconvection in a Newtonian liquid under temperature modulation. International Journal of Engineering Mathematics. 2014;**2014**:1-14

[46] Manjula SH, Kiran P. Nonlinear thermal instability of couple-stress fluids in porous media under thermal modulation. In: Advances in Sustainability Science and Technology. Springer; 2022. DOI: 10.1007/978-981-16-4321-7_31 (2021)

[47] Manjula SH, Kiran P, Narayanamoorthy S. The effect of gravity driven thermal instability in the presence of applied magnetic field and internal heating. AIP Conference Proceedings. 2020;**2261**(1):030042

[48] Kiran P, Manjula SH, Suresh P, Reddy PR. The time periodic solutal effect on oscillatory convection in an electrically conducting lluid layer. AIP

Conference Proceedings. 2020;**2261**(1): 030004

[49] Kiran P, Bhadauria BS, Roslan R. The effect of through flow on weakly nonlinear convection in a viscoelastic saturated porous medium. Journal of Nanofluids. 2020;**9**(1):36-46

[50] Kiran P. Concentration modulation effect on weakly nonlinear thermal instability in a rotating porous medium. Journal of Applied Fluid Mechanics. 2020;**13**(5):1663-1674

[51] Manjula SH, Kiran P, Reddy PR, Bhadauria BS. The complex Ginzburg landau model for an oscillatory convection in a rotating fluid layer. International Journal of Applied Mechanics and Engineering. 2020;**25**(1): 75-91

[52] Kiran P, Manjula SH. Weakly nonlinear double-diffusive oscillatory magneto-convection under gravity modulation. Sensor Letters. 2020;**18**(9): 725-738

[53] Kiran P. Gravity modulation effect on double diffusive oscillatory convection in a viscoelastic fluid layer. Journal of Nanofluids. 2022;**11**:01-13

[54] Manjula SH, Kiran P, Narsimlu G, Roslan R. The effect of modulation on heat transport by a weakly nonlinear thermal instability in the presence of applied magnetic field and internal heating. International Journal of Applied Mechanics and Engineering. 2020;**25**(4): 96-115

[55] Kiran P, Manjula SH, Roslan R. The effect of gravity modulation on double diffusive convection in the presence of applied magnetic field and internal heat source. Advanced Science Engineering and Medicine. 2020;**12**(6):792-805

[56] Manjula SH, Kiran P. Nonlinear thermal instability of couple-stress fluids in porous media under thermal modulation. In: Proc of 4th International Conference on Inventive Material Science. Singapore: Springer; 2022

[57] Kiran P, Bhadauria BS. Weakly nonlinear oscillatory convection in a rotating fluid layer under temperature modulation, Journal of Heat Transfer. Thermo-Rheological Effect on Weak Nonlinear Rayleigh-Benard Convection under Rotation Speed Modulation. 2016; **138**(5):051702-(1-10)

[58] Chandrasekhar S. Hydrodynamic and Hydromagnetic Stability. London: Oxford University Press; 1961

Chapter 12

Pressure Inhomogeneities across Large Samples Using Gas Pressure Media at Low Temperatures

Juscelino Batista Leão

Abstract

In-situ inert gas high pressure vessels for neutron scattering for pressures of up to 1.0 GPa, and temperatures as low as 1.5 K, pose a particular problem due to the $P \times T$ phase diagram of the pressure media. Hydrostatic pressure under constant pressure and volume, passing through the pressure versus temperature phase diagram ($P \times T$) of the gas to achieve low temperatures (1.5 K < T < 30 K) will cause an overall pressure reduction at base temperatures of at most 25% of the pressure read at the $P \times T$. A methodology for pressurization to ensure minimal pressure loss as temperature falls below the pressure media phase change and at the same time minimizing pressure inhomogeneity throughout the length of the sample is presented within this work. The technique proved to reduce the isochore loss of pressure by a factor of 5. Moreover, for the first-time direct experimental quantitative evidence of the reduction in pressure inhomogeneities across large samples is reported here, and the average inhomogeneity reduction in pressure across top and bottom of a 45 mm long sample is better than a factor of 3.

Keywords: pressure, hydrostatic, neutron scattering, helium, pressure gradient

1. Introduction

There has been a constant interest in high-pressure neutron scattering studies involving the *in-situ* application of pressure [1–9]. Techniques vary depending on many factors not limited to type of scattering, pressure vessel construction, and pressure transmitting media (PTM) (see Klotz [1] for a literature review on the subject). Neutron scattering is a powerful tool to investigate both the structure and the dynamics of molecules; however, neutron scattering under pressure poses special challenges associated with the need of large sample volumes to obtain acceptable signal-to-noise ratio in reasonable time; pressure vessels design safety; sample pressure determination; as well as PTM considerations to name a few.

Neutron scattering under pressure (P < 2.5 GPa) typically requires bulky samples as a result of large neutron beam line cross sections. With the exception of neutron experiments involving pressures beyond 2.5 GPa (where other pressure intensifying methods may be better tailor to the measurement at the cost of counting time and signal-to-nose ratio), most experiments using gas as the pressure medium benefit from samples where large amounts are available. Powder samples are easily adaptable to pressure vessels that can provide a maximum in beam cross section. Single crystal samples may employ co-alignment of a number of specimens [10] or take advantage of unusually large crystals [11].

Pressure vessel construction must take into consideration extensive and conservative calculations in order to mitigate risk to personnel and facilities [12–15]. Moreover, a standard operating procedure (SOP) is essential for the safe operation of all high pressure apparatus. It must include engineered and administrative controls for the mitigation of operational risk. Notwithstanding the design and SOP, the high pressure equipment must be submitted to frequent maintenance analysis and procedures.

Not only engineered, but also administrative controls must be in place when determining a pressure vessel operational maximum pressure range. Codes and regulatory bodies differ from country to country and between neutron scattering facilities. The interpretation to these codes also differs between the researcher and their respective safety officers. Therefore, it is in the best interest of the experimental community that, at a minimal, some regulatory guidelines must be followed in the design, testing, and operation of pressure systems. Experiment scheduling problems may rise when a certain pressure apparatus for neutron scattering is to be taken from one facility and used at another. The hazard assessment procedure is a necessary time consuming process intrinsically tied to the safe operation of the system.

Monoblock thick-wall cylindrical vessels are often used in neutron scattering for the 360° available scattering window but pose difficulties in resolving vessel integrity after normal operating temperature and pressure cycling over time. During the manufacturing process in this design, the vessel undergoes autofrettage. The internal pressure is raised above the elastic limit of the construction material without reaching failure and, as a consequence, the inner part of the unloaded cylinder is left under tangential tension while the outer part is left under tangential compression after an initial pressurization well above the final vessel working pressure. Nevertheless, the residual stresses over many temperature and pressure cycles lack quantitative data in general. This issue may be resolved using neutron diffraction (ND) residual stress mapping as a non-invasive long term quality control procedure [16].

Many neutron scattering experiments under pressure focus on the study of first order phase transition properties, and thus require refined precision in the pressure determination. Unquestionably, in the range of pressures reaching 1.5 GPa, noble gases are particularly desirable as PTM. And among these, helium is by far the gas of choice [13]. The use of gases as PTM in such experiments must also take into consideration the pressure media phase transition at low temperatures. Homogeneous hydrostatic conditions are paramount for such experiments, and this is obtained fairly straight forward for measurements where the gas media is in the temperature regime above its melting point in the $P \times T$ diagram (**Figure 1**).

However, for experiments where temperatures and pressures dip into the solid media phase, the problem of pressure inhomogeneity across large samples becomes obvious [11]. While the media present at temperatures above the $P \times T$ are in the gas phase, the media inside the pressure vessel solidifies as its temperature traverses the $P \times T$ curve. In this scenario the pressurization is typically done still in the gas phase by maintaining the temperature of the entire system just above the $P \times T$ melting curve for helium (PT_{He}). Once the target pressure is achieved, while in the gas phase, the temperature of the pressure vessel is allowed to drop passing through the freezing point. This cooling under constant pressure (P_C) attempts to minimize pressure

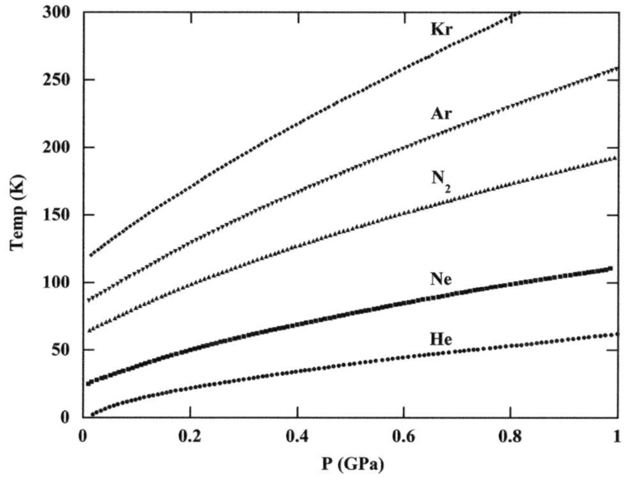

Figure 1.
Melting curves for He, Ne, N_2, Ar, and Kr (adapted from Ref. [4]).

inhomogeneities but it is responsible for pressure losses due to helium gas-solid phase change [13]. Sherman and Stadtmuller reported a pressure loss of 20% at 0.1 GPa [14]. There is mention, in literature, of the possibility for pressure differences across a sample [14]; however, to best knowledge, this has not been documented experimentally elsewhere. Pressure inhomogeneities across large samples are of significant importance for experiments, such as in the pressure-temperature phase diagram of URu_2Si_2 in helium [11] for example, where the mapping of first-order phase changes is significantly sensitive to pressure.

Another consideration for such neutron scattering experiments under pressure is the accuracy and precision of the pressure measurement. Again, in the $P \times T$ case where the PTM is held in the gas phase, the pressure measurement can be achieved fairly easily through the use of various commercially available gauges [12] as long as the hydrostatic condition is unimpeded by a possible "blockage" within the length of the capillary tube that links the pressurizing device to the pressure vessel. The scenario changes once the pressure measurement temperature crosses the solid phase in the media $P \times T$ diagram. The pressure loss as the temperature drops from the melting point for helium is well known and understood [13, 14] in terms of the media isochores. Nevertheless, as mentioned previously, the pressure loss due to the volumetric changes in the PTM must be tabulated with the help of secondary pressure gauges.

The accuracy in pressure measurements depends greatly on the gauge used. There are primary gauges that measure the force applied on a precise area, secondary gauges that depend on the change of a certain property of the gauge that then can be translated into a pressure, and fixed points that propagate the precision of the fixed point measured. The topic of pressure gauges and accuracies is extensive and better suited elsewhere. For the pressure range at hand, the accuracy of the measurements is high [17] across the different gauges; however, the reliance on a single technique for measurement has its drawbacks.

For the case where it is not possible to use optical pressure gauges and measurements other than room temperature, the use of fixed point calibration provides reliable data with small loss in accuracy. Due to its high degree of structural anisotropy, HOPG is a good choice for pressure calibration in neutron scattering experiments using an autofrettage monoblock vessel for pressures up to 15 GPa. The carbon atoms within the two-dimensional hexagonal lattice structure of HOPG are bound by strong covalent bonds, in contrast to much weaker bonding between adjacent planes [18].

Here it is discussed a methodology for pressurization using He as PTM in neutron scattering experiments to minimize the pressure loss in experiments that require temperatures below the pressure media phase change. At the same time minimizing pressure inhomogeneity throughout the length of the sample. Direct evidence of stacking pressure inhomogeneities presented here have not been addressed nor quantified until now. Pressure measurements for first-order phenomena are greatly influenced by these pressure differences since neutron scattering measurements provide an illumination area average [19]. Pressure difference across large samples were significantly reduced following the novel procedure here demonstrated.

2. Experimental details

A systematic approach to resolving the issue of pressure inhomogeneity requires replicating and quantifying the pressure differences across a sample under typical conditions. There is mention, in literature, of the possibility for pressure differences across a sample due to the irregular formation of helium crystals [14]; however, this has not been documented experimentally. Once quantified, a procedure to minimize these average pressures is demonstrated.

A typical neutron scattering experiments under gas pressure consists of a gas pressure intensifier (or "booster") capable of increase the pressure of a gas cylinder to orders of magnitude. Connecting such "booster" to a pressure vessel requires a pressure gas line robust enough, and yet offering low thermal sink from room temperature. Moreover, the pressure vessel and a small portion of the pressure line must also be inserted onto a cryostat that can provide the heat removal necessary to cool the pressure vessel to the desired temperature.

The apparatus consisted of a monoblock aluminum alloy Al 7075-T6 pressure vessels (**Figure 2**) [20]. The vessel was designed with a working pressure of 0.65 GPa, 1.5 cm^3 sample volume, and having a 69% neutron transmission at 2 Å. *In-situ* pressurization was achieved using helium as the pressure media. High pressure was achieved using a commercially available two-stage helium intensifier [21]. The pressure vessel was connected to the intensifier through a high-pressure capillary brazed to a pressure plug and contained a line heater (LH) wound through the entire length of the cryostat sample well down to the immediate sample vessel connector. A commercially available cryostat, or top-loading closed cycle refrigerator (TLCCR), was used to cool the sample down from room temperature to 4 K [20].

The sample measured consisted of three uniformly shaped pyrolytic graphite (HOPG) single crystals each measuring 10 mm × 5 mm × 3 mm and placed onto an aluminum sample holder designed as to hold each of the crystals with the c-direction normal to the neutron beam while maintaining each with a rotation off-set of 10° from one another. The sample holder containing the samples was loaded into the pressure vessel and sealed. The pressure vessel was flushed with helium and connected to the cryostat sample center stick containing the pressure capillary connection form the intensifier. Once the sample centering stick was lowered into the cryostat, a small amount of helium gas (~15 MPa) was then allowed to fill the system from the intensifier to the sample vessel. The sample was cooled to 80 K overnight and pressurized to 0.65 GPa.

Single crystal ND measurements performed in the $\hat{c}$ scattering plane were conducted at the NIST Center for Neutron Research, using the BT-4 triple-axis

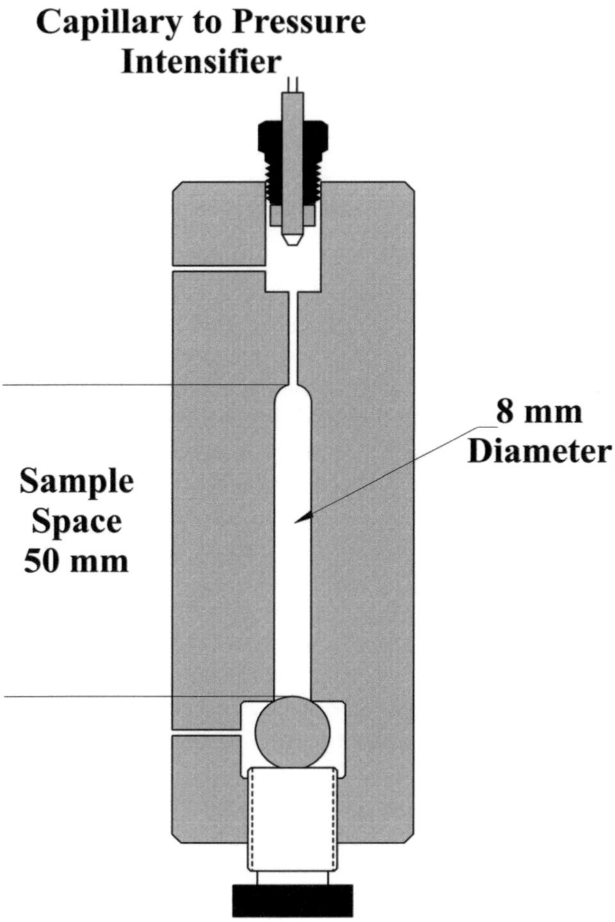

Figure 2.
Schematic of monoblock construction aluminum pressure vessel showing the capillary connection, sample space, and the check-ball closure.

spectrometer [22] operated in diffraction mode, employing an incident wavelength of $\lambda = 2.35$ Å (14.7 meV) defined by pyrolytic graphite PG(002) crystals at both the monochromator and analyzer positions.

ND data for each of the HOPG crystals were obtained by rotating the TLCCR between each 10° off-sets without disturbing neither the temperature, nor the pressure in the vessel. Thus, three measurements were conducted for each temperature in both the cooling under constant volume (V_C) and cooling under constant pressure (P_C) conditions.

3. Discussion

A modified version of the Murnaghan equation [23] (Eq. (1)) was used to explore the relation between the pressure (P) and the lattice constant (r) along one of the HOPG crystal axes. For any given temperature.

$$P = \left(\frac{\beta_0}{\beta'}\right)\left[\left(\frac{r}{r_0}\right)^{-\beta'} - 1\right] \tag{1}$$

Where β_0^{-1} = 37.1(5) GPa is the linear compressibility at ambient pressure, and β' = 14.4(7) [24] is the pressure derivative of the linear compressibility. The axial compression coefficients of graphite vary greatly among references [25–30], this is partially due to the methodology used in determining the modulus for graphite. The measurement techniques that depend on pressure measurement themselves tend to be highly susceptible to propagation in errors. Using explosive-generated pressures these moduli were determined through empirical shock wave and particle velocity experiments by Coleburn [24]. This was particularly interesting since the results were, according to the author, in close agreement with other pressure-gauge independent results. Moreover, good agreement was observed while determining the vessel pressure at temperatures ≥70 K using the pressure gauge reading and Eq. (1) (**Figure 3**).

The results for the room temperature pressurizing intensifier gauge reading and the calculated pressure from the $\hat{c}$ -direction lattice parameters for HOPG were in excellent agreement (±2 MPa). After cooling the pressure vessel to 80 K (still above the PT_{He}), the system was then pressurized to 0.65 GPa under constant volume. The vessel temperature was allowed to equilibrate for 30 minutes before each diffraction pattern was collected. The data in **Figure 3** for pressure measurements above the PT_{He} and their calculations from the ND spectra are in close agreement with the pressure readings while the sample was under hydrostatic conditions. The $\hat{c}$ -direction lattice parameters were measured for each of the HOPG crystals at ambient pressure (P_0) and maximum pressure ($P_{0.65}$) for temperatures above and below the PT_{He}. For the temperature range below the PT_{He}, the vessel was pressurized at a temperature above and close to the melting curve of He. Pressure and diffraction data were collected for V_C and P_C conditions.

Figure 3 also shows the results for all temperatures measured from 80 K down to 4 K. The pressure intensifier gauge (P_{Gauge}) monitored the hydrostatic pressure

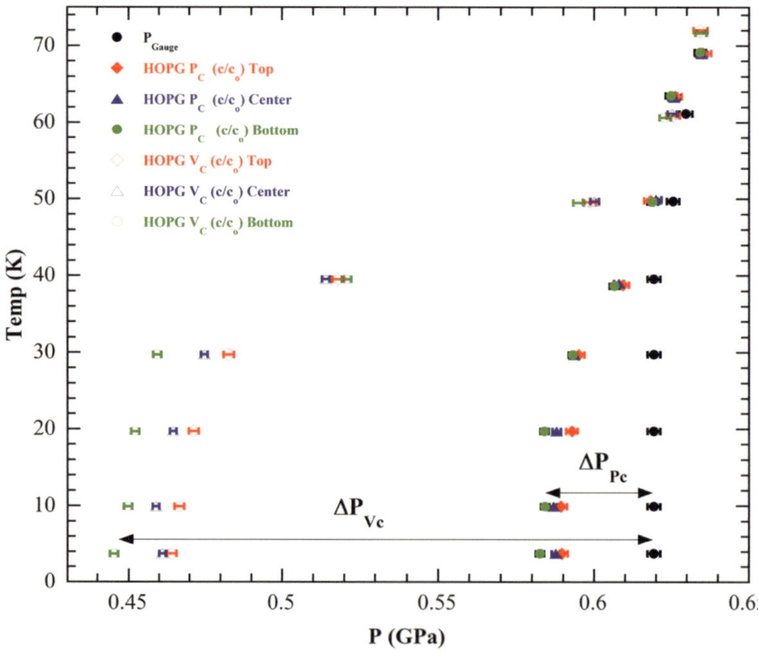

Figure 3.
Pressure vs. temperature data for HOPG under helium as the pressure media. The straight arrow lines are guides to the eye for pressure gradient for cooling under constant pressure (ΔP_{Pc}) and under constant volume (ΔP_{Vc}).

transmitted through the connecting pressure capillary. Once the pressure LH was turned off, P_{Gauge} as expected demonstrated the behavior consistent of a PTM "ice" blockage; that is, the pressure stabilized at a reading within the pressure immediately above the PT_{He} down to the base temperature.

Results for pressures below PT_{He} are shown to be distinct (**Figure 3**). While pressurizing and cooling under constant volume (V_C) the pressures calculated using the ND data confirm the behavior described by Sherman and Stadtmuller [14] and show a significant overall drop in pressure staring at 40 K and steadily increasing down to the lowest temperature (4 K). Readily it is observed that as the vessel is cooled from 60 to 40 K there is a divergence in pressures within the vessel partially due to crossing the PT_{He} (this is also observed in the pressure intensifier gauge reading). For the case of cooling under constant volume (ΔP_{VC}) the pressure gradient is of the order of 25% and substantiates the isochore correction calculations done by Spain and Segall [13].

To minimize the isochore loss of pressure, subsequent pressurization was done under cooling at constant pressure (P_c) by following a rigorous procedure to ensure that during pressurization the helium solidifying at the bottom of the vessel will be formed at a pressure as close as possible to the solid helium nearest the "He ice" blockage in the system. In this case of cooling under constant pressure (ΔP_{PC}) the data shows that the overall pressure loss on solidification of the helium within the sample space was reduced to 5% overall (**Figure 3**).

Finally, results of the direct evidence of pressure inhomogeneities across the vessel's sample space are shown and quantified in **Figure 3** and **Table 1**, respectively. For the first-time direct experimental quantitative evidence of the pressure inhomogeneities across large samples is reported. The average inhomogeneity reduction in pressure between top and bottom of a 45 mm long sample is better than a factor of 3.

Neutron scattering proved to be the defining probe technique to unravel these results. By mounting three HOPG samples off-set in degrees of rotation, the pressures within the vessel could be tabulated across the three distinct regions of the sample space by using Eq. (1).

The total pressure inhomogeneity for cooling at constant pressure (TPI_{PC}) at a given temperature is compared with the total pressure inhomogeneity for cooling at constant volume (TPI_{VC}) across the three HOPG single crystals (top, center, and bottom).

Inhomogeneities between the top and the bottom of a sample are, for the first time, quantified showing the importance of proper pressurization procedure.

A procedure to minimize this pressure drop caused by the phase change in the PTM (assuming helium) is then: (i) Firstly the pressurizing apparatus must contain a properly heated capillary line throughout the entire length of the sample stick to remove the risk of creating a helium ice plug during the procedure and monitor the capillary temperature maintaining it at least 30 K above the PT_{He} for the pressure

Temp (K)	TPI_{PC} (GPa)	TPI_{VC} (GPa)
4	0.018(7)	0.006(9)
10	0.016(9)	0.005(1)
20	0.019(3)	0.008(8)
30	0.023(3)	0.001(8)

Table 1.
Pressure inhomogeneities shown for the range $4\ K \leq T \leq 30\ K$.

target. Monitor three temperature sensors (one for the capillary line, one for the sample vessel and one for the cryostat temperature). (ii) Cool the sample to a temperature at most 10 K above the PT_{He}. (iii) Pressurize the vessel to the target pressure (it is recommended to exceed the target pressure by 10% to allow for the pressure loss as the helium cools to the freezing temperature) to just above the PT_{He} to ensure the helium remains fluid. (iv) Cool the cryostat (while maintaining the capillary line temperature above PT_{He}) under constant pressure to the PT_{He}. It might be necessary to increase the temperature of the capillary at this point to maintain it at least 30 K above the PT_{He}. (v) Freeze the helium slowly under constant pressure, and (approximately) constant volume conditions. While freezing the helium, allow the cryostat temperature to drop well below the PT_{He} and maintain the capillary LH providing heat to the system. This way the vessel begins cooling below the PT_{He} from the bottom and the top of the vessel connected to the capillary is held above PT_{He}. This step is important to guarantee that helium ice is forming gradually inside the vessel and immersing the sample throughout its length more or less uniformly, thus minimizing the inhomogeneities in pressure as the helium ice forms. (vi) When the cryostat temperature reaches at least 15 K below the PT_{He} the capillary temperature is lowered at a rate of 0.5 K/min and continuing until freezing is complete. (vii) Cool slowly to the required temperature under constant volume conditions.

4. Conclusion

For the first time pressure inhomogeneities across large samples in neutron scattering experiments at low temperature are quantified and documented experimentally. A methodology was demonstrated to minimize these pressure inhomogeneities across large samples under pressure at low temperatures, when using helium as PTM, in low temperature neutron scattering experiments. The technique employed here using HOPG to determine pressure measurements corroborate, in the helium gas phase, with previously published calculations of helium isochores to good approximation and provided means to reduce the isochore loss of pressure by a factor of 5. As of the date of publication, this is the first-time direct experimental quantitative evidence of the reduction in pressure inhomogeneities across large samples is reported, and the average inhomogeneity reduction in pressure between top and bottom of a 45 mm long sample is better than a factor of 3.

Author details

Juscelino Batista Leão
NIST Center for Neutron Research, National Institute of Standards and Technology, Gaithersburg, Maryland, USA

*Address all correspondence to: juscelino.leao@nist.gov

References

[1] Klotz S. Techniques in High Pressure Neutron Scattering. Boca Raton, Florida: CRC Press; 2013

[2] Bridgman PW. The effect of homogeneous mechanical stress on the electrical resistance of crystals. Physical Review. 1932;**42**:858. DOI: 10.1103/PhysRev.42.858

[3] Lechner R. High pressure vessel for measurement of pressure induced phonon energy shifts by inelastic scattering of neutrons. The Review of Scientific Instruments. 1966;**37**:1534. DOI: 10.1063/1.1720036

[4] McWhan DB, Bloch D, Parisot G. Apparatus for neutron-diffraction at high-pressure. The Review of Scientific Instruments. 1974;**45**:643. DOI: 10.1063/1.1686704

[5] Paureau J, Vettier C. New high-pressure cell for neutron-scattering at very low-temperatures. The Review of Scientific Instruments. 1975;**46**:1484. DOI: 10.1063/1.1134083

[6] Fujiwara H, Kadomatsu H, Tohma K. Simple clamp pressure cell up to 30-kbar. The Review of Scientific Instruments. 1980;**51**:1345. DOI: 10.1063/1.1136061

[7] Wu AY, Whalley E, Dolling G. Neutron-diffraction of liquids to 20 kbar—An aluminum-alloy pressure-vessel. The Review of Scientific Instruments. 1985;**56**:1409. DOI: 10.1063/1.1138494

[8] Onodera A et al. High-pressure-cell for neutron-scattering. Japanese Journal of Applied Physics. 1987;**26**:152. DOI: 10.1143/JJAP.26.152

[9] Bao W, Broholm C, Trevino SF. Simple high-pressure-cell for neutron-scattering. The Review of Scientific Instruments. 1995;**66**:1260. DOI: 10.1063/1.1146019

[10] Kreyssig A et al. Pressure-induced volume-collapsed tetragonal phase of $CaFe_2As_2$ as seen via neutron scattering. Physical Review B. 2008;**78**:184517. DOI: 10.1103/PhysRevB.78.184517

[11] Butch NP, Jeffries JR, Chi S, Leão JB, Lynn JW, Maple MB. Antiferromagnetic critical pressure in URu_2Si_2 under hydrostatic conditions. Physical Review B. 2010;**82**:060408(R). DOI: 10.1103/PhysRevB.82.060408

[12] Dawson VCD. In: Spain I, Paauwe J, editors. High Pressure Technology: Equipment Design, Materials, and Properties. Vol. 1. Chapter 3. New York: Marcel Dekker; 1977

[13] Spain IL, Segall S. Equation of state of solid helium—Pressure scale to 20-kbar for high pressure measurements at low temperature. Cryogenics. 1971;**11**(1):26. DOI: 10.1016/0011-2275(71)90006-3

[14] Sherman WF, Stadtmuller AA. Experimental Techniques in High-Pressure Research. Chichester, West Sussex: Wiley; 1987

[15] BPVC Section VIII-Rules for Construction of Pressure Vessels Division 3-Alternative Rules for Construction of High Pressure Vessels BPVC-VIII-3—2021. ASME; 2021. p. 384. ISBN: 9780791874257

[16] Leão JB, Milner J. In-situ stress measurement comparison of aluminum pressure vessels under high working pressures. Journal of Neutron Research. 2017;**19**(1-2):65-72. DOI: 10.3233/JNR-170040

[17] Bean VE. In: Vodar B, Marteau PH, editors. Proceedings of the VII International Conference on High Pressure Engineering, pg. 231. Le Creusot, France. July 30th, 1979. Oxford, UK: Pergamon; 1980

[18] Hanfland M, Beister H, Syassen K. Graphite under pressure: Equation of state and first-order Raman modes. Physical Review B. 1989;**39**:12598. DOI: 10.1103/PhysRevB.39.12598

[19] Squires GL. Introduction to the Theory of Thermal Neutron Scattering. Mineola New York: Dover Publications; 1978

[20] NIST Center for Neutron Research list of Sample Environment Equipment. Available from: https://www.nist.gov/ncnr/sample-environment/equipment/high-pressure/hw-02

[21] Harwood Engineering Company, Inc. Available from: http://www.harwoodeng.com/

[22] NIST Center for Neutron Research BT-4 Triple Axis Spectrometer. Available from: https://www.nist.gov/ncnr/bt-4-triple-axis-spectrometer

[23] Murnaghan FD. The compressibility of media under extreme pressures. Proceedings of the National Academy of Sciences. 1944;**30**(9):244-247. DOI: 10.1073/pnas.30.9.244

[24] Coleburn NL. Compressibility of pyrolytic graphite. The Journal of Chemical Physics. 1964;**40**:71. DOI: 10.1063/1.1724896

[25] Seldin EJ, Nezbeda CW. Elastic constants and electron-microscope observations of neutron-irradiated compression-annealed pyrolytic and single crystal graphite. Journal of Applied Physiology. 1970;**41**:3389. DOI: 10.1063/1.1659430

[26] Blakslee OL, Proctor DG, Seldin EJ, Spence GB, Weng T. Elastic constants of compression-annealed pyrolytic graphite. Journal of Applied Physics. 1970;**41**:3373. DOI: 10.1063/1.1659428

[27] Lynch RW, Drickamer HG. Effect of high pressure on the lattice parameters of diamond, graphite, and hexagonal boron nitride. The Journal of Chemical Physics. 1966;**44**:181

[28] Kabalkina SS, Vereshchagin LF. X-ray study of linear compression of graphite at pressures up to 16000 kg/cmÂ^2. Soviet Physics – Doklady. 1960;**5**:373

[29] Wada N, Clarke R, Solin SA. Solid State Communications. 1980;**35**:675. DOI: 10.1016/0038-1098(80)90872-8

[30] Alzyab B, Perry CH, Zahopoulos C, Pringle OA, Nicklow RM. Physical Review B. 1988;**38**:1544. DOI: 10.1103/PhysRevB.38.1544